The Singing Neanderthals

The Singing Neanderthals

The Origins of Music, Language, Mind and Body

Steven Mithen

Harvard University Press
Cambridge, Massachusetts

Printed in the United States of America

First Harvard University Press paperback edition, 2007

First published in 2005 by Weidenfeld &
Nicolson Ltd, London

Library of Congress Cataloging-in-Publication Data

Mithen, Steven J.
The singing neanderthals : the origins of music, language, mind, and body / Steven Mithen.
p. cm.
Includes bibliographical references (p.) and index.
ISBN-13 978-0-674-02192-1 (cloth: alk. paper)
ISBN-10 0-674-02192-4 (cloth: alk. paper)
ISBN-13 978-0-674-02559-2 (pbk.)
ISBN-10 0-674-02559-8 (pbk.)
1. Music—Origin. 2. Music—Psychological aspects.
3. Human evolution. I. Title.

ML3800.M73 2006
780.9'01—dc22 2005030187

Contents

Preface

The propensity to make music is the most mysterious, wonderful and neglected feature of humankind. I began writing this book to describe my theory as to why we should be so compelled to make and listen to music. I wished to draw together and explain the connections between the evidence that has recently emerged from a variety of disciplines including archaeology, anthropology, psychology, neuroscience and, of course, musicology. It was only after I had begun that I came to appreciate that it was not only music I was addressing but also language: it is impossible to explain one without the other. As the work progressed, I also realized that their evolution could not be accounted for without being thoroughly embedded in the evolution of the human body and mind. And so the result is necessarily an ambitious work, but one that I hope will be of interest to academics in a wide range of fields and accessible to general readers – indeed, to anyone who has an interest in the human condition, of which music is an indelible part.

When I mentioned to my family that I intended to write a book about music there was a moment's silence and then laughter. Although I frequently listen to music, I can neither sing in tune nor clap in rhythm. I am unable to play any musical instrument. Sue, my wife, sings; she and my three children play the piano. Heather, my youngest daughter, also plays the violin. It saddens me that I will never be able to join one of them in a piano duet or accompany them with a song. I've tried the latter and it is a deeply unpleasant experience for all involved.

Writing this book has been an attempt to compensate for my musical limitations. When my children were valiantly struggling with piano practice in one room, I would be doing the same with the theory of music in another. It's been the closest we will ever get to making music together. If they hadn't worked so hard, neither would I, and this book would never have been written. So I am most grateful for their efforts; I am both proud and envious of their musical achievements.

Nicholas Bannan, my University of Reading colleague, helped to stimulate my longstanding but latent interest in how music evolved by his infectious enthusiasm for this topic and his spontaneous singing demonstrations.

I began my research vicariously by supervising student dissertations on relevant subjects, notably by Iain Morley and Hannah Deacon. These further excited my interest and I immersed myself in the relevant literature. To my immense pleasure, I then found that I could justify long hours simply listening to music while digesting what I had read and thinking about what it might mean.

During the course of my research I have benefited greatly from the help and advice of many friends, colleagues and academic acquaintances. My greatest academic debt is to Alison Wray. Her work has been the most influential on my own thinking. She kindly commented on the first draft of my manuscript and provided copies of her unpublished papers. Kenny Smith also kindly read and commented on the whole draft manuscript, while Jill Bowie, Dick Byrne, Anna Machin, Isabelle Peretz and Ilona Roth did the same for selected chapters. I am most grateful for their advice and support.

I would also like to thank Brent Berlin, Margaret Clegg, Ian Cross, Valerie Curtis, Franceso D'Errico, Robin Dunbar, Christopher Henshilwood, Maggie Tallerman and Andy Whiten. They pointed me to publications, discussed ideas, answered email queries and provided me with copies of unpublished manuscripts.

Nicholas Bannan and I co-organized the 'Music, Language and Human Evolution' workshop that took place at the University of Reading in October 2004, kindly funded by the European Science Foundation. This occurred after I had completed the first draft of this book and it gave me the opportunity to meet and discuss with many of the academics whose work I had cited but whom I had not previously met. I am most grateful to all the participants for that most stimulating event, which provided me with further information, ideas and inspiration to complete the book: Margaret Clegg, Ian Cross, Pedro Espi-Sanchis, Tecumseh Fitch, Robert Foley, Clive Gamble, Tran Quang Hai, Björn Merker, Iain Morley, Isabelle Peretz, Ilona Roth, Johan Sundberg, Elizabeth Tolbert and Sandra Trehub.

While I have benefited enormously from discussion and debate with all of those mentioned above, any errors or fuzzy thinking within this book are unquestionably my responsibility alone.

During the course of my writing I have been fortunate to have had clerical support from Chris Jones. She eagerly checked bibliographic references, undertook Internet searches and photocopying, and provided all manner of miscellaneous help that made writing this book possible in the gaps in the inundation of meetings and paperwork that comes with university jobs today. I am also grateful to Tom Wharton, my editor at Weidenfeld & Nicolson, who provided considerable advice on my draft manuscript, to the book's immense benefit.

The most intense musical experience of my life was on a winter's evening in 1983 when Sue took me to hear Fauré's Requiem in a candlelit York Minster. It has taken me the last twenty-one years of studying human evolution and the last year of writing this book to understand why I and so many others were moved and inspired by the music we heard that night. That was the first time I had heard Fauré's Requiem. Ever since, Sue has been introducing me to new music, often through the concerts of the Reading Festival Chorus, with which she sings. This has truly enhanced my life. And so it is with my gratitude that I dedicate this book to her.

31 December 2004

1 The mystery of music

The need for an evolutionary history of music

It has been said that the adagio from Schubert's String Quintet in C Major is a perfect piece of music.[1] I have it playing at this very moment and might easily agree. Bach's Cello Suites could also take that accolade, as could 'Kind of Blue' by Miles Davis. That, at least, is my view. Yours will be different. In this book I am not concerned with the specific music that we like but with the fact that we like it at all – that we spend a great deal of time, effort and often money to listen to it, that many people practise so hard to perform it, and that we admire and often idolize those who do so with expertise, originality and flair.

The explanation has to be more profound than merely invoking our upbringing and the society in which we live, although these may largely account for our specific musical tastes. The appreciation of music is a universal feature of humankind; music-making is found in all societies and it is normal for everyone to participate in some manner; the modern-day West is quite unusual in having significant numbers of people who do not actively participate and may even claim to be unmusical. Rather than looking at sociological or historical factors, we can only explain the human propensity to make and listen to music by recognizing that it has been encoded into the human genome during the evolutionary history of our species. How, when and why are the mysteries that I intend to resolve.

While several other universal attributes of the human mind have recently been examined and debated at length, notably our capacities for language and creative thought, music has been sorely neglected. Accordingly, a fundamental aspect of the human condition has been ignored and we have gained no more than a partial understanding of what it means to be human.

The contrast with language is striking. As part of its founding statutes in 1866, the Société de Linguistique de Paris banned all discussions about the origin of language.[2] This position held sway with academics for more than a century, before a surge of research began during the last decade of the twentieth century. Linguists, psychologists, philosophers, neuroscientists, anthropologists and archaeologists now frequently debate the origin and evolutionary history of language, and have published many articles and books on the topic.[3]

Even though the Société de Linguistique de Paris had nothing to say about studying the origin of music, academics appear to have observed their own, self-imposed ban, which has effectively continued up to the present day. There have, of course, been some notable exceptions which should be acknowledged. One was Charles Darwin, who devoted a few pages in his 1871 book, *The Descent of Man*, to the evolution of music. More recently, there was the renowned ethnomusicologist John Blacking whose 1973 book, *How Musical Is Man?*, broached the idea that music is an inherent and universal human quality.[4]

Not only does the origin of music deserve as much attention as that of language, but we should not treat one without the other. Indeed, in spite of the recent activity, only limited progress has been made in understanding the evolution of language. This can be explained partly by neglect of the archaeological and fossil evidence, and partly by the neglect of music. Those writers who had evidently annoyed the Société de Linguistique de Paris were aware that music and language have an evolutionary relationship. Jean-Jacques Rousseau's *Essai sur l'origine des langues* (1781) was a reflection on both music and language.[5] More than a century later, Otto Jespersen, one of the greatest language scholars of the nineteenth and twentieth centuries, concluded his 1895 book, *Progress in Language*, by stating that 'language ... began with half-musical unanalysed expressions for individual beings and events'.[6] Such insights appear to have been forgotten, for music has hardly received any mention at all in the recent surge of publications concerning the origin of language.[7]

This might be simply because music is so much harder to address. Whereas language has a self-evident function – the transmission of information[8] – and can be readily accepted as a product of evolution even if its specific evolutionary history remains unclear, what is the point of music?[9]

That question leads us to a further neglected area: emotion. If music is about anything, it is about expressing and inducing emotion. But while archaeologists have put significant effort into examining the intellectual capacities of our ancestors, their emotional lives have remained as neglected as their music. This, too, has constrained our understanding of how the capacity for language evolved.

Two views of proto-language

Language is a particularly complex system of communication. It has to have evolved gradually in a succession of ever more complex communication systems used by the ancestors and relatives of modern humans.[10] Academics refer to these communication systems by the catch-all term 'proto-language'. Identifying the correct nature of proto-language is the most important task facing anyone attempting to understand how language evolved.

Because defining the nature of language is such a challenging task – one that I will address in the next chapter – it will come as no surprise that academics vehemently disagree about the nature of proto-language. Their theories fall into two camps: those who believe that proto-language was 'compositional' in character, and those who believe it was 'holistic'.[11]

The essence of the compositional theories is that proto-language consisted of words, with limited, if any, grammar. The champion of this view is Derek Bickerton, a linguist who has had a profound influence on the debates about how language evolved with a succession of books and articles over the past decade.[12]

According to Bickerton, human ancestors and relatives such as the Neanderthals may have had a relatively large lexicon of words, each of which related to a mental concept such as 'meat', 'fire', 'hunt' and so forth. They were able to string such words together but could do so only in a near-arbitrary fashion. Bickerton recognizes that this could result in some ambiguity. For instance, would 'man killed bear' have meant that a man has killed a bear or that a bear has killed a man? Ray Jackendoff, a cognitive scientist who has written about both music and language, suggests that simple rules such as 'agent-first' (that is, the man killed the bear) might have reduced the potential ambiguity.[13] Nevertheless, the number and complexity of potential utterances would have been severely limited. The transformation of such proto-language into language required the evolution of grammar – rules that define the order in which a finite number of words can be strung together to create an infinite number of utterances, each with a specific meaning.

Compositional theories of proto-language have dominated studies of language evolution for the past decade; they have been highly influential but have, I believe, led us in the wrong direction for understanding the earliest stages of language evolution. Alternative views have recently emerged that fall into the category of 'holistic' theories. Their champion is a less well-known linguist who has, I believe, identified the true nature of proto-language. Her name is Alison Wray and she is a linguist at the University of Cardiff.[14] By using the term 'holistic', she means that the precursor to language was a communication system composed of 'messages' rather than words; each hominid utterance was uniquely associated with an arbitrary meaning – as are the words of modern language and, indeed, those of a Bickertonian proto-language. But in Wray's proto-language, hominid multisyllabic utterances were not composed out of smaller units of meaning (that is to say, words) which could be combined together either in an arbitrary fashion or by using rules to produce emergent meanings. In her view, modern language only evolved when holistic utterances were 'segmented' to produce words which could then be composed together to create statements with novel meanings. Hence, while Bickerton believes that words were present in the early stages of language evolution, Wray believes that they only appeared during its later stages.

Neglect and dismissal

While considerable attention has been paid to the nature of proto-language, its musical counterpart has been virtually ignored, especially by palaeo-anthropologists, who interpret the skeletal and artefactual remains of human ancestors. I am guilty myself, having failed to consider music in my 1996 book, *The Prehistory of the Mind*. That proposed an evolutionary scenario for the human mind involving the switch from a 'domain-specific' to a 'cognitively fluid' mentality, the latter being attributed to *Homo sapiens* alone. Cognitive fluidity refers to the combination of knowledge and ways of thinking from different mental modules, which enables the use of metaphor and produces creative imagination. It provides the basis for science, religion and art. Although *The Prehistory of the Mind* recognized language as a vehicle for cognitive fluidity, it paid insufficient attention to the actual evolution of language and did not satisfactorily address the thorny issue of how pre-modern humans, such as the Neanderthals, communicated in its assumed absence.

I became acutely aware of my own neglect of music when writing my most recent book, *After the Ice*. That work included reconstruction scenarios for many of the prehistoric hunter-gatherer and early farming communities that existed between 20,000 and 5000 BCE. Few of these felt realistic to me until I imagined music: people singing to themselves as they undertook some mundane task, communal singing and dancing at times of celebration and mourning, mothers humming to their babies, children playing musical games. The communities examined in *After the Ice* were all relatively recent, but the same is true of early *Homo sapiens*, the Neanderthals, and even older ancestral species such as *Homo heidelbergensis* and *Homo ergaster*. Without music, the prehistoric past is just too quiet to be believed.

When music has not been ignored, it has been explained away as no more than a spin-off from the human capacity for language.[15] In his ambitious and in many ways quite brilliant 1997 book, *How the Mind Works*, the linguist and Darwinist Steven Pinker devoted eleven of his 660 pages to music. In fact, he dismissed the idea that music was in any way central to human mentality. For Pinker, music is derivative from other evolved propensities, something that humans have invented merely as entertainment: 'As far as biological cause and effect are concerned, music is useless ... music is quite different from language ... it is a technology, not an adaptation.'[16]

For those who devote their academic lives to the study of music, its dismissal by Pinker as nothing more than auditory cheesecake and 'the making of plinking noises'[17] was bound to raise hackles. The most eloquent response came from the Cambridge-based musicologist Ian Cross. Confessing to a personal motivation in defending the value of music as a human activity, he claimed that music is not only deeply rooted in human biology but also critical to the cognitive development of the child.[18]

Ian Cross is one of a few academics who have begun to address the evolution of musical abilities – a development that had begun while Pinker was writing his dismissal.[19] Elizabeth Tolbert of the Peabody Conservatory at Johns Hopkins University also reacted against Pinker's proposal. She stressed the relationship of music to symbolism and bodily movement, proposing that it must have coevolved with language.[20] Nicholas Bannan, a specialist in music education at Reading University, argued that the 'instinct to sing' is just as powerful as the 'instinct to speak' championed by Pinker.[21] In his 2004 book, *The Human Story*, Robin Dunbar, the distinguished evolutionary psychologist from Liverpool University, proposed that language went through a musical phase during the course of its evolution.[22]

All of these academics, as well as myself, are writing in the wake of not only Jean-Jacques Rousseau and Otto Jespersen but also John Blacking. In one of his last essays, written ten years after *How Musical Is Man?*, he proposed there had been a 'nonverbal, prelinguistic, "musical" mode of thought and action'.[23] It is that proposal that I wish to explore and ultimately to vindicate in this book.

This book

I am embarrassed by my own previous neglect of music, persuaded by Alison Wray's theory of a holistic proto-language, ambitious to understand how our prehistoric ancestors communicated, and convinced that the evolution of music must also hold the key to language. This book sets out my own ideas about how music and language evolved, and evaluates the proposals of others by exposing them to the archaeological and fossil evidence.

The book is in two parts. The first, chapters 2–7, is concerned with music and language today. It describes the character of music and language, and what we understand about their relationship, covering those features that I intend to explain in terms of the evolutionary history provided in the book's second part, chapters 8–17.

I begin by considering the similarities and differences between music and language (chapter 2), and then focus on three topics: how music and language are constituted in the brain (chapters 3–5), how we communicate with prelinguistic infants (chapter 6), and the relationship between music and emotion (chapter 7).

These are the topics that both require an evolutionary explanation and may provide some clues to the origins of music and language. There are, of course, many other aspects of music and language that also require explanation, most notably the diversity and distribution of languages and musical styles in the world today. These are not addressed, because their explanations lie principally in the historical development of human societies and dispersals of populations that took place long after the universally shared capacities for language and music had evolved.[24]

Having established what requires explanation, Part Two begins by remaining in the present but addressing the communication systems of monkeys and apes (chapter 8). These are, I believe, similar to those of our early ancestors, from which our own capacities for music and language evolved. I do not consider the communication systems of other animals, even though these are often described as having musical or even linguistic qualities, notably in the cases of birds and whales. There are, indeed, striking similarities between these and human music.[25] However, owing to the evolutionary distance between humans, bird and whales, such similarities must have arisen by convergent evolution – that is to say, natural selection coming up with similar solutions to similar problems, notably the defence of territory and the attraction of mates. My concern is with the specific evolutionary history of language and music within the human lineage and I have no room to explore such convergent evolution.

In chapter 9, I begin that evolutionary history by considering our early hominid ancestors, which evolved in Africa between 6 and 2 million years ago. Chapters 10–14 examine the evolution and societies of Early Humans, species such as *Homo ergaster*, *Homo erectus* and *Homo heidelbergensis*, addressing the significance for the evolution of communication of bipedalism, hunting and gathering, social relations, life history and cooperation. Each chapter provides some key palaeoanthropological evidence and introduces further material from disciplines such as primate studies, psychology and linguistics. Chapter 15 considers the lifestyles and communication systems of the Neanderthals; chapters 16 and 17 examine the final steps in the origins of language and music respectively, identifying these with the appearance of *Homo sapiens* in Africa soon after 200,000 years ago.

In these concluding chapters I return to the present-day characteristics of language and music as described in Part One and show how these are

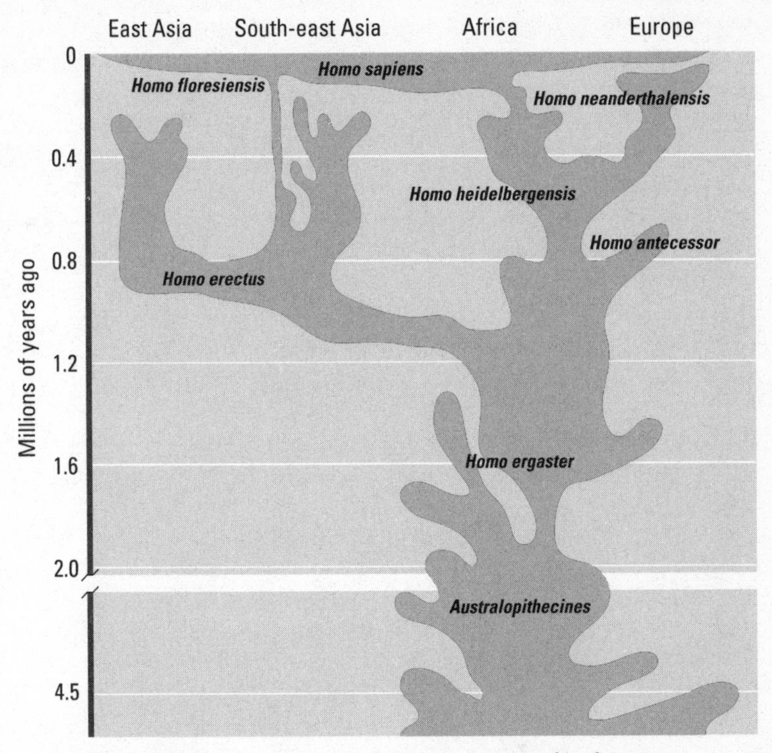

Figure 1 Hominid phylogeny, referring to the species mentioned in the text.

explained by the evolutionary history I propose. The result is a complete account not only of how music and language evolved but also of how they relate to the evolution of the human mind, body and society. It will explain why we enjoy music, whether one's taste is for Bach, the blues or Britney.

Part One
The Present

2 More than cheesecake?

The similarities and differences between music and language

Is music no more than auditory cheesecake, as Pinker would have us believe? Is it simply an evolutionary spin-off from language – a lucky break for humankind, providing song and dance as a relief from the tedium of survival and reproduction? Or is this view itself simply a knee-jerk reaction to contrary claims that music is adaptive and as deeply rooted in our biology as language? Can either of these views be justified? How and why, if at all, are language and music related? The starting point for an answer to such questions is a careful consideration of their similarities and differences.

The issue is not simply the manifest nature of language and music, but the extent to which they rely on the same computational processes in the brain.[1] Our ultimate concern is with the evolved physical and psychological propensities that provide the capacities for language and music in our species, *Homo sapiens*. When dealing with such capacities, as much attention must be given to those for listening as to those for producing.[2]

I will begin this chapter with some uncontentious similarities, before addressing whether the three key features of language – symbols, grammar, and information transmission – are also found in music.

Problematic definitions

Bruno Nettl, the distinguished ethnomusicologist, defined music as 'human sound communication outside the scope of language'.[3] That is perhaps as good a definition as we can get. When I refer to 'music' I mean something that we, as members of a modern Western society, would recognize as music. But we must immediately appreciate that the concept of music is culturally variable; some languages lack words or phrases to encompass music as the total phenomenon that we understand in the West. Instead, they may have words for specific types of musical activities, such as religious song, secular song, dance, and the playing of particular instruments.[4] Even in the West, of course, music is partly in the ear of the beholder – although this statement itself runs into trouble with John Cage's 1952 composition, *4′ 33″*.[5]

The definition of language is, perhaps, more straightforward: a communication system consisting of a lexicon – a collection of words with agreed meanings – and a grammar – a set of rules for how words are combined to form utterances. But even this definition is contentious. Alison Wray, the champion of holistic proto-language, has argued that a considerable element of spoken language consists of 'formulaic' utterances – prefabricated phrases that are learnt and used as a whole. Idioms are the most obvious example, such as 'straight from the horse's mouth' or 'a pig in a poke'.[6] Unlike the other sentences within this paragraph, the meaning of such phrases cannot be understood by knowledge of the English lexicon and grammatical rules.

Wray and certain other linguists argue that the 'words and rules' definition of language places undue emphasis on the analysis of written sentences and pays insufficient attention to the everyday use of spontaneous speech, which often contains very little corresponding to a grammatically correct sentence.[7] Peter Auer and the co-authors of the 1999 book, *Language in Time*, would agree with this. They argue that traditional linguistics has neglected to study the rhythms and tempos of verbal interaction – the manner in which we synchronize our utterances when having a conversation. This is a fundamental and universal feature of our language use, and has an evident link with communal music-making.

Some cultures have forms of vocal expression that fit neither our category of music nor that of language. The most evident are the mantras recited within Indian religions.[8] These are lengthy speech acts and often sound very much like spoken language, but they lack any meaning or grammatical structure. They are transmitted from master to pupil and require not only correct pronunciation but also the correct rhythm, melody and bodily posture. Buddhist mantras therefore constitute a repertoire of fixed expressions that have not changed for centuries. They have characteristics of both language and music, but can be defined as neither.

Universal, but culturally diverse

Both music and language are known to exist in all extant human societies and all those that have been historically documented; archaeologists are confident that both music and language were present in all prehistoric societies of *Homo sapiens*.[9] While the concept of music may vary, all cultures have song and dance, and make some form of internal repetition and variation in their musical utterances; they use rhythmic structures based on distinctions between note lengths and dynamic stresses.[10]

The contexts in which music is used and the function it appears to play in societies are also highly variable, with entertainment, the validation of social institutions, and the fostering of social bonds being particularly widespread. But the most prominent and perhaps the only universal context is that of religion: music is used everywhere to communicate with, glorify and/or serve the divinities identified within any particular culture.[11]

Another form of universality is that found at the level of the individual rather than of the society or culture: with the exception of those who suffer from a cognitive deficit, all individuals have a capacity to acquire language and are born with an inherent appreciation of music. This is a claim I will substantiate and explore in chapter 6, where infant development is examined. Bruno Nettl summarizes the global situation: 'Evidently humanity has decided not only to make music but, despite the vast amount of variation among the musics of the world, to make it in a particular way.'[12]

The universality of music is, perhaps, more contentious than that of language because we place greater emphasis on production than listening, with many individuals declaring themselves to be unmusical. In this regard, John Blacking's comments, made in the 1970s, on the contradiction between theory and practice in the middle-class, Western society in which he grew up (High Church Anglicanism, public school, Cambridge), remain pertinent today. Music was and remains all around us: we hear it when we eat and try to talk in restaurants and airport lounges; it is played all day long on the radio; in fact, there are few occasions when someone is not trying to fill moments of potential silence with music. Blacking remarked that 'society claims that only a limited number of people are musical, and yet it behaves as if all possessed the basic capacity without which no musical tradition can exist – the capacity to listen and distinguish patterns of sound'.[13] He favoured the idea that there was no such thing as an unmusical human being, and noted that the existence of a Bach or a Beethoven was only possible because of the presence of a discriminating audience.[14]

While both language and music are found in all societies, and share some common features, the history of their study has been largely dominated by attempts to document and explain their diversity.[15] More than six thousand languages are spoken in the world today, this being a small fraction of the total number of languages ever spoken. The number of musics in the world is likely to be even greater and to display considerably more diversity.

Just as we have English, Chinese and Arabic languages, we have jazz, Blackfoot Indian and Tibetan chant musics. Like languages, musics have stylistic, geographical and social boundaries. They can be grouped into families; patterns of descent, blending and development can be traced. Such diversity and patterning in both languages and musics arises from the processes of cultural transmission from one generation to the next and from one society to another. This can make identifying the boundaries between languages and musics difficult, whether in a historical sense (when did Old English become Middle English, or Classical music become Romantic?) or in the contemporary world. Where, for instance, are the boundaries between folk, blues, gospel, country and jazz, let alone between old skool, electro, nu-electro, hard house, funky house, deep house and progressive house.[16]

There is a contrast between musics and languages regarding the extent to which they can be translated from one cultural form to another. If I listen to someone speaking a language with which I am unfamiliar I will have very little, if any, idea of what they are saying – especially if their language comes from an entirely different language family from my own, such as Japanese or one of the African so-called 'click' languages. I may pick up some idea of mood from facial expression, intonation and gesture, but I will have hardly any knowledge of content. If, however, I have a translator present, then those mysterious utterances I hear can be rendered in English for me to understand.

Music is quite different. Whereas we can translate Japanese not only into English but into any other language spoken in the world, though recognizing that we may lose much of the subtlety of the original in the process, it makes no sense to attempt to translate the music used by one culture into that of another, and there is no reason to do so.[17] As John Blacking noted back in 1973, this appears to suggest that while there is the possibility of a universal theory of linguistic behaviour, no such possibility exists for music. Blacking went on to discuss the Venda, people from South Africa with whom he lived during the 1950s, expressing views that match my own rationale for writing this book and which deserve quoting at length:

> Music can transcend time and culture. Music that was exciting to the contemporaries of Mozart and Beethoven is still exciting, although we do not share their culture. The early Beatles' songs are still exciting although the Beatles have unfortunately broken up. Similarly, some Venda songs that must have been composed hundreds of years ago still excite me. Many of us are thrilled by Koto music from Japan, sitar music from India, Chopi xylophone music, and so on. I do not say that we receive music in exactly the same way as the players (and I have already suggested that even members of a single society do not receive their own music in the same way), but our own experiences suggest that there are some possibilities of cross-cultural communication. I am convinced that the explanation for this is to be found in the fact that at the level of deep structures in music there are elements that are common to the human psyche, although they may not appear in the surface structures.[18]

Brains and bodies as well as voices

Language and music share three modes of expression: they can be vocal, as in speech and song; they can be gestural, as in sign language and dance; and they can be written down. In each case, they have a biological basis in the brain; cognitive pathologies can lead to aphasia, the loss of language, and/or amusia, the loss of music (see chapters 3 to 5). The first modality, vocalization, will be my prime concern throughout this book, while the third, writing, is a product of human history outside my chronological scope. The use of the body as an element of musical and linguistic communication is, however, of key importance. Indeed, one of my concerns is the relationship between music, language and the evolution of the modern human physique.

One of the most striking and under-researched aspects of music is bodily entrainment: why we start tapping our fingers and toes, and sometimes move our whole bodies, when listening to music. Indeed, to separate rhythmic and melodic sound from rhythmic and melodic movement – song from dance – is quite artificial, and throughout this book I will use the word 'music' to encompass both sound and movement. The easiest way to appreciate this is simply to recognize that song itself is no more than a product of movement, made by the various parts of the vocal tract reaching from the diaphragm to the lips.

Even when sign language is taken out of the equation, many would argue that it is equally artificial to separate language from gesture. Movements of the hands or the whole body very frequently accompany spoken utterances; indeed, several anthropologists have argued that spoken language evolved from gesture.[19] As with music, although some gestures that occur with spoken language are intentional, the majority are quite spontaneous. Speakers are often unaware that they are gesticulating and many find it difficult to inhibit such movements – in a manner similar to people's inability to stop moving their bodies when they hear music (as I will further consider in chapter 10).[20]

Acquisition and competence

The cultural transmission of a specific language from one generation to another seemingly happens by passive childhood acquisition: they just listen and learn. I say 'seemingly' because children certainly need to practise; they do not become competent in language unless they engage in meaningful exchanges with experienced language users. By the age of four they will have a lexicon of several thousand words and a suite of grammatical rules at their disposal, allowing them to compose their words into a huge range of utterances.[21] Moreover, irrespective of what language children are learning, they appear to pass through the same stages of language acquisition. While children gain knowledge of musical traditions in a similar fashion, it might seem that expertise in music is far more difficult to acquire – as those who have suffered piano, violin or singing lessons for years will testify (and their parents will confirm). We must, however, be cautious, because our perceptions are dominated by modern Western culture.

In 'traditional' societies, song is often far more pervasive in everyday life, and hence infant acquisition of musical knowledge may be far easier than it is in Western society. Indeed, if we were to place the emphasis on listening and informal singing and dancing,[22] rather than on the technically demanding playing of musical instruments, which is perhaps more akin to writing than to speaking, then the development of musicality might appear as natural as that of language. John Blacking's studies of music in cultures throughout the world led him to conclude: 'It seems to me that what is ultimately of most importance in music cannot be learned like other cultural skills; it is there in the body, waiting to be brought out and developed, like the basic principles of language.'[23]

Although many adults are bilingual, and some are fluent in many languages, the majority are primarily familiar with just one language, in which they are as competent in speaking as in listening. At least, that is our common perception, which is perhaps strongly influenced by the presence of large nation states unified by a single language. Bilingualism and even multilingualism may well have been more frequent in the past, and might be the current norm outside of the industrialized West. Even if this is the case, though, there is still a marked contrast with levels of competence in music: the majority of people will be familiar with a variety of musical styles, but will be far more limited when it comes to producing rather than listening. Few can compose a musical score and many (myself included) cannot hold a tune.[24] Yet this again may be a product of current Western society rather than of the human condition at large: it may reflect the relative unimportance of music in Western educational systems and the rather elitist and formalized attitudes towards music that have arisen in consequence.

Hierarchical structure, recursion and rhythm

Both language and music have a hierarchical structure, being constituted by acoustic elements (words or tones) that are combined into phrases (utterances or melodies), which can be further combined to make language or musical events. They can both be described as 'combinatorial systems'.

The manner in which such combinations are made often leads to recursion – the embedding of a linguistic or musical phrase within a phrase of a similar type, such as a clause within a clause. This enables a potentially infinite range of expressions to be generated from a suite of finite elements.

A recent review of human language in the prestigious journal *Science*, by the equally prestigious Harvard University team of Marc Hauser, Noam Chomsky and Tecumseh Fitch, concluded that recursion is the only attribute of language that lacks any parallel within animal communication systems, and that it therefore appears to have evolved recently – that is to say, in the *Homo* lineage after the time of separation from the common human–chimpanzee ancestor at 6 million years ago.[25] But Chomsky and his colleagues doubt whether recursion originated as part of a linguistic system. They suggest that it evolved as a means to solve computational problems relating to navigation, number manipulation or social interaction, and then became part of the language system as an evolutionary spin-off. What they do not notice is that recursion is one of the most critical features of music.[26] As Ian Cross has noted, by stressing the role of recursion, Chomsky's definition of language comes to be equally applicable to music.[27]

Conversely, rhythm is acknowledged as a key feature of music while little, if any, significance is usually attached to the rhythmic nature of language use. Peter Auer and his colleagues have argued that by neglecting the rhythmic interaction of speakers within a conversation we fail to address one of the most important features of language. They have analysed the temporal and rhythmic structures of many conversations, held in English, Italian and German, examining how it is possible for multiple speakers to engage in conversations with smooth transitions from one 'floor-holder' to another. They conclude that the initial speaker establishes a clear rhythmic pattern, and then 'unproblematic turn taking . . . is not simply a matter of the absence of overlapped (simultaneous) talk and/or "gaps" but, rather, depends on a sense of rhythmicity that makes it possible to predict when the first stressed syllable of the next speaker's turn is due'.[28]

Symbols and meaning

Although both music and language are hierarchical systems constructed from discrete units (words and tones), the nature of these units is fundamentally different: those of language are symbols; those of music are not.[29] By calling them symbols, I simply mean that the large majority of words have an arbitrary association with the entity to which they refer; the word 'dog' stands for, or refers to, a particular type of hairy mammal, but it no more looks or sounds like a dog than the words 'tree' or 'man' sound like a large woody-stemmed plant or the male form of the human species. Not all words are of this nature. Onomatopoeias are important in all languages, and later in this book I will consider sound synaesthesia – the manner in which some words appear to sound similar to the way their referents look, move or feel.[30] Nevertheless, one of the most remarkable achievements of infants is that they are able to learn the arbitrary meaning of a vast number of words within such a short space of time.

Musical notes lack referential meanings. Middle C is middle C and nothing else: it has no referent; it is not a symbol. High-pitched loud tones are often used to attract attention and suggest danger, as in fire alarms or police sirens, but this derives as much from the context of the sound as the tone itself; if heard in a concert hall they would have a quite different impact. Although some notes, and, more particularly, some sequences of notes, may create similar emotional responses in different individuals (as I will examine in chapter 7), there is no agreed convention describing how notes refer to emotions as there is for words and their meanings.

If meanings are attached to pieces of music, they are usually of a highly individual nature, perhaps relating to memories of past times when the same music was heard – the 'they're playing our tune, darling' syndrome. As Ian Cross has recently stated, 'one and the same piece of music can bear quite different meanings for performer and listener, or for two different listeners; it may even bear multiple disparate meanings for a single listener or participant at a particular time'.[31] This is most evidently the case with religious music; I am an atheist, so when I listen to the great eighteenth-century choral works that were written to glorify God, such as Handel's *Messiah* or Bach's *St Matthew Passion*, they 'mean' something quite different to me than they would to someone of religious faith. As John Blacking has explained, the music of Handel and Bach cannot be fully understood without reference to the eighteenth-century view of the world – just as northern Indian music cannot be fully understood outside the context of Hindu culture. So I, with limited knowledge either of eighteenth-century European culture or of Hindu culture, will inevitably gain something quite different from those musics from people listening to them within those respective cultures.[32]

The theme tunes for soap operas illustrate how some pieces of music can have a shared meaning within a single community. That for the *Archers*, a long-running BBC Radio 4 serial, will be instantly recognizable to a great many people throughout the UK and will be understood as a symbol for that radio programme – although the term 'sign' or 'index' is actually more appropriate.[33] They will not be able to listen to the tune without thinking of the *Archers*. But most people (I guess) from elsewhere in the world would have little inkling of this association, and would listen to the tune as a piece of music without any symbolic association.

This type of meaning derives from the piece of music as a whole, or at least a significant chunk of the repeated melody, rather than being constructed from individual meanings attached to the separate notes. Some of the meanings attributed to spoken utterances are the same. These are the prefabricated phrases referred to at the start of this chapter, such as 'don't count your chickens before they're hatched', which are understood in a holistic manner and are just as meaningless when decomposed into individual words or speech sounds as is the *Archers* theme tune when decomposed into individual notes.

It is also the case that we might come to understand the meaning of an utterance spoken to us in a foreign language by processing it holistically rather than compositionally. For instance, I have learnt that the Welsh phrase *mae'n flin 'da fi* means 'I'm sorry', but I do not understand its compositional structure – which sound means 'I am' and which means 'sorry' – in the way a Welsh-speaker would. These formulaic aspects of language suggest a greater similarity with music than might initially be apparent.[34]

Grammar in language and music

Language is compositional when words are combined into spoken utterances or written sentences by the use of rules – the grammar of the language concerned. To be more precise, this is achieved by the use of syntax, which is one of the three rule systems that constitute grammar. The two others are morphology – the manner in which words and part-words are combined to make complex words – and phonology – the rules that govern the sound system of a language.

Grammatical rules are special because they provide meaning to the phrase beyond that contained within the symbols themselves; they map word meanings into sentence meanings. Hence 'man bites dog' has a different meaning from 'dog bites man' because in English grammar the subject precedes the verb. Grammatical rules also provide the means by which the finite number of words within a language can be used to create an infinite number of sentences, partly by delivering the feature of recursion, as described above.

The acquisition of grammatical rules by children is remarkable, because the rules have to be abstracted from the utterances they hear. Indeed, when writing in the 1950s, Noam Chomsky found it inconceivable that children could learn language in the same manner as they learn to draw, play a musical instrument or ride a bicycle. He argued that the quantity and character of the utterances children hear from their parents, siblings and others would simply provide insufficient information for them to extract the rules of grammar so that they could produce their own unique but meaningful utterances. This was referred to as the 'poverty of the stimulus' argument. It led to the idea of a 'Universal Grammar' – an innate set of predispositions that is universal among all *Homo sapiens* and provides a form of head start in acquiring grammar. This implies that there are some rules of grammar that are shared by all languages and are simply manifest in different forms. When an infant listens to its parents and other language users, certain 'parameters' of 'Universal Grammar' are set in the manner appropriate for the particular language being acquired.

After Chomsky made his proposals about 'Universal Grammar', musicologists began searching for the musical equivalent.[35] They did not get very far. Grammars were produced for specific musics, such as Swedish folk songs, and were able to generate new pieces of music with the same essential structure. Indeed, by definition, all musical styles accord to a set of rules for what is acceptable within that style. If the rules of a musical style are not followed, then the piece of music may sound as awkward as a spoken utterance that breaks the rules of grammar. Most of us in the West, for instance, have developed an intuitive and shared 'tonal knowledge' of music and can tell when a suite of musical tones do not conform; they will simply sound 'wrong', as will be explained in chapter 4. The description of such tonal knowledge rules (see below, pp. 51–2) is as obtuse to most readers as would be a description of the linguistic grammar that we equally effortlessly acquire and apply.

The most ambitious attempt at elucidating the grammar of music was that of Fred Lerdahl and Ray Jackendoff in their 1983 book, *A Generative Theory of Tonal Music*.[36] They argued that knowledge of tonal music relies on a musical competence that is innate – a musical equivalent of 'Universal Grammar'. But few musicologists have accepted that this musical competence is equivalent to a grammar like that of language. As Douglas Dempster, a philosopher of music, has recently explained, many systems are governed by rules (for example, the game of chess, or cooking recipes) and by routines (as when undertaking various computer tasks, or setting a video recorder), but we gain little, if anything, by describing these as 'grammars'.[37] The same applies to music.

The rules of a musical style and the rules of a language are profoundly different. Those of music do not provide meaning in the same way that linguistic grammar provides meaning to language. While 'man bites dog' is fundamentally different from 'dog bites man', reversing the order of three notes is of far less significance to a piece of music. It may make the piece sound awkward, but the reversal cannot be said to change its meaning, because there was none to change in the first place.

Change through time

A second major difference between the rules of a musical style and the grammar of a language is the rate of change. Dempster suggests that because we use language to communicate and need to understand a speaker's intentions, there is enormous pressure on the grammatical rules of a language to remain stable. The pressures on the rules of music are quite different, because we value deviation within musical styles – although, once again, this may be a view that is biased by the experience of modern Western music and would be contradicted by the degrees of conformity and stability found in other musical traditions. Nevertheless, Dempster argues that there are enormous aesthetic pressures for the rules of musical styles to remain comparatively unstable in comparison with those of languages.[38]

Languages do, of course, change through time. Pronunciations evolve, new words are invented or imported from other languages, the meanings of words drift (epitomized by the way in which the word 'gay' has changed its meaning within the last two or three decades). Language is used as a means of asserting social identity – by the adoption of a particular slang or jargon, for example – and hence as the structure of society changes so will the character of its language. Such change, however, is relatively slow, because effective communication between generations and social groups must be maintained. It most probably takes as much as a thousand years for a language to evolve into an unintelligible form.

The speed at which music can evolve is readily apparent from the development since the early 1970s of rap, part of a more general subculture of hip hop which includes break-dancing, 'cutting and scratching' records and graffiti. It originated in the Bronx and is often identified with the arrival of the Jamaican DJ known as Kool Herc, who began to chant over the instrumental sections of the day's popular songs. From that point, rap underwent swift development because there were very few, if any, rules other than rhyming in time to the music; one could rap about anything, with a premium on being original and injecting one's own personality. In the mid-1980s rap, along with hip hop in general, became mainstream, while the 1990s saw the rise of political and then gangsta rap. Artists have been prepared to borrow sounds from many disparate sources including folk music, jazz and television news broadcasts, and with the advent of the Internet rap is now a truly global musical phenomenon. And what is particularly interesting to note here is that the slang of hip hop culture has influenced language at large, becoming part of the standard vocabulary among many groups of various ethnic origins.

Information transmission, reference, manipulation

Language, whether spoken, written or gestural, is principally used as an intentional means to communicate ideas or knowledge to one or more other individuals, with a clearly defined 'sender' and one or more 'receiver(s)' of the message. Perhaps the most remarkable thing about language is how suited it is to this role; language is used to communicate about virtually everything, from our most intimate feelings, through daily and routine events, to theories about how the universe originated.[39]

In this regard, language is fundamentally different from music. According to Ian Cross, a great deal of music lacks any clear communicative function: 'the members of a recreational choir or of an amateur rock band may rarely, if ever, fulfil the role of performer; for them, music may be more a medium for participatory interaction where all are equally and simultaneously performers and listeners ... similarly if we look to non-western musical practices many seem to have as their raison d'être not the transmission of musical information from the active performer to the passive listener but collective engagement in the synchronous production and perception of complex patterns of sounds and movement'.[40]

John Sloboda, a distinguished music psychologist, has explained that 'we use language to make assertions or ask questions about the real world and the objects and relationships in it. If music has any subject matter at all, then it is certainly not the same as normal language.'[41] This does not exclude the possibilities that music can be used to tell stories or make reference to the real world.[42] Sloboda cites the case of someone who has been asked for his response to a recently heard symphony. It is likely to contain not only some description ('it was loud') and some evaluation ('I liked it') but also some characterization of the music through metaphor ('it had the feeling of a long heroic struggle triumphantly resolved').[43] People, Sloboda explains, nearly always have some comment to offer, even though they may arrive at quite different characterizations for the same piece of music.

Such responses to music may have been inspired by the use of mimicry – woodwind 'birdsong' to evoke a pastoral scene, or glissandi violins to suggest the howling wind of a storm. The reference is, therefore, achieved by using musical notes and phrases not as symbols but as signs or even icons. In the rare instances when symbolic reference is found, this tends to be to events outside of the music itself and to be reliant on linguistic and/or visual input. Sloboda provides the example of an opera where a particular musical theme is associated with the appearance of the hero. It might then be used in his absence to signify that, say, the heroine is thinking of him. This could be described as symbolism, although the music is really acting as no more than an index or sign of the hero's appearance.

In summary, music is a non-referential system of communication. Yet although a piece of music does not 'tell' us anything about the world, it can have a profound impact on our emotions – for instance, by making us feel happy or sad (as I will consider below and in chapter 7). It can also make us move, by the phenomenon of entrainment. We can therefore describe music as 'manipulative' rather than referential in character. Language is often thought to be principally referential because it tells us things about the world; often, however, when we are told something we are driven to action, and so language, too, might be portrayed as being manipulative.

The cognitive challenge

Language can be characterized as a means of expressing thoughts. Such thoughts are generally believed to exist independently of language and could be entertained by a non-linguistic or even a prelinguistic human. Once again, we must be cautious, because some aspects of human thought, especially those concerned with abstract entities, might be dependent upon the possession and use of language,[44] rather than the other way round (as we will explore later, in chapter 16). Also, there is a great risk of attributing human-like thoughts to other animals simply because we have difficulty in imagining minds without such thoughts.

Because most utterances are expressions of thoughts, it is those thoughts, rather than the words themselves, that a listener wishes to understand. This often involves considerable inference and listeners will employ a great deal of additional information in order to garner the meaning of an utterance – such as what they know about the person speaking, the entities being referred to, and the intonation of the utterance. When asked to repeat what someone has said, a listener will rarely do this verbatim, but will instead express what they understood to be the thought behind the utterance.[45] In light of this, language appears to be dependent upon another cognitive skill, that known as 'theory of mind'. This is the ability to understand that another individual may have beliefs and desires that are different from one's own. Apes may lack this ability, which may constrain their ability to acquire language – as is the case with very young children before their theory of mind abilities have developed.[46]

Music appears to make fewer cognitive demands. Listeners are not required to infer the thoughts and intentions of the performer or composer – although they may attempt to do so, and this may enhance their experience. You can simply sit back and let the music 'wash over' your body and mind without having to concentrate or actively listen at all. Indeed, we often have music playing in the background while we make conversation at dinner or engage in other activities (I have J. S. Bach's Cello Suites playing in the background as I write today; yesterday it was Bob Dylan). Nevertheless, even such background music has a cognitive and physiological impact: just as we automatically infer a speaker's thoughts when listening to a spoken utterance, and to some extent come to share those thoughts, so we automatically have emotions aroused within ourselves while listening to music. In fact, music often manipulates our mood.

The 'language' of emotion

Both music and language have the quality of expressive phrasing. This refers to how the acoustic properties of both spoken utterances and musical phrases can be modulated to convey emphasis and emotion.[47] It can apply either to a whole utterance or phrase, or to selected parts. The word 'prosody' refers to the melodic and rhythmical nature of spoken utterances; when the prosody is intense, speech sounds highly musical. Prosody plays a major role in the speech directed towards infants; indeed, whether the utterances 'spoken' to very young babies should be considered as language or as music is contentious – I will explore this overlap in chapter 6.

Although the content of language can be used to express emotion, it is subservient to the prosody. I can, for instance, state that 'I am feeling sad'. The words alone, however, may be unconvincing. If I say that 'I am feeling sad' in a really happy voice, priority will be given to the intonation and the inference drawn that I am, for some unknown reason, being ironic when I claim to be feeling sad.

Emotional expression is more central to music than to language. If I listen to the song-like cry of a mother who has lost her child, I can more easily appreciate her grief than if she simply tells me that her child has died and that she feels distraught. I may well be moved to tears by her dirge. I can, of course, listen to exactly the same song performed by an actress in a film and experience the same feeling of empathetic grief while knowing that the actress is not really experiencing such an emotion at all. Similarly, I can listen to a requiem mass on a CD and come to feel mournful, even though I would not necessarily attribute the same emotions to either the musicians or the composer, all of whom may be long dead, let alone the CD player from which the music emanates.

The psychologist Roger Watt believes that music itself acts as a virtual person to whom listeners attribute not only emotional states but also a gender and personality characteristics such as good or evil.[48] John Sloboda has made a similar proposal. He argues that when we experience music we gain an impression of tension and resolution, of anticipation, growth and decay. Such 'dynamic awareness', as he describes it, 'involves reading the music as an embodiment of something else . . . [which is] broadly, the physical world in motion, including that very special sub-class of moving objects, the living organism'.[49] The most obvious example of this is the manner in which so much music sounds – and has often been deliberately composed and/or performed to sound – as if a human conversation is taking place.

Music is remarkably good at expressing emotion and arousing emotion in its listeners – a fact captured in the popular notion that music is the 'language of the emotions'. The emotional power of music has long been the subject of academic study, marked by two seminal books of the 1950s, Leonard Meyer's *Emotion and Meaning in Music* (1956) and Deryck Cooke's *The Language of Music* (1959).[50] While Meyer explained how music might represent the emotions, Cooke proposed relationships, which are universal across all Western tonal music, between certain musical phrases and certain emotions. There has been a great deal more recent research on music and emotion, some of which will be examined, along with Cooke's proposals, in chapter 7.

This feature of music appears incompatible with Pinker's view that music is no more than a spin-off from language and has no biological value. We don't have emotions for free or for fun: they are critical to human thought and behaviour, and have a long evolutionary history. As I will explore later, our basic emotions – anger, joy, disgust and sadness – are most probably shared with our primate relatives and are closely related to the character of their vocalizations. Emotions are deeply entwined with the functioning of human cognition and physiology; they are a control system for body and mind. It is most unlikely, therefore, that our deepest emotions would be so easily and profoundly stirred by music if it were no more than a recent human invention. And neither would our bodies, as they are when we automatically begin tapping our fingers and toes while listening to music. In fact, even when we sit still, the motor areas of our brain are activated by music.[51]

Summary: shared or independent evolutionary histories?

Music and language are universal features of human society. They can be manifest vocally, physically and in writing; they are hierarchical, combinatorial systems which involve expressive phrasing and are reliant on rules that provide recursion and generate an infinite number of expressions from a finite set of elements. Both communication systems involve gesture and body movement. In all of these regards, they may well share what Douglas Dempster calls some 'basic cognitive stuff'.

Yet the differences are profound. Spoken language transmits information because it is constituted by symbols, which are given their full meaning by grammatical rules; notwithstanding formulaic phrases, linguistic utterances are compositional. On the other hand, musical phrases, gestures and body language are holistic: their 'meaning' derives from the whole phrase as a single entity. Spoken language is both referential and manipulative; some utterances refer to things in the world, while others make the hearer think and behave in particular ways. Music, on the other hand, is principally manipulative because it induces emotional states and physical movement by entrainment.

So where does this leave us with regard to the relationship between music and language? Is music no more than auditory cheesecake, as Steven Pinker would have us believe? The answer must be probably not, because the material I have discussed in this chapter indicates that music is too different from language to be adequately explained as an evolutionary spin-off. What, then, are the alternatives?[52]

Well, the relationship might be the precise converse to that which Pinker suggests: language might be derivative of music. This hypothesis also appears unlikely, for precisely the same reason: it does not explain the unique properties of language. Another alternative is that music and language evolved in parallel to each other, as completely separate communication systems. This is also unconvincing: while music and language have their own unique properties, they still share more features than one would expect from entirely independent evolutionary histories.

The remaining possibility is that there was a single precursor for both music and language: a communication system that had the characteristics that are now shared by music and language, but that split into two systems at some date in our evolutionary history.

In a recent essay, the musicologist Stephen Brown favours this view and describes the single precursor as 'musilanguage'.[53] He believes that this formed a type of ancient communication system used by human ancestors, although he makes no reference to human evolution and provides no indication of when he believes that 'musilanguage' might have existed. According to Brown, at some unspecified date in the past, 'musilanguage' became differentiated into two separate and specialized systems, each of which then acquired additional unique properties. One became music while the other became language.

Stephen Brown's notion of 'musilanguage' is, I believe, essentially the same as Alison Wray's notion of holistic proto-language, mentioned in my introductory chapter – an example of convergent intellectual evolution in the fields of musicology and linguistics. From my reading of their work, neither Brown nor Wray has fully appreciated the true significance of their separate insights. It is my task in this book to reveal that significance – not only to explain the origin of music and language, but also to provide a more accurate picture of the life and thought of our human ancestors. With the idea of a single precursor for music and language we do, in fact, return to the ideas of Jean-Jacques Rousseau in the *Essai sur l'originie des langues*, where he reconstructed the first language as a kind of song.[54]

Neither 'holistic proto-language' nor 'musilanguage' is a user-friendly or particularly informative term; indeed, the use of the part-word 'language' within these terms is misleading. So I will use an alternative term to refer to the single precursor to music and language: 'Hmmmmm'. This is, in fact, an acronym, and its meaning will become apparent as my book develops. My next task is to continue exploring the relationship between music and language by another means. Rather than looking at what people say, sing and do, it is time to look at what is happening within the brain.

3 Music without language

The brain, aphasia and musical savants

The human brain is the most complex organ known to biology and the second most complex entity in the universe – the first being the universe itself. A human brain can be held within the palm of one hand and has the consistency of uncooked egg. And yet it contains a hundred billion nerve cells, called neurons. Each neuron is connected to many other neurons via tentacle-like extensions, known as axons and dendrites. The contact between an axon from one neuron and a dendrite from another is called a synapse. Scientists estimate that there are between ten and one hundred trillion synapses within the brain – that's ten followed by either nineteen or twenty zeros – and six hundred million synapses per cubic millimetre.[1]

When a neuron is activated it generates an electrical signal which passes down its axon and triggers the release of a chemical from its end, known as a neural transmitter. This diffuses across the gap, called the synaptic cleft, that separates the axon from the dendrite of an adjacent neuron. That dendrite has chemical receptors for the neural transmitter, and when the transmitter reaches them the dendrite becomes activated and an electrical signal spreads along its own neuron.

We are only able to speak and sing because of the manner in which neurons are connected to each other and the brain activity they promote – although quite how the firing of neurons and the release of chemicals within the brain become a thought or a sensation remains unclear. Sets of connected neurons relating to specific activities are termed neural networks. The key question that we must address here is whether the same neural networks are used for language and for music. Perhaps there is a partial overlap, perhaps none at all. If there is an overlap, did those shared neural networks originate for language, for music, or for something else entirely?

Inside the brain

These neural networks are located in the cerebral cortex, the most recently

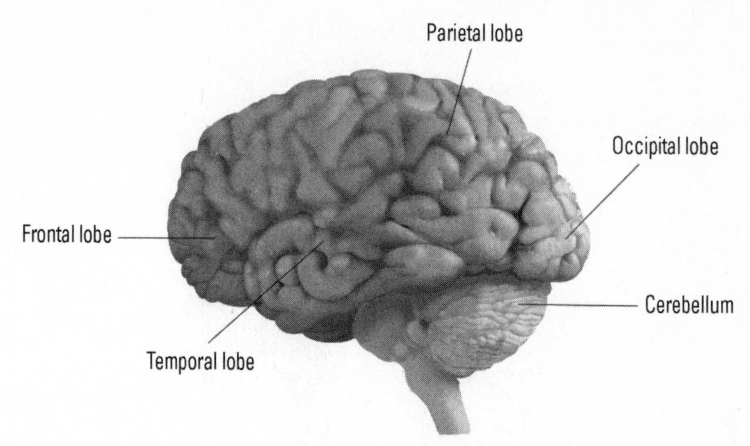

Figure 2 The four lobes of the human brain and the cerebellum.

evolved part of the brain, which expanded dramatically during human evolution.[2] The brain consists of a suite of distinct anatomical elements, all of which are present in other mammals, although their relative size and significance varies markedly. It is dominated by the cerebrum, which is divided into the left and right cerebral hemispheres, which have heavily convoluted or folded surfaces. Each of the 'hills' of the folds is referred to as a gyrus (plural gyri), and each of the 'valleys' is called a sulcus (plural sulci). The cerebrum overlies the stalk-like brainstem which connects the brain to the central nervous system, which extends throughout the body.

The two cerebral hemispheres are composed of an outer layer of grey matter, known as the cortex, and an inner bulk of white matter. The grey matter is dense and consists of neuron cell bodies – the brain cells. Their axons are often covered in a white sheath, composed of a substance called myelin, and they are the main constituent of the white matter. About 90 per cent of the bulk of the human cerebrum is contained in its outer layer, which is called the neocortex; the word literally means 'new bark'. This outer layer is between 1.5 and 4 millimetres thick and is composed of a mere six layers of cells; yet because it is so heavily folded the surface area of the neocortex is substantial and it contains between ten and fifteen billion neurons.

The cortex is divided into four areas known as lobes. Specific regions of each lobe are designated by the terms 'superior', 'inferior', 'anterior' and 'posterior', which refer to their proximity to the crown of the skull (rather than to their significance in brain activity). The lobe nearest the front of the skull is called the frontal lobe, and the foremost (anterior) area of it is called the prefrontal cortex. This is known to be important for tasks such as planning

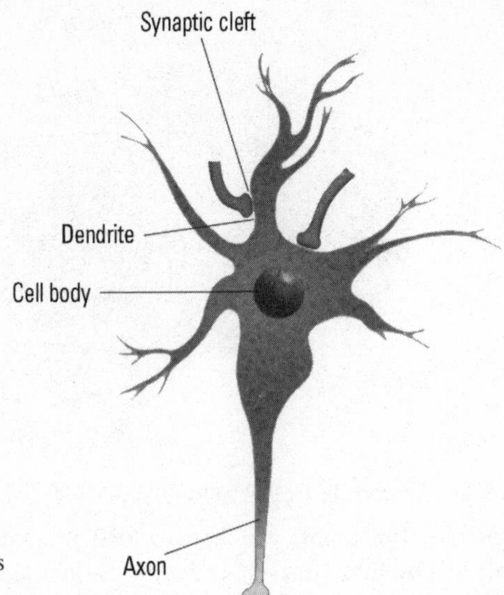

Figure 3 A neuron cell, each of which can be connected with thousands of others.

and problem-solving – tasks that require the integration of different types of information. The rear (posterior) part of the frontal lobe consists of the motor area of the brain, the place where neurons that control movement are clustered.

The frontal lobe is demarcated from the smaller parietal lobe, which lies directly behind it, by the central fissure, a particularly deep sulcus. The parietal lobe is divided between the left and right sides of the brain, and contains the primary sensory cortex which controls sensations such as touch. The parietal lobe is in turn separated from the temporal lobes, one on each side of the brain, by another deep sulcus known as the Sylvian fissure. The temporal lobes are found at about the level of the ears and are involved in short-term memory and, of most significance for our study, in the processing of auditory stimuli. Finally, at the rear of the brain, but not demarcated by any major fissure, is the occipital lobe. This is where the brain processes visual information.

Studying the brain

There are two main ways in which the brain is studied.[3] One is brain imaging. Since the 1970s it has been possible to use various methods to study both brain anatomy and the patterns of brain activation that arise when specified tasks are undertaken. The methods have rather daunting names: computerized tomography (CT), positron emission tomography (PET), functional magnetic resonance imagining (fMRI), electroencephalography (EEG), and magnetoencephalography (MEG). Although such methods lie at the cutting edge of scientific research, their underlying principles are quite straightforward. CT is like a sophisticated X-ray; it records what the brain would look like if it were surgically removed, cut into slices and laid out on a table. PET effectively measures blood flow in the brain and hence identifies which parts of the brain are currently active by making use of the fact that active neurons require more oxygen. fMRI also measures blood flow but does so by exploiting minute changes in the magnetic properties of blood haemoglobin molecules as they release oxygen into the brain. Active neurons also generate electrical signals, which were once termed 'brainwaves', and these provide the basis for EEG; by placing electrodes on the scalp it is possible to detect the location within the brain from which the electrical signals are emanating. These active areas in the brain also produce tiny magnetic fields, which can be measured by MEG.

These methods have already revolutionized our understanding of the brain; one suspects that further advances will soon be made. From them we are beginning to acquire a detailed understanding of which parts of the brain are used for which activities – for language, music, vision and so forth. We are also beginning to understand the plasticity of the brain – how new neural networks can develop to replace those that have been damaged and thus restore lost functions to the brain.

The fact that different functions are to some extent localized in different parts of the brain has been known for more than a hundred years, thanks to the second method of brain science, lesion analysis. A lesion is an abnormal disruption of tissue produced by injury, stroke or disease. If the lesion is survived – they often prove fatal – it is possible to monitor what types of sensory, cognitive or motor activities have been lost and then to form hypotheses about the role of the particular part of the brain that has become inactive due to the lesion. Before the days of brain imaging, scientists had to wait for the patient to die and then perform an autopsy in order to discover where the lesion had occurred. Today, brain-imaging techniques can locate the lesion while a battery of cleverly designed experimental tests is used to identify the specific deficit that has arisen. It is this combination of lesion analysis and brain imaging that has provided the greatest insights into the neural bases for music and language.[4]

Lesions and language

The very first use of lesion analysis identified a key locality for language processing within the brain. Paul Broca, born in 1824, was a distinguished professor of surgical anatomy in Paris, and secretary of the Anthropological Society of Paris between its establishment in 1859 and his death in 1880. In

April 1861 he examined a fifty-one-year-old man, named Leborgne, who had spent twenty-one years in a hospice. This patient was also known by the name 'Tan', as this was the only word he could speak, even though he appeared to understand all that was said to him and to be of normal intelligence. When Tan died, a mere six days later, Broca removed his brain, stored it in alcohol and subjected it to detailed examination. He concluded that Tan's language deficits had arisen from damage to the third convolution of the frontal lobe of the left cerebral hemisphere – henceforth known as Broca's area. The presentation of his conclusions towards the end of 1861 began the modern investigation of the biological basis of language. Today, Tan's brain remains preserved in the anatomy museum of the École de Médecine of Paris, and further studies of lesions, as well as brain imaging, have confirmed that Broca's area is indeed essential for the production of speech.

Tan's language deficit – an understanding of language but an inability to speak – is known as Broca's aphasia. A second type of language deficit is Wernicke's aphasia, identified in 1874 by Carl Wernicke, a twenty-six-year-old neurologist. This is quite different: sufferers are able to speak fluently but their utterances are meaningless, a jumbled assortment of words. Moreover, they lack comprehension of any words spoken to them beyond the simplest of instructions. Wernicke discovered that this deficit arose from damage to a part of the left cerebral hemisphere quite separate from Broca's area, the superior surface of the anterior left temporal lobe – now known as Wernicke's area. In those of us with normal linguistic abilities, there must be a web of neural connections between Wernicke's and Broca's areas to allow us to comprehend a question and then make an appropriate verbal response.

There are numerous ways in which lesions can impact on language. Some people suffer such extensive lesions to their brains that they lose the whole of their language abilities, and they are referred to as suffering from global aphasia. At the other end of the scale some experience more specific difficulties, such as not being able to find the appropriate words when speaking, a condition known as anomic aphasia. Despite this variability, as research progressed throughout the late nineteenth and twentieth centuries, a consistent finding was that language deficits arose from lesions to the left cerebral hemisphere. So it was within this part of the brain that the neural networks for language appeared to be located, most notably in those areas named after Broca and Wernicke. This, however, is a simplification of a more complex reality, as we now know that the language processing system is in fact widely distributed throughout the brain.

Medical case studies of people who have suffered from aphasia provide us with the perfect opportunity to examine the neural relationships between music and language. If, for example, music was derivative of language, or

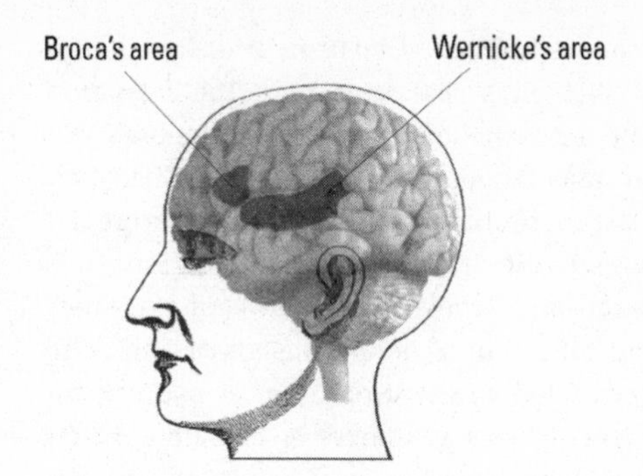

Figure 4 The language areas discovered by Paul Broca and Carl Wernicke.

vice versa, then the loss of musical ability ought to be an automatic consequence of the loss of language. Alternatively, if music and language rely on entirely independent neural networks, then the loss of one should have no impact upon the other. We now turn, therefore, to consider the impact of aphasia on musical ability.

'A brilliant creative work'

We begin with a musician, someone who was in the first division, if not the premier league, of twentieth-century composers: Vissarion Yakovlevich Shebalin. Born in 1902, he composed his first symphony while still a schoolboy and went on to be elected professor of the Moscow Conservatoire at the age of forty, becoming tutor to many well-known Russian composers. He taught composition and wrote symphonies, pieces for the pianoforte and operas, one of which was performed at the Bolshoi Theatre in Moscow.

In 1953, at the age of fifty-one, Shebalin suffered a mild stroke in his left temporal lobe which impaired his right hand, the right side of his face and disturbed his speech. Shebalin recovered from these symptoms within a few weeks, continued his work and progressed to become director of the Moscow Conservatoire. Then, on 9 October 1959, he suffered a second and more severe stroke. He lost consciousness for thirty-six hours; when he recovered his right side was partially paralysed and his linguistic capabilities had almost entirely disappeared. Shebalin's physical abilities returned over the next six months, but he continued to have problems talking and understanding what was said to him. After experiencing two epileptic fits, he died from a third stroke on 29 May 1963. A few months prior to that he had completed his fifth symphony, described by Dmitri Shostakovich as 'a brilliant creative work, filled with highest emotions, optimistic and full of life'.

Professor A. Luria and his colleagues from the Department of Psychology of Moscow University reported Shebalin's case history in the *Journal of Neurological Science* in 1965.[5] To them it provided evidence that the brain has quite separate systems for music and language. The extent of Shebalin's aphasia while he composed his last orchestral works was indeed severe. He was quite unable to repeat sentences, while straightforward instructions such as 'point to your nose' had to be repeated several times and kept extremely simple if they were to have any effect. In general, Shebalin's ability to comprehend speech had been devastated by his stroke. Just as bad was his own speech; he was left unable to construct sentences or to name objects when there were more than two of them present, even if prompted with the beginnings of the words. He could still read and write short words, although when he was tired even these proved too much for him.

Shebalin was aware of his difficulties. Luria quotes him as saying, 'The words ... do I really hear them? But I am sure ... not so clear ... I can't grasp them ... Sometimes – yes ... But I can't grasp the meaning. I don't know what it is.' And yet, despite such severe afflictions, Shebalin continued to compose and to teach his pupils by listening to their compositions, analysing and correcting them. He completed musical pieces that he had begun before his illness and created a series of new works described as being equal to anything he had produced in his earlier years. Indeed, Luria cites no less than eleven major pieces published between 1959 and 1963, including sonatas, quartets and songs, in addition to that fifth symphony which was so much admired by Shostakovich.

'For whom the spell extols'

One could argue that professional musicians such as Shebalin are quite untypical of humankind. The vast number of hours spent listening to, playing and composing music, especially during childhood, might have led to the development of neural networks devoted to music that would otherwise be quite absent. Indeed, it is known that the left planum temporale region of the brain, located in the posterior superior temporal gyrus, is larger in musicians than in non-musicians.[6] It is also the case, however, that musicians who received training before the age of twelve generally have a better memory for spoken words than non-musicians, which makes Shebalin's loss of language abilities even more striking.[7] Nevertheless, for those of us who are non-musicians the 'music as cheesecake' theory might still be correct: our musical abilities may simply be dependent upon neural circuits that evolved for language.

In this light, the case of NS (it is, by the way, usual practice for medical patients to be referred to by their initials alone), a sixty-eight-year-old male stroke victim, is particularly interesting.[8] His stroke occurred while under anaesthetic during heart surgery. When he awoke, he was unable to understand what people were saying, explaining that they seemed to be 'speaking too fast or in Chinese'. His own speech, however, had not been affected and he could still read and write. MRI indicated a lesion in the right temporoparietal region along the superior temporal gyrus.

NS's inability to comprehend speech was severe and permanent. His wife resorted to writing him notes because he could not follow her conversation; very soon he was reliant on writing for all communication. Twelve years after his stroke he underwent formal examination by Professor Mario Mendez of the Department of Neurology and Psychiatry at the University of California, Los Angeles. NS was highly talkative but quite unable to understand simple spoken requests such as 'touch your chin', although he could immediately understand when they were written down. He could repeat single words but struggled with phrases. For example, 'for whom the spell extols' was his best effort at repeating 'for whom the bell tolls'.

An added complication was that NS had also lost his ability to identify environmental sounds – those that we would intuitively think of as lacking any musical qualities.[9] These had become indistinct and difficult to understand, so that he would mistake an alarm clock for a man snoring, a fire engine for church bells, or a choir for a children's playground. With such difficulties, NS's quality of life had seriously declined. But some compensation was forthcoming: his appreciation of music had been enhanced. Before his stroke music had been of little interest to him, but in the twelve years between the stroke and Mendez's examination NS and his wife had become avid attenders of concerts and other musical performances. In fact, listening to music had become NS's main activity. During his interviews with Mendez, he often broke into song to endorse his new-found love of music.

While NS had lost his ability to understand simple phrases and environmental noises, he could still recognize melodies. When asked to identify ten common melodies, such as 'Happy Birthday', 'Auld Lang Syne' and the Christmas carol 'Silent Night', he often got the names wrong but was able to sing them all and to point to pictures illustrating the appropriate settings for the melodies. Similarly, he was able to discriminate between different rhythms; he could identify whether successively played examples were the same or different. He was also quite able to hum the rhythms and tap them out on a table.

NS's is a particularly interesting case for several reasons. First, his exposure to music as a child and young man had been limited. It had certainly been no more, and perhaps a little less, than that of the average person, so there is no reason to suppose that he would have developed unusually specialized musical circuits in his brain which might have preserved musical ability after the loss of language. Secondly, his appreciation of music appears actually to have been enhanced when his auditory capacities for speech and environmental sounds became severely disrupted. Thirdly, while a right temporal lesion has often been recorded as impairing the ability to comprehend environmental sounds, NS's was the first reported case in which this had also caused speech aphasia. Unlike the vast majority of people, he must have been reliant on a neural pathway for language in the right rather than the left cerebral hemisphere. But whatever neural circuits had been damaged, these had evidently been quite separate from those used for recognizing melody and rhythm.

Environmental sounds, foreign languages and prosody

Shebalin's and NS's case histories are unusual but not unique, as the next studies will illustrate. These introduce a couple of extra complicating factors: foreign languages and prosody.

NS had lost his capacity to identify environmental sounds such as ringing bells and barking dogs; whether this was also the case for Shebalin was not reported. It was not so for the three people in the following studies, although they were otherwise similar to NS in having preserved musical and inhibited language abilities following brain damage. They concern native speakers of French, Arabic and Japanese respectively, who varied in the extent to which they could recognize languages other than their own after their affliction. They also varied in their abilities to process the prosody of sentences, to identify their emotional moods, and to distinguish whether they were questions, exclamations or statements on the basis of the intonation of particular words. None of the subjects has been identified by either name or initials in the scientific literature.

On 16 September 1980, a twenty-four-year-old woman was admitted to the Clinique Neurologique of the Hospices Civils in Strasbourg following the sudden onset of difficulties with language.[10] Her speech had become garbled and she was quite unable to understand that of others, although she could lip-read isolated words and remained able to read simple sentences. Within a few days her abilities to speak, read and write returned almost to normal, but her inability to understand speech persisted.[11]

During the week after the woman's admission, Drs Marie-Noelle Metz-Lutz and Evelyne Dahl examined her auditory abilities. She proved quite able to recognize non-verbal sounds, correctly associating each tape-recorded sound with an appropriate picture. She was also able to recognize musical instruments from their sounds and to identify familiar melodies. Her success was greater when the melodies were hummed rather than sung, suggesting that the words hindered her recognition. But one aspect of her musical abilities was particularly affected, that of rhythm. When asked to tap a pencil against the table, she could only imitate very simple and short rhythms.[12]

This patient's language deficits showed an interesting pattern. Although she could no longer understand words – she was completely word-deaf – some of her language-processing abilities remained. When asked to identify whether each of twenty short sentences had been spoken in French (her native language), German, English or Spanish, she did so correctly in every case, without hesitation. She was able to identify on the basis of intonation whether a sentence had been spoken as a question, a statement or an exclamation, even though she could not understand the sentence itself. She could discriminate nonsense words from real words, although she found the former much more difficult to repeat, only correctly repeating twelve out of fifty nonsense words in contrast to thirty-six out of fifty real words.

Drs Metz-Lutz and Dahl explained the patient's particular suite of retained and lost language abilities in the face of complete word-deafness by the location of her brain lesion in the left cerebral hemisphere. EEG had suggested that the anterior and middle temporal lobe was impaired in its function, while a CT scan made one year after the onset of the patient's difficulties indicated left temporal hypodensity and a vascular lesion in the left middle cerebral artery. The processing of intonation, Metz-Lutz and Dahl argued, takes place in the right cerebral hemisphere and was therefore not inhibited. Her recognition of foreign languages may have relied as much on the typical intonation of the language as on the words themselves. The right hemisphere must also, they argued, be able to undertake 'lexical decisions' – that is, discriminating between words and non-words.

The next case was reported by Basim Yaqub of the King Khalid University Hospital in Riyadh, Saudi Arabia.[13] A thirty-eight-year-old Arabic-speaking Syrian male was admitted to the coronary care unit. Brain scans showed a lesion in the left cerebral hemisphere, in the posterior superior temporal gyrus, and a more extensive lesion in the right cerebral hemisphere, in the posterior superior and midtemporal gyrus.

On the third day after admission, it was noticed that he could not respond appropriately to verbal commands, although he himself spoke fluently and sensibly. He explained that he had a bad cold, which was interfering with his hearing. However, when asked questions, his answers were grammatically correct but entirely irrelevant. Yaqub and his colleagues formally tested the man's language and found that he had entirely lost his ability to comprehend speech.

Unlike the Frenchwoman described above, this Arabic man was no longer able to discriminate between his own and other spoken languages, nor could he tell whether a real word had been uttered or just a nonsensical collection of syllables. But all other elements of language remained intact; he could not only speak fluently but also read and write. His appreciation of intonation and prosody, too, remained intact; although he could not understand its meaning, he could tell whether a sentence had been delivered as a question, a command or an exclamation. He could recognize when someone's voice was emotionally neutral, happy or angry. And he could discriminate between laughing and crying.

The man remained able to recognize non-verbal sounds, such as a key jingling, a telephone ringing, rustling paper or animal calls. He could identify the sounds of different musical instruments and was adept at recognizing melodies and songs, although he was, of course, unable to understand the lyrics. He was also still able to identify spoken vowels. One hundred Arabic vowel stimuli were spoken to him and after each he was asked to point to the letter on a card; he scored 97 per cent correct. Only when they were combined in increasingly complex ways and he had to infer their meaning did his ability to comprehend them become impaired.

The third case is of a fifty-five-year-old Japanese-speaking man. He was admitted to the Chiba Emergency Medical Centre, in Teikyo, Japan, on 17 October 1982, following the onset of a severe headache.[14] He was not fully conscious and was suffering from global aphasia – a complete loss of his capacity for language. He had suffered a haematoma which had caused damage to his left cerebral hemisphere, specifically in the left thalamus and the white matter of the left temporal and parietal lobes.

Full consciousness returned and the man regained his ability to speak, which was fluent, with normal speed and intonation. He could also read and write without difficulty. But he, too, had entirely lost his ability to understand what others were saying.

When a team led by Professor Nobuyoshi Takahashi undertook an examination, the man was found to be able to discriminate between speech and other environmental sounds, and to identify the number of words in a sentence and the number of syllables in a word. When he was spoken to slowly, or when the words were repeated, he gained a little understanding, and this improved when he began to lip-read. Unlike the Frenchwoman or the Arabic man, this Japanese man had lost not only the ability to understand the meaning of words but also his ability to appreciate prosody. When asked to identify the mood of fourteen short sentences he failed in every case.

He scored full marks in tests of environmental sound recognition and was also quite able to identify human voices. Similarly, his musical abilities appeared quite normal, with the possible exception of his appreciation of rhythm. While he had no difficulties discriminating between pitches and melodies, identifying melodies, appreciating the mood of a piece of music, or recognizing the kinds of instruments on which it was played, he was able to identify whether two consecutively played rhythms were the same or different in only one out of four attempts.[15]

Differential recovery rates for musical, verbal and environmental sound recognition

NS and the French, Syrian and Japanese patients all suffered lesions that led to loss of language-processing abilities but preservation of musical abilities. They varied, however, with regard to their remaining capacity to process environmental sounds, foreign languages and prosody. We have also seen that some language abilities can return. This must happen either because new neural networks are formed to replace those that have been damaged or because existing networks take on new functions. Our next case study is an interesting example of such recovery.

The subject was a fifty-eight-year-old French-speaking woman who suffered a stroke and whose case was reported by Olivier Godefroy and his colleagues from the Department of Neurology, University of Lille, in 1995.[16] This woman had initially experienced a mild right-sided sensory loss; twenty-four hours later the same weakness occurred on her left side. An ambulance was called and while she was being taken to hospital she entered an entirely silent world; she had become totally deaf. People's lip movements provided some understanding of what was being said to her, but she soon began asking them to write down their questions. CT and PET brain scans showed two deep hemorrhages, one on either side of the brain, within the temporal lobes. These had caused complete deafness – but not complete aphasia, as she was still quite able to speak and read.

Within two weeks her hearing had returned, but she complained that sounds remained difficult to identify. The medical team undertook a series of tests to examine whether this applied to all sounds or just to those of a particular type. Having established that her general intellectual and memory abilities were normal, they evaluated her accuracy at identifying environmental and musical sounds and speech against that of seven healthy controls. For the environmental noise recognition test, a series of forty-eight sounds were played, which the patient and controls were asked to match to drawings. The music recognition test involved identifying thirty-three well-known tunes. The speech test required thirty-four consonant–vowel pairs heard through headphones to be repeated and then matched with thirty-four written syllables. The tests were repeated at regular intervals after the patient's hearing had returned, and they showed that within two months she was as competent as the controls at identifying all three types of sounds.

But the different abilities had not all returned at the same rate.

When tested on the eighth day after her hearing had returned, the patient was still unable to recognize environmental and speech sounds, achieving correct scores of only 11 and 21 per cent respectively, compared with an average of 88 and 95 per cent for the controls (with standard deviations of 3.1 and 4.5 per cent respectively). Her ability to recognize musical melody was a little better, with 21 per cent correct responses, while the average control score was 49 per cent (with a standard deviation of 11 per cent). Ten days later she had improved across the board; her environmental and speech recognition were still significantly impaired (48 and 32 per cent correct responses), but her music recognition had returned to normal (64 per cent of the melodies identified correctly). It was not until the sixtieth day after her hearing had returned that her abilities at recognizing environmental noises and speech were equivalent to those of the controls.

Eddie and other musical savants

The cases of Shebalin, NS and the other subjects described above show a clear separation between linguistic and musical capacities in the brain, as well as some degree of partitioning within the linguistic capacity itself. But they also hint at some areas of overlap. We are unquestionably dealing with an immensely complex issue and should not expect there to be any simple answers.

One further complicating factor is that these case studies have dealt with mature adults. While music may have some independence from language within an adult brain, it may have required the language networks for its development in childhood. Having been built on those for language, the neural networks for music might later become independent and then survive when the networks for language are destroyed by brain damage. If so, the idea that music is derivative of language would remain intact.

So-called 'musical savants' contradict such claims because their musical abilities develop in the absence of language, or at least the complete package of abilities that constitute language.[17] As the following case histories illustrate, all musical savants appear to be very sensitive to sound and to appreciate how sounds can be combined together into sequences, as in speech. Such abilities are indeed essential to language, but there is no necessary reason to suppose that they evolved for language rather than for perceiving environmental sounds or, indeed, music itself.

I will focus on one particular musical savant, known as Eddie, whose case history has been recounted by Leon Miller, a psychologist specializing in childhood disability at the University of Illinois. In 1985, Miller was invited by a teacher at a day school for the multiply handicapped to meet Eddie, a five-year-old boy who had just begun attending the school and was able to play the piano surprisingly well.[18] Eddie appeared as a fragile child, bony, thin and small for his age; his speech was echolalic – when spoken to he would frequently simply repeat what he had heard. He wore thick glasses and walked hesitantly with his feet splayed, but when the teacher mentioned the word 'piano', Eddie smiled excitedly and went purposefully to the music room. He climbed onto the stool, found the keys and began to play, his head tilted back at the ceiling with an intent expression on his face.

Eddie played the Christmas carol 'Silent Night' and Miller was impressed: the melody was well articulated, the tempo appropriate and there was a nice rolling broken chord figure in the bass. Although Eddie had difficulty holding a pencil, his hands were at home on the keyboard. As his fourth and fifth fingers on each hand were weak, he supported these with other fingers so that he could strike the keys firmly. When his rendition was finished, Miller and the teacher clapped enthusiastically. Eddie smiled in appreciation.

This was Miller's first encounter with Eddie, one that he later described as 'unexpected and dramatic'. On his next visit, he tested Eddie's musical abilities, suspecting that he might simply have learnt 'Silent Night' by intensive practice. Miller played 'Twinkle, Twinkle, Little Star', a melody with which Eddie was familiar. But rather than playing it in the key of C, Miller played it in G, A flat and F sharp. In each case Eddie was asked to try the melody on his own, and his playing reflected the transposition to the new key with no hesitation. After Miller had once again played 'Twinkle, Twinkle' in the key of C, Eddie was no longer content just to repeat the melody. Instead, he added several left-hand chords and transposed the piece into a minor key with several unexpected modulations in its harmonic structure, using minor thirds rather than the major thirds of the original. In effect, Eddie – a five-year-old boy with severe sight, hearing and learning difficulties, who had hardly any linguistic abilities and was physically underdeveloped and disabled – had generated an entirely new version of the piece.

Miller spent several years studying Eddie and published an account of this remarkable child in his 1989 book entitled *Musical Savants: Exceptional Skill in the Mentally Retarded*. The subtitle was appropriate not only for Eddie but for a number of other children that Miller described and tried to understand. One of these was Thomas Wiggins, otherwise known as 'Blind Tom', who was the earliest known case of a musical savant. He was born a slave on a plantation in the American South in 1849. Blind from birth, he was exceptionally sensitive to the sounds of nature, accurately imitating the calls of animals and birds. By the age of four he was picking out melodies on the piano in the house of the plantation owner; by the age of eight he was giving concerts and by ten he was touring the South. A long performing career followed, throughout the USA and in Europe. Tom had an extensive repertoire, including his own compositions, and a talent that has been compared to that of other concert pianists of the period.

Tom's exceptional musical talent was accompanied by severe mental deficits. He had the type of physical mannerisms offstage that are often associated with autistic disorders, such as rapid flicking of his fingers, grimaces, body twirling and rocking. His earliest speech was, like that of Eddie, echolalic. When older, he apparently spoke to himself in unintelligible jargon and his conversation with others was minimal and monosyllabic.

The nature of Tom's musical ability and the extent of his deficits are difficult to establish owing to conflicting nineteenth-century reports. Some may have been deliberately misleading, written by those wishing either to exaggerate or to minimize the contrast between his musical and other abilities. He was, for instance, called an 'untutored musical genius' even though he had received extensive piano lessons. Fortunately, Miller was able to locate more rigorous documentation about eleven other musical savants, several of whom he was able to meet and work with.

One of these was CA.[19] He, too, was blind at birth, and by the age of six he had been diagnosed as severely mentally retarded; his language was, and remained up to the date of Miller's book, almost entirely echolalic. CA was placed in a residential institution; he was aggressive and, until the age of eleven, had to be fed, dressed and toileted by the staff. It was during those years that his special sensitivity to sound became apparent. If he heard someone tapping a sound with a spoon on a table, CA would tap objects until he found the correct match for what he had heard. He was attracted to an accordion that one of the supervisors brought into the ward. The supervisor took on the task of teaching him to play – an extraordinarily difficult task since there could be virtually no verbal instruction. But once CA had learnt the required fingering to produce different sounds, his progress was rapid. He began lessons at the supervisor's house, some of which lasted whole days. At the age of fourteen he appeared in his first public concert and he then began to make regular performances, often with his teacher. He moved into a private home and began to learn the piano; he favoured waltzes and polkas, having learnt those during his first few years of playing.

Eddie and CA are just two of the thirteen musical savants that Leon Miller describes in his book. Other cases are known, such as an autistic young man named Noel who could perform near-perfect renditions of classical pieces that he had just heard.[20] Miller compared the musical savants he studied both with one another and with other musical savants, as well as with groups of adults and children who showed exceptional musical talent but, with a few exceptions, were cognitively unimpaired. The exceptions were either blind or had some language deficits, although less severe than those of the musical savants. Several of the adults were professional pianists. Miller's study was rigorous and all his subjects undertook the same types of tests of pitch and rhythm recognition as were used on the patients who had suffered brain damage. These tests provided a considerable challenge, especially when the subjects had a tendency towards echolalia.[21]

Miller's most striking finding was the existence of strong similarities in the specific deficits and abilities of the musical savants. All thirteen were identified as having perfect pitch (also known as absolute pitch) – the ability to identify or to sing a note of a specific pitch to order. Although this may be widespread and perhaps even universal in very young children, only one adult in every ten thousand has perfect pitch. None of Miller's control group, which included professional musicians, had perfect pitch. Both Miller and Beate Hermelin, another psychologist specializing in savants, have concluded that the possession of perfect pitch is most likely a necessary, although not sufficient, condition for the development of musical abilities in cognitively impaired children because it provides the basic building blocks for understanding more complex musical structures.[22]

A second similarity among the musical savants was that at least eight of them had a tendency towards echolalia when young; in several cases this persisted into adulthood. People with echolalia appear unable to understand that words have meanings and/or that words relate to experiences shared between the speaker and the listener. Unlike other types of language deficits, echolalia requires sensitivity to sound and an appreciation of how sounds are combined together to make sequences.[23]

A further common feature among the musical savants was gender: ten out of the thirteen documented by Miller were male. A similar gender bias is found in the case of autistic children and adults, among whom there is also a high incidence of perfect pitch.[24] This bias may be a reflection of a general difference between male and female cognition which leaves the former particularly susceptible to deficits relating to language and the ability to empathize with other people's feelings.[25]

The most interesting finding came when Miller compared the musical abilities of savants with those of his control group. Contrary to some people's expectations, they did not play by rote and were not limited to a fixed number of pieces. In contrast, they displayed a level of musical understanding similar to that possessed by professional musicians. The extent to which they practised and had lessons varied hugely. Some, like Blind Tom and CA, had had long periods of intensive tutoring, repeatedly playing the same piece until they had mastered it; others appear to have been more spontaneous. Eddie rarely practised the piano. In every case, however, their musical knowledge was extensive, with no particular holes or deficits. It included sensitivity to the rules of musical composition, enabling them to improvise and to capture immediately the essence of a piece of music. Miller describes how one Christmas he introduced Eddie to the full orchestral version of Tchaikovsky's *Nutcracker* suite: 'After each segment Eddie ran to the piano, giving his rendition. These were always faithful to the style and ambience of the original. The Sugar Plum Fairy was light and delicate, the Cossack Dance exuberant.'[26] Eddie was thus able to capture the gist of the pieces he played, showing an understanding of their general, abstract qualities as well as their particulars.

With their seemingly complete mastery of the domain of music, musical savants appear to contrast with other types of savants. 'Calendar calculators' are those who are able to provide the day of the week for any given calendar date with apparently minimal effort. This is a remarkable mathematical feat and, like musical savants, they must achieve it by having an implicit understanding of the rules underlying how days are defined rather than a reliance on rote memory. But the mathematical talent of calendar calculators rarely extends to other areas of mathematics, in which they often appear inept. Hyperlexic savants appear to be similarly limited. Although able to read rapidly and then recall accurately long passages of text, they appear unable to extract the overall meaning by identifying and linking together the ideas contained within each passage. Unlike Eddie with his music, they cannot extract the gist of a story.[27]

We will return to Eddie and the other musical savants later in this book as we explore further the significance of perfect pitch for the relationship between music and language; indeed, the relationship between language and the development of musical talent is one of the most important issues arising from Miller's work, and will be explored in chapter 6. But before we leave them, I must cite one further passage from Miller's book. This describes a walk with Eddie when he was nine years old, by which time his language, personality and physical abilities had improved markedly since Miller's first encounter with him when he was five. He had begun attending a new school and now had a music teacher who took him to recitals, where Eddie could sometimes play with the professional musicians. This teacher's experience of working with Eddie provides a fascinating appendix to Miller's own studies, and among her recollections is the following:

> I found that a walk with Eddie is a journey through a panorama of sounds. He runs his hand along metal gates to hear the rattle; he bangs on every lamp post and names the pitch if it has good tone; he stops to hear a car stereo; he looks into the sky to track airplanes and helicopters; he imitates the birds chirping;

he points out the trucks rumbling down the street. We go into a small store and I hardly notice the radio in the background but he reports to me, 'That man is singing in Spanish.' If it is aural Eddie is alert to it, and through the aural he is alert to so much more.[28]

4 Language without music

Acquired and congenital amusia

Shebalin, NS, Eddie and those others we have so far considered all demonstrate that the capacity for music can exist within the brain in the absence of language. But is the converse also true? Can language exist without music? If so, then we can refer to music and language as having a 'double dissociation' in the brain, which is often taken as a sign of both developmental and evolutionary independence. This double dissociation does indeed exist, as will become apparent from the following accounts of people who either lost their musical abilities while retaining those for language, or who simply never developed any musical ability at all. The word for this condition is amusia, the musical equivalent of aphasia.

Although reference will be made to the work of numerous scientists as we examine various case histories of amusia, that of Dr Isabelle Peretz of Montreal University is of most significance. She has undertaken extensive studies of amusia and discovered many important aspects of this condition. In particular, she has found compelling evidence that the capacity for music is not a single entity in the brain but is constituted by multiple components, some of which can be lost while others remain intact. Moreover, some appear to be dedicated to music while others are evidently shared by the language system. We must start with another professional musician and a case that is neither strictly one of amusia nor of aphasia.

'This opera is in my head, I hear it, but I will never write it'

These were the words of the French composer Maurice Ravel in November 1933 when confiding to one of his friends about his new opera, *Jeanne d'Arc*, four years before he died. His last years were blighted by a brain degeneration that caused mild aphasia and entirely removed Ravel's ability to write down the music he composed. We can, perhaps, imagine how devastating this must have been by thinking about our own frustration when we have a word 'on the tip of the tongue' but can't quite say it. Imagine a whole opera sitting there. The inability to write is called agraphia.

The first signs of Ravel's illness appeared early in 1927 when he began to make 'blunders in writing'.[1] In November 1928 he lost his place when playing his *Sonatine* in Madrid, jumping straight from the first movement to the finale. Ravel began to suffer intense frustration, flying into rages when he could not find the words he wanted to say. By 1933 his friends were helping to write down his compositions, notably the three songs of *Don Quichotte à Dulcinée* and the orchestration for *Ronsard à son âme*. However, Ravel's agraphia soon became severe and was effectively complete by the end of that year, after which he never signed his name again. His last public appearance was in November, conducting *Boléro*, but, as was noted at the time, 'without doubt the orchestra managed on its own'. Ravel's speech also suffered, causing him to struggle to find proper names. Unlike Shebalin, however, he remained able to express himself and to understand speech. He died acutely aware of how his illness had left him able to compose music in his head but quite unable to write it down.

Dr Théophile Alajouanine, a neurologist, cared for and examined Ravel during the last two years of his life.[2] He found that Ravel was still able to play major and minor scales on the piano, and that he could write down some of his compositions from memory, although with much hesitation and some errors. When listening to these, Ravel could detect any departures from what he had originally written, indicating that his hearing remained intact. But sight-reading, writing down new works, and playing from a score were entirely lost. His key deficit was in translating musical representations from one modality – hearing them inside his head – to another – writing them down on paper or playing them on a piano.

Alajouanine did not identify the precise cause of Ravel's illness, and there was no post-mortem examination of Ravel's brain. Although discussion of his case in recent medical and science journals has been informed by advances in our understanding of the brain, no firm conclusions can now be drawn. The best guess is that Ravel suffered from a degeneration of the posterior region of the left cerebral hemisphere, in the superior temporal gyrus and the inferior parietal lobe. Disturbance to these areas is known to cause amusia, but the loss of selective musical abilities may have arisen from degeneration of other brain areas.

Fingers losing their way

A similar deficit has been found in other cases of amusia. 'I know what I want to play but I can no longer transfer the music from my head into my hands' was the heartfelt cry of a sixty-seven-year-old Australian known in the medical literature as HJ.[3] He had been an amateur musician, although he had never learnt to read or write music, and had composed nothing more than ditties. HJ suffered a stroke in 1993, resulting in severe amusia while leaving his abilities for language, reasoning, writing, concentration and memory unaffected. He also reported that he felt the sound of his singing had changed for the worse and that he had forgotten how to play his clarinet and harmonica. He seems to have suffered much the same as Ravel – an inability to express the music that he continued to experience within his head.

HJ was examined by Dr Sarah Wilson from the Department of Psychology at the University of Melbourne. He was an enthusiastic subject, keen to understand his condition, but often became angry and distressed at the loss of his musical abilities. Such feelings arose when he attempted to perform by either singing or playing instruments: he was perfectly able to recognize the errors he made but quite unable to rectify them.

Before his stroke, HJ had been a keen pianist and had received lessons throughout his childhood. When asked to play some familiar songs from his repertoire and some major scales, Wilson noted how he would seat himself and position his hands at the piano quite correctly. But when he began to play he made repeated mistakes with the fingering of his right hand, which lacked fluency. His left hand was even worse, appearing clumsy and often entirely unable to find the correct finger patterns. In fact, his hands appeared to play independently of each other, so any elements of melody, rhythm and harmony he could produce were dissociated from one another. He would begin a song by correctly playing the opening but then he appeared unable to recall the chord progressions, sometimes omitting them and skipping from the opening line to a later, similar phrase. On other occasions his two hands would become entirely uncoordinated, fingering a simple melody with the right hand while repeatedly playing the same chord with the left. Ravel might have been suffering from something similar when he jumped from the first movement to the finale in his Madrid performance of *Sonatine* in November 1928.

HJ's problems did not arise from a deficiency in motor coordination per se because he could still play one piece of music with an impressive fluency, providing listeners with a sense of his previous 'musical flair'. This was 'Kalinka', a Russian folk song that he had learnt at an early age. Wilson suggests that this had become the most 'automatic, over-learnt piece in his music repertoire'. HJ's performance of 'Kalinka' contrasted with his loud and clumsy playing of other melodies. These often ended abruptly, lacking musical phrasing and generally suffering from a blurring of the notes by an overuse of the sustain pedal. He suffered from difficulty in translating his auditory representations into the correct sequence of finger movements. This was most likely caused by damage to his right inferior parietal lobe, as revealed by a CT scan, which would explain why his left rather than his right hand was most seriously affected.

Unsurprisingly, HJ showed his greatest deficit when attempting to play new pieces. When asked to play on the piano some very simple melodic passages for one-handed sight-reading, he was able to achieve this for the right hand but not the left, even though he could correctly identify the notes. He found the effort of trying to transpose the notes of the music from his mind to the fingers of his left hand so distressing that he refused to complete the tasks. Wilson commented that his fingers seemed totally unable to find their way around the piano keys.

When singing sounds like shouting

In December 1997 a twenty-year-old man was referred to Professor Massimo Piccirilli and his colleagues at the Medical School and Institute of Neurology of the University of Perugia.[4] The patient – neither a name nor initials are given in the medical report – had experienced a sudden and violent headache, followed by nausea, vomiting and the onset of a language disorder. An emergency brain scan showed a haematoma – a tumour filled with blood – in the left temporal region. Drainage and closure of the tumour, a process known as 'clipping', prevented any further brain damage beyond that which had already occurred in the left superior temporal gyrus and the temporal lobe.

Although immediate tests revealed that the patient was experiencing substantial difficulties with comprehension, writing and reading, a rapid and spontaneous improvement had begun within a few days. One month after the event linguistic errors had become rare, and after three months the patient's language abilities had effectively returned to normal. But his perception of music had changed: the patient was now unable to sing or play his guitar and he reported that sounds had become 'empty and cold', while singing 'sounds like shouting'. 'When I hear a song,' the patient explained, 'it sounds familiar at first but I can't recognize it. I think I've heard it correctly but then I lose its musicality.' HJ, the Australian man whose amusia was similar to Ravel's, also complained that music now sounded 'like awful noise'.[5]

For both these sufferers, listening to music had become an unpleasant experience. This was tragic: Piccirilli's patient had been an avid music lover and a competent guitarist. Although he had not learnt to read music, he had practised the guitar since the age of fourteen and formed a band with his friends in which he both played and sang. Similarly, HJ had been able to play the piano, clarinet, drums and harmonica, was often asked to play at social functions and particularly enjoyed improvising on the piano.

Massimo Piccirilli and his colleagues carried out a wide range of tests to examine their patient's linguistic and other cognitive functions, such as memory, perceptual recognition and learning abilities. All proved absolutely normal. He was able correctly to identify environmental sounds and to tell whether they emanated from people (for example, coughing or laughing), animals, nature (for instance, running water), vehicles or machines. He was also able to recognize the prosodic tone of a spoken phrase, correctly identifying the mood of the speaker. But something had certainly happened to his musical abilities – or at least to some of them.

The patient could not identify familiar melodies, those that are commonly hummed or well known such as the national anthem. Neither could he recognize what had once been his favourite musical pieces. But he could recognize the intensity of a piece of music and its timbre (which varies with the instrument being played), as well as familiar rhythms such as the waltz or tango. A whole battery of detailed tests confirmed that while the patient had entirely lost his ability to process melody, he remained competent at processing rhythm and timbre. This contrasts with HJ's situation: while his language abilities could also be considered normal, his amusia was more complete, leaving him unable to maintain a steady pulse in response to an external tone, to clap and dance in time to music, or to discriminate between and to reproduce rhythms. So, as we found with the loss of language in cases of aphasia, the extent to which the various components that make up our musical abilities are lost or preserved varies considerably in cases of amusia.

Losing tonal knowledge

Our next case focuses on a fifty-one-year-old businessman from Quebec, known as GL, and provides further evidence of how the brain's capacity for music is constituted by multiple components which have some degree of independence from one another. Although a non-musician, GL was a keen listener to popular and classical music, and frequently attended concerts. In 1980 he suffered an aneurysm of the right middle cerebral artery. After this was clipped he recovered and was able to return to work, with no apparent symptoms. But one year later, GL suffered a mirror aneurysm on the left side, which was also clipped but left him with severe aphasia. Two years of speech therapy enabled him to recover his language abilities, and his lifestyle was now similar to that of any other well-off retired person – with one substantial exception. The second aneurysm had caused permanent amusia, and GL was no longer able to enjoy music. A CT scan in 1990 confirmed lesions in the left temporal lobe and the right frontal opercular region of the brain.

Ten years after the first aneurysm, Isabelle Peretz examined GL's amusia. During the course of a year she performed an extensive series of tests in an attempt to identify its precise nature. First, she assessed his cognitive and motor abilities in general, such as memory, language and visual discrimination. These were all quite normal and reflected his high educational attainment. Moreover, he was also found to be quite normal at recognizing animal cries, environmental noises and the sounds of specific musical instruments. But his amusia was severe. He failed to identify a single one out of 140 musical excerpts that were judged to be familiar to residents of Quebec at the time and which he would undoubtedly have been able to identify prior to his brain damage.

Interestingly, when Peretz examined each separate component of GL's musical ability in isolation, she found that most had remained intact. For instance, when played successive notes on the piano, he could state whether they were of the same or different pitches. His ability at pitch recognition was, in fact, no different from that of five controls – normal subjects of the same sex and similar age, and with equivalent educational and musical experience. Similarly, he was able to recognize whether pieces of music had the same or different rhythms, and was able to make use of contours – the manner in which pieces of music rise and fall in pitch as they are played – in discrimination tasks.

Where GL failed was on those tests that required the use of tonal knowledge – those implicit rules we possess about how music should be structured. There is a consensus among musicologists that we acquire such knowledge very early in childhood, without explicit tutoring, and that it is essential for musical experience. As I noted in chapter 2, one might compare it with the grammatical knowledge that we also automatically acquire at an early age, although the two are not strictly equivalent. Tonal knowledge enables us to identify whether a piece of music sounds 'right', in the same way that our implicit knowledge of grammar informs us whether or not a sentence makes sense.

Peretz provides a succinct summary of tonal knowledge:

> In our [i.e. the Western] musical idiom, the set of musical tones consists of a finite ensemble of pitches, roughly corresponding to the tones of a piano keyboard. Moreover, only a small subset of these pitches is generally used in a given piece, i.e. those from a particular musical scale. The most common scale used in popular Western music is the diatonic scale, which contains seven tones, repeated at octave intervals. The structure of the scale is fixed and

> asymmetrical in terms of pitch distance. It is built of five whole steps and two half steps. Scale tones are not equivalent and are organized around a central tone, called the tonic. Usually a piece starts and ends on the tonic. Among the other scale or diatonic tones, there is a hierarchy of importance or stability, with the fifth scale tone – often substituting for the tonic – and the third scale tone being more closely related to the tonic than the other scale tones. Together, the tonic, the third and the fifth form what is referred to as a major triad chord, which provides a strong cue for the sense of key. The remaining scale tones are less related to the tonic, and the non-scale tones are the least related; the latter often sound like 'foreign' tones.[6]

While the specific character of these rules relates to Western music, we should note that the use of scales is universal in music. The majority of scales have some common properties: most make use of pitches an octave apart, and are organized around five to seven focal pitches. This relatively small number might, it has been suggested, relate to cognitive constraints on human memory.

By having implicitly acquired tonal knowledge, listeners can readily detect tones that depart from the scale of a melody. Moreover, Western listeners have a strong aesthetic preference for listening to music that conforms to their tonal system. There can be little doubt that, prior to his brain damage, GL possessed such implicit tonal knowledge; this would have provided the basis for his immense enjoyment of music.

Peretz played GL and control subjects a set of melodies and asked them to pick out those that sounded complete. All the melodies ended with a descending pitch contour, but only some of them had the tonic at the end and so would normally be recognized as complete. The control subjects identified these with a success rate of about 90 per cent; GL's success rate was 44 per cent. In another test, Peretz examined whether GL had a preference for tonal or atonal melodies. As one might expect, the control subjects had a strong preference for the former, but GL could not even tell them apart. He complained that the test did not make any sense to him; he could hear the differences but did not know how to interpret them. HJ, the sufferer who, like Ravel, could still hear music within his head, was also unable to distinguish between tonal and atonal melodies, and he failed, too, on other tasks relating to tonal knowledge that he knew he would have found quite easy prior to his stroke.[7]

The rest of GL's musical abilities – his appreciation of rhythm, discrimination of pitch, recognition of melodic contour and identification of timbre – had either remained entirely intact or had become only moderately impaired. Peretz concluded that GL's amusia had arisen from a very specific deficit – an impairment of tonal knowledge, which she otherwise describes as the tonal encoding of pitch. When this is lacking, Peretz argues, an inevitable consequence will be an inability to recognize melodies.

The significance of words

The next case concerns another dissociation, this time between song and melody. Singing is an important topic because it involves the integration of both melody and speech – music and language. Whether the words and melodies of songs are stored in memory independently of each other or in some integrated form remains a point of contention among neuropsychologists, although there has been limited research into the issue. This makes the case of KB, a Canadian man who suffered a stroke at the age of sixty-four, all the more interesting because it provides some clues as to how the lyrics and the melody of the same song are stored within the brain.[8]

KB was an amateur musician. He had played trumpet and drums at school, and spent ten years singing in a barbershop quartet and in amateur operettas. He sang frequently at home and regularly listened to his large collection of jazz and classical records. In July 1994, KB was admitted to hospital after he suffered a stroke causing left-sided paralysis and speech difficulties. A series of CT scans showed focal damage in the right fronto-parietal area and to a lesser extent in the right cerebellum.

After a period of convalescence, KB was subjected to a battery of psychological tests. With the exception of some minor deficits, his speech had recovered and his memory was unimpaired. But he had suffered a slight decline in his general intellectual functioning, experiencing difficulties in tasks such as sequencing symbols and alternating between alphabetic and numeric symbols.

KB's hearing itself had not been affected; when tested he was able to identify environmental sounds and correctly recognized thirteen out of seventeen musical instruments played to him. Most of his mistakes were not unusual, such as confusing an oboe with a bassoon. But some musical abilities had been lost, and tests led to the diagnosis of amusia. KB himself recognized this, finding that he sounded flat when singing aloud and that music no longer had any interest for him. After his stroke, he no longer listened to his record collection and avoided musical performances or activities. Interestingly, although his personality became rather more emotional and his behaviour more impulsive, his speech now lacked any prosodic element – it sounded monotonous and unemotional.

Professor Willi Steinke of Queen's University in Kingston, Canada, and his colleagues studied KB's amusia. As in other case studies, they compared his performance in a series of tests against a set of controls, people of similar age and musical background to KB but with quite normal musical abilities. KB's perception of pitch and rhythm was found to be severely impaired, and he could no longer recognize well-known melodies, such as Ravel's *Boléro* or the opening theme from Beethoven's Fifth Symphony. However, those melodies are purely instrumental, and when KB's ability to recognize song melodies played without their lyrics was tested, a very different result was found – he was as good as any of the controls. Steinke also found that KB could identify some instrumental melodies – those for which he had at one time learned some lyrics, such as the Wedding March and the William Tell Overture.

Further sets of tests were undertaken to confirm and explore this surprising finding. It was shown to be very robust. It was even found that KB could learn new melodies if they were presented in the context of songs with lyrics, although in such cases his learning ability was limited. He showed a deficit when asked to identify a song melody while the lyrics to a quite different song were being played; the controls had a significantly higher success rate at this task than KB.

Steinke and his colleagues tested, and then rejected, the idea that song melodies have a different structure from instrumental melodies, which might make them easier to identify. They also rejected the notion that KB might have been generally more familiar with song than with instrumental melodies; his musical background was quite the opposite. In any case, that would not explain his ability to recognize those instrumental melodies for which he had once learnt some lyrics, or his continuing ability, albeit limited, to learn new melodies when they were presented with lyrics.

Steinke and his colleagues concluded that the specific form of KB's amusia had to derive from the particular manner in which songs are stored in our memories. They proposed that melody and lyrics are actually stored separately in the brain but nevertheless remain connected so that one can act as a means to recall the other. Consequently, every time we hear a familiar song two interconnected neural systems are involved. One, the 'melody analysis system', activates the stored memory of the tune, while the other, the 'speech analysis system', does the same for the lyrics. Repeated listening to a song builds strong neural links between these two systems, so that activation of one will automatically stimulate the other: when the melody of a song is played one will recall the lyrics, and vice versa.

Support for the hypothesis of a separation between memory for tune and memory for lyrics comes from the case of GL, the Quebec businessman who lost tonal knowledge, and that of another sufferer, known as CN, who in 1986, at the age of thirty-five, suffered brain damage that caused amusia but left her language capabilities intact. Both GL and CN were able to recognize the lyrics of familiar melodies when they were sung to them, but were unable to recognize the corresponding tunes.[9] The same was found with HJ, the Australian who, like Ravel, could still hear music in his head but was unable to express it. He could identify the choruses from songs but was quite unable to identify melodies that had never had an association with lyrics.[10]

This memory structure explains another incidental finding from Steinke's study of KB – that the control subjects were better at recognizing songs than purely instrumental melodies. The memorizing of songs establishes a more extensive and elaborate network of information in the brain because both the speech memory and the melody memory systems are involved. This makes the recognition of a song melody easier, because more neural circuits are stimulated when the music is played than is the case when a melody is purely instrumental in nature. With regard to KB, Steinke suggested that a sufficient amount of his severely damaged melody analysis system had survived for it to activate his speech analysis system when he heard a familiar song melody. This enabled him to identify the melody, even though he was not hearing the words that would normally accompany it. When listening to an instrumental melody, there was no stimulation of neural circuits beyond his damaged melody analysis system and so he failed at the recognition task.

This would also explain why KB had some limited ability to learn new songs but not instrumental melodies: the former activated the speech analysis system as well as the remnants of his melody analysis system, building up a relatively elaborate neural network which facilitated memory formation. We can also understand why KB struggled to recall song melodies when they were played with the lyrics to a different melody. These would have caused interference in KB's speech analysis system, which was being activated both by the words he was played and by the links to his melody analysis system, which was 'recalling' quite different lyrics. If his melody analysis system had been intact, then its activation would have been sufficient to overcome the interference, allowing recognition of both the melody and the lyrics, as was the case with the controls.

Prosody and melodic contour

KB's stroke had one significant impact on his speech: it lost the variations in intonation – dynamics, speed, timbre and so forth – that infuse speech with emotional content and often influence its meaning. Prosody, as this is called, can sound very music-like, especially when exaggerated, as in the speech used to address young children. And it has a musical equivalent in melodic contour – the way pitch rises and falls as a piece of music is played out. We have already seen that prosody has been variously maintained or lost in those who have suffered from either aphasia or amusia. This has, however, been reported as little more than an incidental observation. Consequently, a study published in 1998 by Isabelle Peretz and her colleagues is particularly important because it explicitly attempts to identify whether the same neural network within the brain processes sentence prosody and melodic contour, or whether independent systems are used.[11]

This study examined two individuals who were suffering from amusia but who appeared to differ in their abilities to perceive prosody in speech and melodic contour in music: CN (mentioned above) and IR. Like CN, IR was a French-speaking woman, and she had a similar history of brain damage in both left and right cerebral hemispheres. When examined by Peretz, they were forty and thirty-eight years old respectively, and had experienced their brain injuries seven and nine years previously. Both suffered from amusia, and by that time neither had sufficiently significant speech deficits to warrant the diagnosis of aphasia.

CN had bilateral lesions on the superior temporal gyrus. She had normal intellectual and memory abilities for her age and education, but auditory deficits that inhibited her processing of music to a much greater extent than speech and environmental sounds. Previous studies by Peretz had established CN's main deficits as long-term musical memory and, as with GL, the perception of tonality. Although she could understand speech, initial tests had suggested that CN was impaired in judging intonation and interpreting pause locations in sentences. IR had more extensive damage to both sides of the brain and had initially suffered from aphasia as well as amusia. Her understanding of speech had returned; she, too, had normal intellectual and memory abilities, and she could recognize environmental sounds. But IR was entirely unable to sing even an isolated single pitch.

Peretz and her colleagues designed a very clever set of tests to examine CN's and IR's relative abilities at processing melodic contour and speech prosody. They began with sixty-eight spoken sentences, recorded as pairs, that were lexically identical but differed in their prosody and hence their meaning. For instance, the sentence 'he wants to leave *now*?' was spoken first as a question, by stressing the final word, and then as a statement, 'he wants to *leave* now'. These are known as 'question–statement' pairs. 'Focus-shift' pairs were also used, in which the emphasis of the sentence was altered – for example, 'take the *train* to Bruges, Anne' was paired with 'take the train to *Bruges*, Anne'. The acoustics of each pair of sentences were very carefully manipulated using a computer program in order to ensure that they were identical in their loudness and timing. Sentences of a third type formed 'timing-shift' pairs, where the location of a pause was varied so as to alter the meaning. For instance, 'Henry, the child, eats a lot' was paired with 'Henry, the child eats a lot'. These were also controlled for their loudness, but variation in pitch between the sentences of a pair could not be removed without them sounding quite unnatural. CN, IR and eight female control subjects of similar age, musical background and educational history as CN and IR but with no history of neurological trauma, were played pairs of these sentences, some of which were different and some identical.

The clever part of the test was then to translate each sentence into a melodic phrase. Since the question–statement pairs and the focus-shift pairs had already been standardized for loudness and timing, each pair of sentences formed a pair of melodies that differed only in pitch contour. In terms of length, loudness and timing, they were identical to the spoken sentences. The essential difference was that the lexical content had been removed so that 'speech' had been translated into 'music'.

With this experimental design, the question that Peretz and her colleagues were seeking to answer was whether CN and IR would prove equally successful at distinguishing whether the sentence and the melody pairs were the same or different. If so, the implication would be that a single neural network is used to process the prosody of sentences and the pitch contour of melodies. If they were able to recognize the prosody of sentences, by successfully identifying whether they were the same or different, but unable to process pitch contour, or vice versa, the implication would be that different neural networks are used for these aspects of language and music. Of all the experimental procedures designated to investigate aphasia and amusia of which I am aware, this is undoubtedly one of the most meticulously designed and thoroughly executed.

The results were very striking indeed. CN performed as well as the control subjects at identifying whether paired sentences and paired melodies were the same or different, for each of the three classes of sentence (statement–question, focus-shift and timing-shift). She was, in other words, quite normal as regards the processing of both speech prosody and pitch contour. This highlighted the specificity of the deficits causing her amusia – long-term memory for melody and perception of tonality.

IR's results were also consistent: she showed a marked inability to answer correctly in any of the trials, but her performance on the statement–question pairs was significantly better than with the focus-shift and timing-shift pairs. In contrast to CN and the control subjects, IR could process neither speech prosody nor pitch contour. Additional tests showed that this did not arise from an inability to detect pitch. IR was able to discriminate between questions and statements when they were presented either as sentences or as melodic analogues. She could also detect which word of a sentence or phrase of a melody had an accentuated pitch, and where pauses were located. What she appeared to be able to do was to hold patterns of speech prosody and patterns of pitch contour in her short-term memory.

As CN was able to process both speech prosody and pitch contour, while IR was able to process neither, Peretz and her colleagues concluded that there is indeed a stage at which the processing of language and of melody utilize a single, shared neural network. From IR's specific deficits, they concluded that this network is used for holding pitch and temporal patterns in short-term memory.

Born without music

In the previous chapter we considered musical savants – children who are deficient in their language capacities but appear to be quite normal or even to excel with regard to music. Such children demonstrate that some degree of separation between language and music exists in the brain from the early stages of development, rather than this being a property of the adult brain alone. We can now undertake the complementary study, those who are born with severe musical deficiencies but are nevertheless quite normal with regard to language.

In 1878 an article was published in the philosophy and psychology journal *Mind* that described a man of thirty who 'does not know one note from another'.[12] Grant Allen, its author, proposed the term 'note-deafness' for this condition, believing it to be analogous to 'colour-blindness'. Unlike the case histories of amusia I have so far discussed, this man (who remained unnamed) had not suffered any brain damage: he had been deficient in musical ability from birth. This was not due to lack of effort on his part, for he had taken singing and piano lessons, both of which had proved entirely unsuccessful. When Allen undertook a series of tests on him, he was found to be quite unable to distinguish between any two adjacent notes on the piano. When asked to sing 'God Save the Queen', Allen thought that the only notes he sang correctly had come about by accident.

The subject's hearing was quite normal; he was described as being able to recognize non-musical sounds with a high degree of accuracy – indeed, as being particularly perceptive in relation to non-musical sounds. When listening to a violin, he became distracted from the notes by the 'scraping and twanging noises' which necessarily accompany the tones, while he described the sound of a piano as 'a musical tone, plus a thud and a sound of wire-works'.

Some ability to perceive rhythm was present, because he could identify some tunes by their timing alone. He could also recognize the general character of a piece – whether it was lively, bright, tender, solemn or majestic – by relying on its loudness and rhythms. But to him any music that depended on what Grant Allen described as a 'delicate distribution of harmonies' was 'an absolute blank'. He was described as speaking in a rather monotonous voice, little modulated by emotional tones.

Grant Allen's description of his subject was an outstanding piece of work for its day, but does not meet the rigorous experimental standards required in modern-day science. His 1878 paper now reads as an interesting but anecdotal report rather than a systematic study, and this is also the case with further reports of supposedly note-deaf or tone-deaf people that appeared over the next hundred years. Even two large-scale studies published in 1948 and 1980, which concluded that 4–5 per cent of the British population was tone-deaf, have been deemed unreliable.[13] Such was the lack of systematic study that the whole condition came into question; in 1985 an article was published in a leading music journal entitled 'The myth of tone deafness'.[14]

However, that was itself shown to be a myth in 2002, when the first fully documented and rigorously studied case was published of someone who had been tone-deaf since birth, a condition now described as 'congenital amusia'. The word 'congenital' distinguished this person's amusia from that arising from brain damage, which is referred to as 'acquired amusia'. The leading scientist of the study was Isabelle Peretz, who simply advertised for volunteers who believed that they had been born tone-deaf. A large number of responses was received and 100 people were interviewed. From these, a smaller number were selected and tested, of whom twenty-two showed unambiguous signs of amusia and fulfilled four necessary criteria for it to be considered a congenital condition. First, they had to have achieved a high level of educational attainment, in order to guard against their amusia being a consequence of general learning difficulties; secondly, they had to have had music lessons in childhood, so as to ensure that their amusia had not arisen from limited exposure to music. Thirdly, they were required to have suffered from amusia for as long as they could remember, thus increasing the likelihood that it had been present from birth. Finally, they had to have had no history of neurological or psychiatric illness. Of the twenty-two subjects who met these criteria, the most clear-cut case was that of Monica, a French-speaking woman in her early forties.[15]

Monica was a well-educated woman who had worked for many years as a nurse and was taking a master's degree in mental health when she responded to the advertisement. Music had always sounded like noise to Monica and she had never been able to sing or dance. When social pressure forced her to join a church choir and school band she experienced great stress and embarrassment.

Peretz and her colleagues subjected Monica to the same battery of tests that they used on their brain-damaged patients, measuring her performance against other women of a similar age and education but of normal musical ability. The majority of tests were based around identifying whether pairs of melodies were the same or different, with the investigators manipulating some of them in order to explore whether Monica could recognize changes in pitch contour and intervals. She could not. Neither could she perceive changes in rhythm. She did show some ability to recognize different metres when she was asked to discriminate between melodies that were waltzes and marches, but even here she was only correct in nineteen out of thirty attempts.

As with the other amusia sufferers described above, Monica was unable to identify familiar melodies. Since she was quite adept at identifying familiar speakers from their voices, this could not be explained by poor hearing, memory or inattentiveness.[16] The same methods used with CN and IR demonstrated that although Monica was unable to distinguish musical pitch, she could recognize speech intonation.

Monica's ability to appreciate speech intonation while being unable to monitor pitch contour is one of the most interesting findings to emerge from Peretz's study. This finding differs from the cases of IR, who could process neither prosody nor pitch contour, and of CN, who could process both. Unlike Monica, however, both IR and CN could discriminate between pitches (IR's deficit lay in maintaining these in short-term memory). Monica's ability to process prosody might simply reflect the much larger variations in pitch that are present in speech than in melodies. Alternatively, it might result from the combined stimuli of both pitch change and words, just as several of the brain-damaged subjects discussed above were better able to recognize songs than instrumental melodies.

Peretz and her colleagues concluded that the root of Monica's congenital amusia lay in her inability to detect variations in pitch, which caused all music to sound monotonous. They felt that her inability to discriminate melodies by their rhythm was a secondary effect, arising from the failure of her musical system to develop owing to the inability to detect pitch.

But it was not just Monica: ten more of those who had responded to the advertisement and showed the same congenital amusia as Monica were also found to be deficient in their ability to detect variations in pitch, although some maintained a limited ability to appreciate rhythm.[17] All could identify well-known people from their speech, recognize common environmental noises and, with one exception, identify familiar songs on hearing the first lyrics. Like Monica, their disorder was entirely music-specific.

This group study of congenital amusia confirmed that patterns of speech intonation could still be recognized; those with congenital amusia had the same ability to recognize intonation and prosody in spoken language as a set of control subjects. But when the linguistic information was removed, leaving just the acoustic waveform of the sentence, the success rate of the group with amusia was significantly reduced. So their success at the prosody tasks appears to have been dependent upon having the word cues to support discrimination between sentences. This sounds very similar to the aid that lyrics gave to melody recognition in the case of KB.

As with the study of Monica alone, Peretz and her colleagues concluded that the underlying cause of congenital amusia was a defect in the ability to recognize pitch. They acknowledged that most of their subjects suffered from deficiencies in musical abilities seemingly unrelated to pitch, such as memory for melodies, the discrimination of melodies by rhythm, and the ability to keep time. But these were thought to be secondary or 'cascade' effects of an underlying inability to recognize pitch which had inhibited the development of the whole musical system within the brain.[18]

5 The modularity of music and language

Music processing within the brain

The case histories described in the two previous chapters indicate that the neural networks that process language and music have some degree of independence from each other; it is possible to 'lose' or never to develop one of them while remaining quite normal in respect to the other.[1] To use the technical term, they exhibit double dissociation, and this requires us to consider language and music as separate cognitive domains. Moreover, it is apparent that both language and music are constituted by a series of mental modules. These also have a degree of independence from one another, so that one can acquire or be born with a deficit in one specific area of music or language processing but not in others. The separation of the modular music and language systems, however, is not complete, as several modules, such as prosody, appear to be shared between the two systems.

Isabelle Peretz has concluded that the structure of music modules in the brain is as illustrated in Figure 5. Each box represents a processing component, some of which are shared with the language system, which is shown in grey. The arrows represent the pathways of information flow between the modules. A cognitive deficit in the music system might arise from the failure of one or more modules, or from the inhibition of one or more pathways between modules.

In this model, any acoustic stimulus is initially processed in one module and then passed to the modules concerned with speech, music and environmental sounds. Each of these modules then extracts the information to which it can respond. So if, for instance, you hear the song 'Happy Birthday', the language system will extract the lyrics, the music system will extract the melody, and the environmental sound system will separate out additional sounds such as intakes of breath and background noise. The music system is, in fact, constituted by two groups of modules, one concerned with extracting the pitch content (pitch contour, pitch intervals and tonal encoding) and the other with temporal content (rhythm and metre). Both the pitch

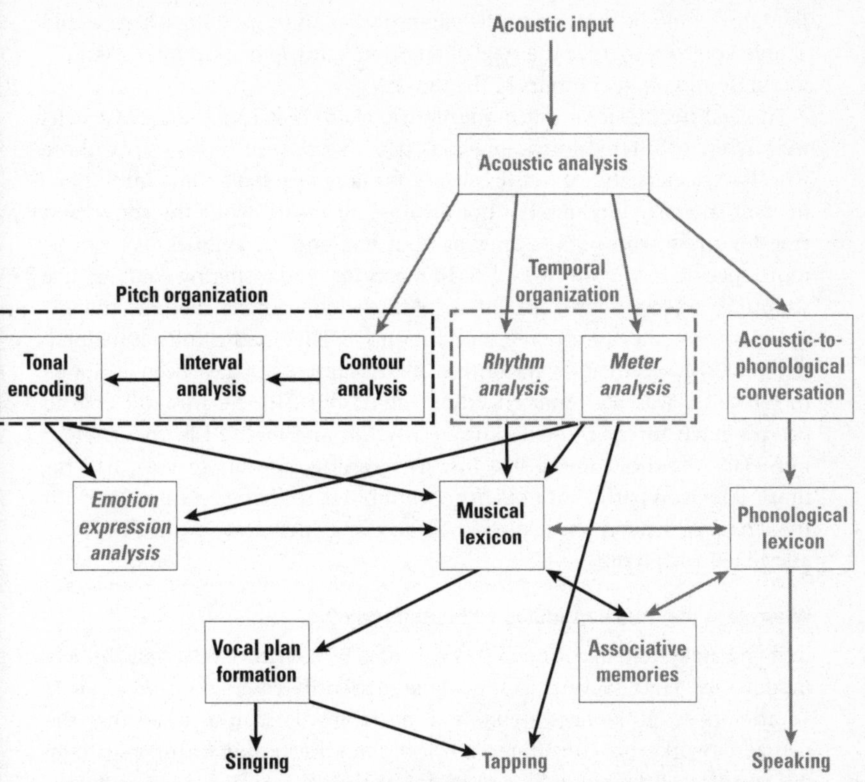

Figure 5 A modular model of music processing. Each box represents a processing component, and the arrows represent pathways of information flow between processing components. A neurological anomaly may either damage a processing component, or interfere with the flow of information between processing components. All components whose domains appear to be specific to music are shown with a dark tint, others in a lighter tint. There are three neurally individuated components in italics – rhythm analysis, meter analysis and emotion expression analysis – whose specificity to music is currently unknown.

and the temporal modules filter their output through the musical lexicon – a memory bank of musical phrases – and the emotional expression analysis module.

If you decide to join in with the song 'Happy Birthday', then the melody will be provided as output from the musical lexicon module, and this will be combined with output from the language lexicon, resulting in singing. If, on the other hand, you were simply required to recall something about the song, the output would be sent to what Peretz describes as the associative memories module. In parallel with this, outputs from the pitch and temporal modules would be directed to the expression analysis module, which would enable you to recognize the type of emotion being expressed by the tempo, loudness and pitch contours of the music.

This architecture for music in the brain explains the diverse range of deficits we have encountered in the previous two chapters. In broad terms, those who have a capacity for music but not for language have some inhibitions in their language modules but not in those for music, while the converse is true for those who possess language but not musical abilities. We can be more specific in certain cases. NS, the man for whom singing sounded like shouting, was evidently inhibited in his pitch content module and his associative memories module, but appears to have had a fully functional temporal content module. In contrast, all that appears to have been inhibited in GL's case was his tonal encoding sub-module, for he was still able to process pitch interval, pitch contour, rhythm and metre. KB, the amateur musician for whom music had lost its meaning, appears to have had his musical lexicon partly but not entirely inhibited, so that its connection with the language lexicon was just sufficient to enable the recall of melodies once associated with lyrics.

Where are the music modules within the brain?

If the music system in the brain is constituted by a series of relatively discrete modules as Peretz suggests, is it possible to identify where they are located? To attempt to do so begs a couple of questions. First, it assumes that the neural networks that constitute a music module are located within a spatially discrete area of the brain. This need not be the case, as the neural networks might be very widely distributed throughout the brain even though they undertake a specific task. Indeed, even if there is a spatially discrete neural module that maps onto a music module, it need not be located in the same place in each individual's brain. Evolution may have 'programmed' the brain to develop neural networks for music without specifying where they are to be located, which might be highly dependent upon the individual concerned. We know that some people, such as NS, have a greater proportion of their neural networks for language on the right side of the brain than on the left, and the extent of brain plasticity for the location of neural networks for music might be considerably greater. Or not. We simply do not know.

Nevertheless, some gross patterns are apparent from the case studies we have examined. Those subjects who suffered from amusia while maintaining their capacity for speech and their abilities to identify environmental noises and voices, tended to have lesions in both their right and left temporal lobes. The key exception was the unnamed patient studied by Massimo Piccirilli and his colleagues, whose lesion was located in the left superior temporal gyrus. He was also exceptional in being left-handed, suggesting that large parts of his brain may have been organized in a different way from the majority of people. The subjects who maintained their musical abilities while losing their other auditory abilities tended to have lesions within their right temporal lobes alone.

Isabelle Peretz herself is rather despondent about the progress that has been made in identifying the location of neural networks for music. In a 2003 article she wrote:

> ... the demonstration of a similar brain organization for music in all humans remains elusive ... In my view, the only consensus that has been reached today about the cerebral organization underlying music concerns pitch contour processing. The vast majority of studies point to the superior temporal gyrus and frontal regions on the right side of the brain as the responsible areas for processing pitch contour information. However, it remains to be determined if this mechanism is music-specific, since the intonation patterns for speech seem to recruit similarly located, if not identical, brain circuitries.[2]

Her final point, about the apparent overlap between the neural networks used for speech and music tasks, is one of the most important, but still unresolved, issues. Aniruddh Patel, of the Neurosciences Institute in San Diego, has worked with Peretz on several studies and also wishes to pinpoint the location of music modules within the brain. It is here, of course, that brain-imaging methods have been of immense value – not just for locating lesions but for examining brain activity while people are actually engaged in making or listening to music.[3] Patel has noted a curious result emerging from these studies: although the lesion studies have shown that music and language capacities can be partially or entirely dissociated, the brain-imaging studies suggest that they do share the same neural networks. This apparent contradiction remains to be resolved.[4]

Broca's area and the processing of musical syntax

To illustrate the rather confusing picture that is emerging from brain imaging we can briefly consider a recent study by Burkhard Maess and his colleagues at the Max Planck Institute of Cognitive Neuroscience in Leipzig.[5] They undertook an MEG study of the way in which the brain processes syntax in speech and music (MEG being the brain-imaging method that relies on detecting the tiny magnetic fields that are produced by active neurons). While no one questions the existence and nature of speech syntax, whether there is an equivalent for music is unclear (as discussed in the previous chapter). Tonal knowledge is probably the closest that music comes to having a syntax – for although the rigid rules apparent in language syntax do not apply, tonal knowledge creates expectations about what pitches will follow each other, especially when a piece of music comes to its end. Maess and his colleagues adopted this approach, interpreting harmonic incongruities in music as the equivalent of grammatical incongruities in speech.

The participants in their experiments, all non-musicians, listened to chord sequences that conformed to the rules of tonal music. As a sequence was played, the chords created expectations in the listeners as to which chords should come next. Maess then introduced unexpected chords, ones that did not conform to the rules of Western music, and used MEG imaging to measure the changes in brain activity that this caused. If one brain area became more active than it had been before, then that was likely to be where musical syntax was being processed. Maess found that this did indeed happen, and that the area was the same as that in which speech syntax is processed – Broca's area.

Since Paul Broca's original lesion study, we have learnt that speech syntax is processed not only in Broca's area, a region of the left frontal lobe, but also in the equivalent area of the right cerebral hemisphere, although this is relatively less important. Musical syntax, as described and studied by Maess, appears to be processed in these same areas, although that in the right rather than the left cerebral hemisphere seems to be the more important. Maess concludes that these brain areas might be involved in the processing of complex rule-based information in general, rather than being restricted to the rules of any single specific activity such as language.

The distributed nature of music processing in the brain

A team led by Lawrence Parsons of the University of Texas has undertaken the most interesting and certainly the most ambitious research programme into music processing in the brain.[6] They have carried out numerous PET studies of people who are either engaged in listening to music, reading musical scores, playing musical instruments, or undertaking two of these activities simultaneously. (PET is the brain-imaging method that measures the flow of blood within the brain, based on the fact that active neurons need an enhanced oxygen supply and hence blood flow.)

One set of experiments involved professional musicians, who played the piano while a PET scan created an image of their brain activity. They were first asked to play a set of musical scales from memory, thus excluding any brain activity that might result from reading a score. Both hands were used, so activity was expected on both sides of the brain. They were then asked to play (again from memory) a piece by Bach, which introduced melody and rhythm into the musical performance while using essentially the same motor processes.

Although both sets of brain images showed activity in the temporal lobes, this was considerably more intense when the Bach was being played. There were also differences in the areas of the temporal lobes that were involved. The scales activated the middle temporal areas of both cerebral hemispheres, but with greater intensity on the left than on the right. In contrast, the Bach activated the superior, middle and inferior temporal areas, with greater emphasis on the right than on the left.

This experiment had a further and quite unexpected result. Whenever the right superior temporal gyrus was activated during the playing of Bach, so, too, was part of the left cerebellum; and activation of the left temporal area was likewise correlated with activity in the right cerebellum, although this finding was less robust. These results are interesting because they show that the cerebellum has roles other than just that of motor control. Moreover, the neural networks for music processing evidently extend beyond the cerebral cortex and into parts of the brain that have had a much longer evolutionary history.

Parsons and his colleagues have also used PET to identify those parts of the brain that process harmony, melody and rhythm. Their studies involved scanning the brains of eight musicians who listened to computer-generated Bach Chorales while reading the corresponding scores. On selected trials, errors were deliberately introduced into the music, affecting the harmony, the rhythm or the melody of the Chorale. Scans were also made of the musicians' brains when reading scores alone, and when listening to error-free music alone. By comparing the sets of brain images made under different conditions, Parsons and his colleagues were able to draw conclusions about which parts of the brain were involved in processing each component of the music.

As one might expect, they found that each test provoked neuronal activity in numerous and distributed parts of the brain. But some differences were detected. While melody activated both cerebral hemispheres to an equal extent, harmony required more activity in the left rather than the right hemisphere, whereas rhythm activated few brain areas outside of the cerebellum. Melody and harmony activated areas of the cerebellum, too, but with significantly less intensity than that produced by rhythm.

Although each musical component activated the frontal cortex, the focus varied: for rhythm it was the superior region of the frontal cortex that was most strongly activated; for melody it was the inferior region; and for harmony it was somewhere in between. Parsons interpreted the spread of such activity as 'translating notes to abstract motor–auditory codes'. There were also differences in the activity that rhythm, harmony and melody generated in the temporal cortex, in terms of the specific regions activated and the relative degrees of left and right activity.

Parsons and his colleagues have conducted further studies, examining the processing of pitch and comparing professional musicians with non-musicians. Pitch processing was found principally to activate the right superior temporal area in non-musicians, but the equivalent area on the left in musicians. When processing rhythm, musicians were found to have a relatively lower degree of brain activity in the temporal lobe and cerebellum than non-musicians.

The overall conclusion drawn by Parsons and his team was that the neural networks for processing music are widely distributed throughout the brain – a finding confirmed by other brain-imaging studies. The implication of the lesion studies that the networks for processing pitch and rhythm are distinct from each other was confirmed, and strong evidence was produced that the location and activity of the neural networks for music differ within the brains of musicians and non-musicians. This last finding is hardly surprising – the brain has a high degree of plasticity, and it must be moulded by the many hours of practice and the skills acquired by a musician. And it is to that moulding process that we must now turn: the development of language and musical abilities during childhood.

6 Talking and singing to baby

Brain maturation, language learning and perfect pitch

Studies of people who have suffered brain lesions have established that the processing of language and music have a degree of independence from each other within the brain. But neither lesion nor brain-imaging studies indicate that one system is necessarily derivative of the other, and hence our progress towards an evolutionary scenario for music and language has been somewhat stymied. There is, however, a source of evidence that suggests that music has a developmental, if not evolutionary, priority over language. This is one of the most delightful sources of evidence that exists and I am envious of those who devote their working lives to its study. Fortunately, practically everybody has some first-hand experience because we generate the evidence ourselves: it is the manner in which we communicate vocally with pre-linguistic children.

Infants are born musicians

'Baby-talk', 'motherese' and 'infant-directed speech' (IDS) are all terms used for the very distinctive manner in which we talk to infants who have not yet acquired full language competence – that is, from birth up until around three years old. The general character of IDS will be well known to all: a higher overall pitch, a wider range of pitch, longer 'hyperarticulated' vowels and pauses, shorter phrases and greater repetition than are found in speech directed to older children and adults.

We talk like this because human infants demonstrate an interest in, and sensitivity to, the rhythms, tempos and melodies of speech long before they are able to understand the meanings of words. In essence, the usual melodic and rhythmic features of spoken language – prosody – are highly exaggerated so that our utterances adopt an explicitly musical character. It is the 'mental machinery' underlying the prosodic elements of language that Steven Pinker claims has been borrowed in order to create our musical ability. But the evidence from IDS suggests that his view of language and music is topsy-turvy: on the basis of child development, it appears that the neural networks for language are built upon or replicate those for music.

During the last decade our scientific understanding of IDS has advanced considerably, thanks to the work of numerous developmental psychologists who have been trying to reveal the function of its rhythms, pauses, pitches and melodies. Dr Anne Fernald of Stanford University is one of the leading academics in this field. She has found that IDS is automatically adopted by men, women and children (even those of pre-school age) when speaking to infants, whether their own offspring or siblings or those of someone else.[1] Subtle differences exist in the types of exaggerated prosody that are found, along with more profound similarities; fathers, for instance, appear not to expand their pitch range as widely as mothers.

On the other hand, those who are new to conversing with babies do it with the same degree of exaggerated prosody as experienced mothers – and the babies enjoy it. Experiments have shown that they much prefer listening to IDS than to normal speech,[2] and that they are far more responsive to intonation of voice than to facial expression. This applies equally to premature infants, who are more frequently calmed by the use of IDS than by other techniques such as stroking. A positive response from the child will then encourage the development of non-linguistic conversation with the infant. Indeed, one function of the prosodic element is to engender the turn-taking that is critical to adult conversation.[3]

The impact of exaggerated prosody is not a function of physical or mental competence; adults who are either sick or elderly will react negatively if spoken to in a similar manner.[4] Conversations between adults do, of course, have a prosodic element, although the extent of this varies between languages. Utterances made in English would lose a great deal of their emotional impact and meaning were it not for the specific intonations that are given to certain words. But while the exaggeration of prosody in IDS may have the same function, it also plays a role in language acquisition itself. When adults are introducing a new word to their infants, they are likely to stress that word with a particularly high pitch and place it at the end of an utterance; but they would not make the same use of pitch when teaching a new word to another adult.[5] Prosody is also used to help children acquire the syntax of language; the placement of pauses is a more reliable cue to the end of one clause and the beginning of another in IDS than it is in adult speech.[6] In general, the exaggerated prosody of IDS helps infants to split up the sound stream they hear, so that individual words and phrases can be identified. In fact, mothers of young children fine-tune the manner in which they use prosody to their infants' current linguistic level.

One would be mistaken, however, to believe that the prosody of IDS is primarily intended to help children accomplish the truly astounding task of learning language.[7]

The four stages of IDS

Anne Fernald has identified four developmental stages of IDS, of which only the last is explicitly about facilitating language acquisition by using pauses and stresses as I have just described. For newborn and very young infants, IDS serves to engage and maintain the child's attention by providing an auditory stimulus to which it responds. Relatively intense sounds will cause an orienting response; sounds with a gently rising pitch may elicit eye opening; while those with an abrupt rising pitch will lead to eye closure and withdrawal.

With slightly older infants, changes to the character of the IDS intuitively used by adults indicate the appearance of its second stage, as it now begins to modulate arousal and emotion. When soothing a distressed infant, an adult is more likely to use low pitch and falling pitch contours; when trying to engage attention and elicit a response, rising pitch contours are more commonly used. If an adult is attempting to maintain a child's gaze, then her speech will most likely display a bell-shaped contour. Occasions when adults need to discourage very young infants are rare; but when these do arise IDS takes on a similar character to the warning signals found in non-human primates – brief and staccato, with steep, high-pitched contours.

As a child ages, IDS enters its third stage and its prosody takes on a more complex function: it now not only arouses the child but also communicates the speaker's feelings and intentions. This marks a major transformation in the child's mental life; in all previous 'conversations' the infant simply enjoyed pleasurable sounds and listened with displeasure to unpleasant sounds – all that was important were the intrinsic acoustic properties of the IDS. Now, however, the melodies and rhythms help the child appreciate the mother's feelings and intentions.

Experiments undertaken by Fernald and her colleagues have shown that, owing to its exaggerated prosody, IDS is a more powerful medium than adult-directed speech for communicating intent to young children.[8] In one such experiment, she recorded samples of IDS directed by mothers towards their children in contexts of approval, prohibition, attention-bidding and comfort. She then recorded analogous samples of adult-directed speech made in equivalent role-playing scenarios. Both samples of speech were 'filtered' of their lexical content – that is, the words themselves were removed – to leave no more than the sound contours of the phrase. When these infant-directed and adult-directed sound phrases were played back to a new sample of adults, all of whom were inexperienced with infants, they were significantly better at identifying the content of the (non-lexical) phrases – that is, whether they had originally been spoken in a context of approval, prohibition, attention-bidding or comfort – when they had originally been infant-directed rather than adult-directed.

Such experiments demonstrate that in IDS 'the melody is the message'[9] – the intention of the speaker can be gained from the prosody alone. Whatever that message is, it appears to be good for the child, because a correlation has been found between the quantity and quality of IDS an infant receives and the rate at which it grows.[10]

As young children begin to understand the meaning of words, further subtle changes to IDS occur which mark the beginning of its fourth stage. In this stage, the specific patterns of intonation and pauses facilitate the acquisition of language itself.

The universality of IDS

The idea that IDS is not primarily about language is supported by the universality of its musical elements. Whatever country we come from and whatever language we speak, we alter our speech patterns in essentially the same way when talking to infants.[11]

Fernald and her colleagues made a cross-linguistic study of IDS in speakers of French, Italian, German, Japanese, British English and American English. They found significant similarities in the use of prosody, suggesting the existence of universals within IDS – the same degrees of heightened pitch, hyperarticulation, repetition and so forth. Some language-specific variations were detected. Most notable was that Japanese-speakers employ a generally lower level of emotional expression in comparison with speakers of other languages, which appears to be a reflection of their culturally influenced patterns of expression in general. Similarly, speakers of American English had the most exaggerated levels of prosody, for probably the same reason.

The universality of IDS was demonstrated by recording the responses of infants to phrases spoken in languages that they would never have heard before. In one such experiment, infants from English-speaking families were sat on their mothers' laps inside a three-sided booth. On the left and right sides of the booth were identical pictures of a woman's face, with a neutral expression, and loudspeakers connected to tape recorders in an adjacent room. The infants then heard a phrase of IDS expressing approval from one loudspeaker, followed by an IDS phrase expressing prohibition from the other. Eight samples of each were played in random order and the responses of the infants to each phrase were recorded on a five-point scale, ranging from a frown (the lowest) to a smile (the highest).

The key feature of this experiment was that the samples of speech were played not only in English but also in four languages that the infants would not have heard before – German, Italian, Greek and Japanese – and then again in English but using nonsense words that maintained the same pattern of prosody. Moreover, increased loudness, the most salient difference between IDS phrases that express approval and prohibition, was filtered out. In effect, the infants were listening to nothing more than melody.

The results were conclusive: the infants responded in the appropriate manner to the type of phrase they were hearing, frowning at the phrases expressing prohibition and smiling at those expressing approval, whatever language was being spoken and even when nonsense words were used. There was, however, one – not unexpected – exception: the infants made no response when the phrases were spoken in Japanese. Fernald explained this in terms of the narrower pitch range used by Japanese mothers than by those speaking European languages. This finding was consistent with other studies that have found Japanese vocal and facial expressions difficult to decode.

Another team of developmental psychologists, led by Mehthild Papousek of the Max Planck Institute for Psychiatry in Munich, made a comparative study of IDS in speakers of Mandarin Chinese and American English in order specifically to compare a tonal and a stress language. Although the American mothers they studied raised their pitch and expanded their pitch ranges more than the Chinese mothers, the same relationships between pitch contours and contexts were found, leading the investigators to support the idea that there are universal elements in IDS. Further studies have found the same patterns of IDS in Xhosa, another tonal language.

It is perhaps not surprising to find the same patterns of prosody in the IDS of native speakers of English, German, French and Italian, because all these languages have the same essential structure owing to their relatively recent common origin, perhaps no more than eight thousand years ago.[12] But that mothers speaking Xhosa, Chinese and Japanese should introduce the same patterns of heightened pitch and expanded pitch range is very significant, even if they are limited in the case of Japanese.[13] The meaning of words in tonal languages such as Xhosa and Chinese varies according to their pitch; whereas in stress languages, such as English, a change in pitch alters only the importance of the word within a particular utterance – its actual meaning is left unchanged. Japanese is known as a 'pitch-accent' language, in which some limited use of contrastive tone is employed.

With tonal languages such as Chinese, the exaggeration of prosody in IDS inevitably risks confusion for the child – is the change of pitch altering the meaning of the word, or merely giving it more emphasis? Experiments have found that Chinese mothers even employ a number of intuitive strategies to outwit the rules of their own language. In some cases, they simply violate those rules, drastically changing the lexical tones of Chinese in favour of the globally universal and emotionally evocative melodies of IDS.

If the exaggerated prosody of IDS were no more than a language-learning device, one would expect to find the IDS of peoples speaking Xhosa, Chinese and Japanese to be quite different from that of those speaking English, German and Italian. That this is not the case strengthens the argument that the mental machinery of IDS belongs originally to a musical ability concerned with regulating social relationships and emotional states. Further evidence for this comes not from IDS but from what we can call 'PDS' – pet-directed speech.

What's new, pussycat?

There is an uncanny relationship between IDS and PDS. When we talk to our cats, dogs and other pets (rabbits, guinea pigs and goldfish, in my case) we also exaggerate the prosodic aspects of our speech. But, unlike when speaking to our children, we are aware that our pets will never acquire language.

Dennis Burnham and his colleagues from the University of Western Sydney decided to explore the similarities and differences in the manner in which mothers spoke to their infants (IDS), pets (PDS), which in this study consisted of cats and dogs, and other adults (adult-directed speech or ADS). Three aspects of speech were measured: pitch, 'affect', a measure of the extent to which intonation and rhythm can be heard but the words not understood; and hyperarticulation of vowels, which measures the extent to which vowel sounds are exaggerated, as is characteristic of IDS.

Pitch in IDS and PDS proved to be statistically the same, and significantly higher than in ADS. The affect rating of IDS was higher than that for PDS, but for both of these it was substantially greater than for ADS. In these two respects, therefore, people speak to their pets in the same manner as they speak to their infants. But when the extent of hyperarticulation of vowels was examined, it was found to be the same in PDS as in ADS; in neither of these situations were vowels exaggerated to the same extent as in IDS.

From this study, Burnham and his colleagues concluded that people intuitively adjust their speech to meet the emotional and linguistic needs of their audience. Like infants, cats and dogs have emotional needs, which may be met by heightening the pitch and affect of utterances. But unlike infants, they lack any linguistic needs and so there is no advantage to hyperarticulation of vowels when addressing pets. This suggests that in IDS we hyperarticulate vowels in order to facilitate the child's acquisition of language, rather than as a further means of attending to its emotional needs.

Burnham's study is fascinating, and one that could usefully be extended to explore further similarities and differences between IDS and PDS. But we must be content with one further observation they recorded: there are no differences between what we might term 'DDS' and 'CDS' – dog-directed speech and cat-directed speech.

Natural-born statisticians

If linguistic acquisition is not the primary function of IDS, how else do infants sift meaning from the words they hear? Jenny Saffran, a developmental psychologist at the University of Wisconsin–Madison, has shown that they are also able to identify statistical regularities in the continuous sound streams they hear.[14] Her work is important for this book because it throws further light on the relative significance of music and language when we are young, and returns us to one of the prevalent characteristics that Leon Miller identified among the musical savants he studied – their possession of perfect pitch.

Saffran began her work by asking a simple question: when infants listen, what do they hear? Few questions are more important for our understanding of how infants acquire language, for they must extract not just the meaning but the very existence of words from the continuous streams of sound they hear from their parents, siblings and others. It is indeed a 'continuous sound stream'; although we leave gaps between the end of one word and the beginning of the next when we write them down, no such gaps exist in our spoken language, so quite how an infant is able to identify discrete words is one of the most fascinating questions in developmental linguistics. We have already seen that prosody helps, but Saffran's work shows that they also deploy another skill: they are natural-born statisticians.

Eight-month-old infants were familiarized for two minutes with a continuous sound string of artificial syllables (for example, *tibudopabikudaropigolatupabiku* ...). Within this stream, some three-syllable sequences were always clustered together (*pabiku*, for instance), whereas other syllable pairs were only occasionally associated. Of critical importance was the elimination of any prosodic cues as to where these regular syllable sequences were located. Once the infants had been familiarized with this sound stream, they were presented with 'words', 'part-words' and 'non-words' to test which of them they found to be of most interest. Each of these consisted of short nonsense syllable strings. The 'words', however, consisted of syllables that had always been found together and in the same order within the continuous sound string. In 'part-words' only the first two syllables had consistently been placed together in the sound string; while in 'non-words' the syllables had never previously been heard in association with each other. The expectation was that if the infants had been able to extract the regularities from the original sound stream, then 'part-words' and especially 'non-words' would be novel and the infants would show greater interest in them, measured by their length of listening time, than in the 'words'.

The results confirmed this expectation: the infants ignored the 'words' but showed interest in the 'part-words' and 'non-words'. This implied that their minds had been busy extracting the statistical regularities, introduced by the recurring syllable sequences, while they were listening to the two-minute syllable stream. And this must be the method they use in 'real' life to identify where words begin and end within the continuous sound streams they hear from their parents, siblings and others – who also help them in this task by the exaggerated prosody of IDS.

One might think that this statistical-learning capability is part of some dedicated mental machinery for acquiring language. But Saffran and her colleagues repeated the experiment with musical tones from the octave above middle C instead of syllables. This time, the infants were played a continuous sound stream of tones (for example, AFBF#A#DEGD#AFB) and then tested as to whether they identified statistical regularities in short tone sequences. The results were equivalent to those for syllables, suggesting that at that age we have a general ability to extract patterning from auditory stimuli. It seems unlikely, owing to the young age of the infants being tested, that this ability originally evolved or developed specifically for language acquisition and was now being used incidentally for music. The converse might be true: a mental adaptation that originated for processing music later came to be used for language. Alternatively, we might be dealing with a very general cognitive ability, as experiments by Saffran's colleagues have suggested that infants can also identify statistical regularities in patterns of light and sequences of physical actions.[15] If that is correct, then this ability must relate to how we get the most general handle on the world at a very young age.

Perfect pitch

Jenny Saffran developed her work by exploring how the infants had been able to recognize the regularly occurring tone sequences within the sound string she had played them. In a study with her colleague Gregory Griepentrog, she devised experiments to test the infants' relative dependence on either perfect or relative pitch when making the required mental computations.[16] They had to be using one or both of these, because there was no prosodic, or pitch contour, element within the sound stream that might otherwise have cued the beginning or end of key tonal sequences.

Perfect (or absolute) pitch is the ability to identify or to produce the pitch of a sound without any reference point. Only about one in every ten thousand adults has this capacity, with the vast majority of people relying solely on relative pitch – the ability to identify the extent of pitch difference between two sounds. There is a higher incidence of perfect pitch among professional musicians than in the public at large, but it remains unclear whether this is a consequence of their musical experiences, especially in early childhood, or the cause of their musical expertise. One of the most intriguing findings in Leon Miller's studies of musical savants, as described in chapter 3, is that almost all of them possessed perfect pitch. This is also a general finding among those suffering from autism.[17]

Saffran and Griepentrog repeated the previous experiment and used it to test whether infants could preferentially recognize 'tone-words' over 'tone-part-words'. In this case, however, the tone-words and tone-part-words were given identical relative pitch sequences so that discrimination had to depend entirely on recognizing and remembering the differences in the absolute pitch of the tones. The infants succeeded at this task: they showed greater interest in listening to the 'tone-part-words', indicating that they had recognized and become familiar with the tone-words in the continuous sound stream. To do so, they had to have relied on recognizing the perfect pitch of the tones.

The next step for Saffran and Griepentrog was to reverse the condition: they devised an experiment to evaluate whether the infants could rely on relative pitch alone to distinguish between 'tone-words' and 'tone-part-words'. The infants now failed to recognize the tone-words, showing that they were unable to use relative pitch as a learning device – or at least that they had a much greater dependency on perfect pitch.

Unsurprisingly, when they repeated the experiments on adults, among whom perfect pitch is rare, the converse was found. It appears, therefore, that when we enter the world, we have perfect pitch but that this ability is replaced by a bias towards relative pitch as we grow older.

Why should this be? In the West, a reliance on perfect pitch is thought to be disadvantageous to learning both language and music, as it prevents generalization. Saffran and Griepentrog suggested that categorizing sounds by their perfect pitch 'would lead to overly specific categories. Infants limited to grouping melodies by perfect pitches would never discover that the songs they hear are the same when sung in different keys or that words spoken at different fundamental frequencies are the same.'[18] They would even be unable to recognize that the same word spoken by a man and by a woman with differently pitched voices is indeed the same word.

Perfect pitch might, however, be an advantage in tonal languages such as Mandarin Chinese, in which some words do have different meanings when spoken at different pitches and therefore need to be separately categorized. Saffran and Griepentrog tested a sample of adult Mandarin-speakers but found that they performed just like the English-speakers with regard to perfect and relative pitch. They suggested three possible explanations: that the Mandarin-speakers all used English as a second language (they were students in the USA); that they spoke several Chinese dialects, which would require a flexible use of tone; or that the pitch abilities for tonal languages might be used for the native language alone. Although Saffran and Griepentrog failed to detect any difference between their speakers of English and Mandarin Chinese, it is known from other studies that Asia has a higher frequency of adults with perfect pitch than is found in the West. This condition exists, however, irrespective of whether the Asian people speak a tonal language such as Mandarin Chinese or an atonal language such as Korean or Japanese, and it appears to reflect a genetic predisposition that is unrelated to language or other cultural factors.[19]

Saffran and Greipentrog's results indicated that a reliance on perfect pitch is most likely disadvantageous for learning language, whether atonal or tonal. However, it is equally the case that a total reliance on relative pitch might also prove problematic because it would provide infants with insufficiently fine-grained information about language – too many of the words they hear would sound the same. Consequently, Saffran and Griepentrog suggest that perfect pitch might be a necessary supplement to the attendance to relative pitch contours during the very first stages of language acquisition. However, because the ability to discriminate between pitch contours is improved by monitoring not only the direction but also the distance of contour changes, perfect pitch comes to be of less significance and eventually a positive hindrance. It becomes, in Saffran and Griepentrog's view, 'unlearned'.

This is a neat idea because it helps to explain why certain experiences in childhood, such as intense music practice, may enable the maintenance of perfect pitch. Indeed, the finding that exposure to music and musical training between the ages of three and six enhances the likelihood that children will maintain perfect pitch into adult life suggests that it is of some value for listening to and performing music.[20] If this were not the case, then one would expect an even lower incidence of perfect pitch among musicians than among non-musicians, because they would have both their language acquisition and their music acquisition acting against its retention. So the possession of perfect pitch must somehow enhance one's musical ability. It may do this by making practice more accurate and facilitating direct connections between physical movements or instrument keys and the desired pitch.[21]

That language acquisition involves the 'unlearning' of perfect pitch (unless it is retained by intense musical activity) explains why those musical savants who have cognitive deficits that prevent their acquisition of language maintain perfect pitch – they have no pressure to unlearn it. Leon Miller believes that it is their possession of perfect pitch that provides the foundation for the development of their musical talents. But this is not sufficient in itself: autistic children who lack language and have no special musical talent also have superior abilities over normal children at identifying and remembering musical pitches.[22] To summarize, in the course of infant development there is an erosion of an inherent musical potential by the demands of language acquisition, unless strenuous efforts are made to maintain perfect pitch.

The evolutionary parallel of this is intriguing: it suggests that prelinguistic hominids may have maintained a bias towards perfect pitch throughout their lives, and hence developed enhanced musical abilities in comparison with those found among language-using hominids, including ourselves. It is, of course, quite impossible to test this proposal. But the presence of a general musical sensitivity in monkeys, which may be based on perfect pitch, is encouraging.[23]

Singing to infants

This chapter has so far concentrated on the musical aspects of IDS, noting its role in the expression of emotion and development of social bonds prior to the eventual use of exaggerated prosody in order to facilitate language acquisition. But parents not only talk to their babies, they sing to them too, and this may have even greater emotional power. Sandra Trehub, a professor of psychology from Toronto University, has undertaken several studies with her colleagues to explore the significance of singing to infants. One set of studies explored lullabies and found a striking degree of cross-cultural uniformity in their melodies, rhythms and tempos. Adult listeners were found to be quite capable of discriminating between lullabies and non-lullabies when tested on songs in an unfamiliar language.[24] Trehub and her colleagues also examined the relative impacts of IDS and song.[25] They found that babies will spend significantly longer periods attending to audio-visual recordings of their mothers when they are singing rather than speaking.[26] Trehub demonstrated that six-month-old, non-distressed infants show a greater physiological response (gauged by the production of salivary cortisol) to their mother's singing than to their speech, indicating the significance of singing as a care-giving tool.[27]

That such responses are not wholly a reflection of socialization is shown by the physiological responses of newborn babies, as demonstrated by work in neonatal intensive care units carried out by Jayne Standley of Florida State University.[28] One of her studies showed that the singing of lullabies by a female vocalist significantly improved the development of sucking abilities in premature infants, and this resulted in measurable impacts on their weight gain. Music was also found to stabilize oxygen saturation levels, which enhances the physical development of premature infants. Premature infants subjected to a combination of music and massage were discharged an average of eleven days earlier than a control group of infants.

Alison Street, at the University of Surrey, has undertaken a particularly interesting study that focused on mothers as much as infants and was concerned with the natural setting of the home, rather than the contrived environment of the laboratory. Street carried out a survey of how frequently mothers sing to their infants and enquired about their attitudes to such singing, basing her work on a carefully controlled sample of 100 mothers, each with a baby of less than one year old. She found that all of the mothers sang to their infants in the privacy of their own home, even though half of her sample claimed that they lacked a singing voice. Frequent responses to the question of why they sang to their infants were 'to soothe, to entertain, to make the infants laugh, smile, gurgle and make sounds themselves'. A particularly interesting and frequent finding was that mothers believed that their singing reassured the infant of their presence. Some mothers also mentioned the fact that singing helped them relax and feel calm, and that it helped recall their own happy childhoods when their own mothers sang to them.

While these results are of considerable academic interest, they also have practical applications, suggesting that health professionals can support mothers by encouraging them to sing to their infants. Street suggests that if mothers could be helped to develop a greater awareness of their infants' musical competencies they would be encouraged to sing even more. This would then help foster a sense of well-being in both mother and infant, confirm their developing companionship, and help empower mothers in supporting their infants' development.[29]

Sandra Trehub is rather more explicit about the biological significance of maternal singing:

> To the extent that maternal singing optimizes infant mood, it could contribute to infant growth and development by facilitating feeding, sleeping, and even learning. Children's extended period of helplessness creates intense selection pressures for parental commitment and for infant behaviours to reward such commitment. Falling asleep to lullabies or entering trance-like states to performance of other songs might be suitable reward for maternal effort. In general, favourable consequences of maternal singing on infant arousal, whether through cry reduction, sleep induction, or positive affect, would contribute to infant well-being while sustaining maternal behaviour. Presumably, the healthy and contented offspring of singing mothers would be

more likely to pass on their genes than would the offspring of non-singing mothers.[30]

'Born with a kind of musical wisdom and appetite' . . .

Much of the material in this chapter can be encapsulated in this phrase, used by Colin Trevarthen, emeritus professor of child psychology at the University of Edinburgh. In those words he has summarized his assessment of the infant's mind, following a lifetime of research on mother–infant interactions.[31] His work has stressed how infants not only listen to the music coming from their carers' voices, but also produce music in their own coos and movements.[32] He, too, is worth quoting at length:

> A mother greets her newborn in ecstatic cries with falling pitch, and by gentle fondling. She is unable to keep her eyes from the baby's face. She touches hands, face, body with rhythmic care, and holds the infant close to her so they can join attention and greet one another. Her speech is a kind of singing, with high gliding gestures of pitch and repetition of regular gentle phrases on a graceful beat, leaving places for the infant to join in with coos, smiles and gestures of the hands and whole body. Every move of the baby elicits a reaction from her. These exchanges are intricately coordinated with a subdued choreography that brings out matching rhythms and phrasing in mother and infant.[33]

. . . and ready to laugh

One of Alison Street's findings was that we sing to our babies with the explicit intention of entertaining them, to make them smile and laugh. We need to spend a few moments on laughter, since it is such a key element of parent–infant interactions in humans. Like all other utterances and gestures, laughter frequently occurs in a turn-taking fashion and is characterized by the same music-like qualities than typify IDS. Infants begin to laugh between fourteen and sixteen weeks of age, and it often regulates the parent–infant interaction; laughter prompts you to repeat whatever you did that made you laugh. Mothers laugh, and have concurrences of speech with laughter, far more frequently when interacting with infants than with older children and other adults.

As a ubiquitous and universal type of human utterance, laughter has been neglected as a research topic, hence its specific role in developing adult–infant social bonds remains unclear. Robert Provine, professor of psychology at the University of Maryland, is one of the few academics to take laughter seriously. He believes that laughter is an ancient form of social signalling, one that is more akin to animal calls and birdsong than to human speech. The fact that laughter is contagious – which TV producers exploit to great effect by adding dubbed-in laughter – suggests that it is a means of social bonding.

We are all aware that laughter can be a powerful social tool: it breaks the ice with strangers, creates bonds, generates goodwill, and reduces aggression and hostility. Businessmen use laughter to increase their rapport with clients; we all use it when flirting. By 'laughing at' rather than 'laughing with' someone else, laughter can be used to create offence, to express or create a power relationship, and to exclude a person from a social group.

Provine's studies have included eavesdropping on groups of adults and children and recording gender differences in the extent of laughter. Whether they are speaking or listening, women laugh a great deal more than men, while both male and female audiences laugh a great deal less when listening to a woman than to a man – hence the lot of the female comic is not an easy one. Provine has also studied how laughter is embedded into speech, and found that it usually comes at the end of an utterance, rather than interrupting it in the middle, suggesting that some neurological priority is given to the speech.

The propensity for laughter must be part of our evolved psyche rather than a purely learned behaviour. Children who are born deaf and blind, and hence have never heard laughter or seen a smile, will nevertheless laugh and smile when tickled. Moreover, laughter-like vocalizations are present in our closest living relatives, the chimpanzees, who produce them when tickled or during play. These sound different from human laughter because they are produced while breathing in and out, whereas human laughter arises from 'chopping up' a single expiration. Chimpanzee laughter also lacks the discrete, vowel-like notes of human laughter (for example, ha, ha, ha). Indeed, if one were simply to listen to chimpanzee laughter without watching the animal itself, one might not recognize it as laughter at all.

While chimpanzee laughter occurs almost exclusively during physical contact or in response to the threat of such contact – for example, during a chase – most adult human laughter occurs during conversation, without any physical contact. My intuitive assumption is that laughter in human babies is likely to be more similar to that of chimpanzees than that of adult humans, in terms of being more closely associated with physical contact. But as far as I can ascertain, the required research has yet to be undertaken.

Some argue that rats, far more distant relatives of humans than chimpanzees, also emit vocalizations when they are tickled that could be interpreted as laughter.[34] Young animals, such as puppies, emit vocalizations during play which appear to signal that play-fights are not serious. It is not surprising, therefore, that neuroscientists believe that human laughter arises from activity in the evolutionarily oldest parts of the brain, notably the limbic system. This contains structures that are also found in the brains not only of other mammals but also of lizards and toads. Cortical activity is also important for human laughter. One experiment conducted on an epileptic woman found that low-intensity electrical stimulation of a specific part of her cortex caused her to smile; when the intensity was increased, she had fits of laughter.

While a great deal about human laughter remains to be explored, this limited evidence suggests that it is an evolved part of the human mind and that its development in early infancy is facilitated by IDS, singing, tickling, facial expressions and play. It benefits the infant because it learns how to use laughter in social interactions that will help it during childhood, adolescence and adulthood. The same conclusions can be drawn about IDS when it is used to facilitate language acquisition – that is, learning not just the meanings of utterances but also how turn-taking is used in communication.

Brain development

To bring this chapter to a close we must return to the brain, because we are ultimately concerned with the impact of IDS, song, laughter and play on the types of neural networks that are formed as a baby matures.[35] We considered the adult brain in terms of its mental modules for language and music, and where these might be located, in the previous chapter. Here, we are concerned with how they might develop and the relative influence of genes and environment on that process.

The brain of a newborn human baby is 25 per cent of its adult size. As virtually all of the neurons of the adult brain are already present at that stage, the increase in size almost entirely arises from the growth of dendrites and axons. Also occurring is the process of myelination – the growth of glial cells. These 'insulate' the axons and increase the speed at which nerve impulses travel within them. Although myelination continues throughout life, it is most intense shortly after birth and plays a critical role in how the brain develops.

The formation of synapses, known as synaptogenesis, arises from the growth of axons and dendrites, and is a major factor in brain development soon after birth. It appears that some neurons release chemicals that act as either attractors or repellents to axons, and the concentrations of these chemicals then influence the direction of growth of axons. Dendritic growth responds to the approach of axons. Millions of new synapses are formed on a daily basis, and these are linked into neural circuits, conditioning the later cognitive abilities of the child.

Since many of the new synapses do not survive into later life, one of the key theories regarding brain development has become known as 'neural

Darwinism'.[36] This argues that there is an initial overproduction of synapses, followed by a selective culling. 'Use it or lose it' seems to be the message: those neural circuits that are most used and stimulated are retained, producing a cognitive ability that is most suitable for the particular environment into which the infant has been born. This has been demonstrated by experiments in which animals are raised in conditions of sensory deprivation and then have their brains examined. Those raised in environments made up solely of horizontal or vertical lines develop brains that are more active in the presence of such lines; similarly, those raised in conditions in which perception of movement is constrained come to be less sensitive to movement. At the most extreme, total visual deprivation can result in the loss of dendritic branching and synapses in that part of the brain responsible for vision, which has simply not been activated by environmental stimuli.

While such evidence supports neural Darwinism there also seems to be a strong case for 'neural constructivism' – the growth of dendrites, axons and synapses in response to environmental stimuli.[37] Animals raised in 'enriched' environments, and which may be subjected to specific training, display differences in brain structure from those raised in sterile laboratory environments. Rats, cats and monkeys have been recorded as showing a more than 20 per cent increase in dendritic and synaptic growth when subjected to environmental stimulation during early development.

We can place the evidence for selective culling of existing neural circuits together with that for the construction of new circuits as a result of environmental stimuli and simply note that the brain displays immense plasticity as it grows. The extent of the environmental stimuli may often be minimal, effectively just a trigger to set neural development on course. This may, for instance, be the case with language learning, in light of the 'poverty of the stimulus' argument: as mentioned earlier, Chomsky and many other linguists have argued that the environmental inputs a child receives – that is to say, the utterances it hears – are just too few and too ambiguous to allow for anything more than the triggering and fine-tuning of what they call a 'Universal Grammar'.

Whether we are dealing with neural Darwinism or neural constructivism, or some combination of the two, it is evident that those who use facial expressions, gestures and utterances to stimulate and communicate with their babies are effectively moulding the infants' brains into the appropriate shape to become effective members of human communities, whether we think of those as families or societies at large. Parents largely do this on an intuitive basis – they do not need to be taught IDS – and use music-like utterances and gestures to develop the emotional capacities of the infant prior to facilitating language acquisition.

7 Music hath charms and can heal

Music, emotion, medicine and intelligence

The seventeenth-century poet William Congreve claimed that 'Music has charms to soothe a savage breast / To soften rocks, or bend a knotted oak'. Few would disagree: music moves us, whatever our taste.[1] It is surprising, therefore, that the scientific study of the relationship between music and emotion has been limited. Although two classic works were published in the later 1950s, Leonard Meyer's *Emotion and Meaning in Music* and Deryck Cooke's *The Language of Music*,[2] the scientific study of music and emotion was neglected in the latter part of the twentieth century.[3] Only recently has the topic begun to receive the attention it deserves, exemplified in the compilation of academic studies edited by Patrik Juslin and John Sloboda entitled *Music and Emotion* (2001).[4]

The popular characterization of music as the 'language of emotion' is one reason why many modern commentators have been unwilling to accord music itself much significance. Traditionally, the emotions have been thought of as the antithesis of human rationality, our most valued cognitive ability.[5] This notion began with Plato in 375 BCE, who argued that our emotions arose from a lower part of the brain and perverted reason. Charles Darwin furthered this idea in his 1872 book, *The Expression of Emotions in Man and Animals*. This was a truly remarkable book, which made the first use of photography as an aid to scientific argument and which began the systematic study of emotion. But Darwin's aim was simply to use the manner in which people express their emotions as a means to support his argument that we evolved from ape-like ancestors. For him, emotional expression was a vestige of our animal past, and one now lacking any functional value because emotion had been surpassed by human reason. Think of the way we describe people as being 'over-emotional' and you'll realize that we still feel that way today.

The last decade, however, has witnessed a sea change in academic attitudes towards human emotions, which has now placed them at the centre of human thought and the evolution of human mentality. As the philosopher Dylan Evans explained in his recent book, *Emotion: The Science of Sentiment*, the ultra-rationality of Mr Spock from *Star Trek*, a creature who has eliminated emotion from his decision-making, could in fact never have evolved because emotion lies at the root of intelligent action in the world.[6]

Emotions as guide to action

The twentieth century saw diverse approaches to the study of emotion and an enormous literature developed around the topic, ranging from the classic works of Sigmund Freud and William James to recent work by the neuroscientists Joseph LeDoux and Antonio Damasio.[7] It was once thought that all emotions are culturally constructed and hence specific to each particular society. This view has now been surpassed by the recognition that some emotions are universal to all members of our species.[8] Such emotions have been wired into the human genome by our evolutionary history, and that in itself suggests that they are more than mere mental frippery.

The precise number and definition of the universal or 'basic' emotions vary according to which academic authority is consulted, but they are always variants of happiness, sadness, anger, fear and disgust. No matter which human culture one visits, these emotions will be found, along with the common facial and bodily expressions that accompany them. One must assume that such emotions and expressions would also be found if one could visit those cultures of prehistoric *Homo sapiens* that no longer exist. Indeed, we can be confident that such emotions existed among all of our ancestors and relatives back to at least 6 million years ago, because they are also possessed by the modern great apes.

Meticulous observations of chimpanzee social life by Jane Goodall and others have described them behaving almost identically to human beings when in highly emotional states: infants having temper tantrums, males acting aggressively, juveniles playing lovingly with each other and with their mothers.[9] Although we have no knowledge of the chimpanzees' subjective experiences – just as we cannot penetrate the feelings of another human being – it would be perverse not to attribute to them such feelings as anger, sadness, happiness and fear.

Whether we should also attribute to apes what psychologists describe as 'complex emotions' is more contentious. These are emotions such as shame, guilt, embarrassment, contempt and love. These are far harder to identify in chimpanzee behaviour – and also more difficult to claim as human universals, because they may be dependent upon the development of one or more basic emotions within a particular cultural and historical setting.

Why have emotions?

But why should humans and chimpanzees have emotions? The argument I wish to follow is that which has been most fully developed by the psychologists Keith Oatley and Philip Johnson-Laird, although its basic elements are shared by numerous other theories.[10] The starting point is to recognize that any living being constantly has to choose between different courses of action: should one continue with a current task or do something else? should one cooperate or compete with another individual? should one flee or fight when faced with a bully or a potential predator? An ultra-rational creature, such as Mr Spock, would solve such problems entirely by logic, carefully measuring the costs, benefits and risks of each alternative and then undertaking some mental mathematics to identify the optimal choice. But Mr Spock does not live in the real world. When we and other animals have to make decisions, we must do so with partial information, conflicting aims and limited time. Under these pressures, it is simply impossible to undertake a logical search through all the possibilities and identify the optimal course of action.

Oatley and Johnson-Laird argue that emotions guide action in situations of imperfect knowledge and multiple, conflicting goals – a state that is otherwise known as 'bounded rationality'. Emotions alter our brain states and make available repertoires of actions that have previously been useful in similar circumstances. When undertaking any task, we set ourselves goals and sub-goals, either consciously or subconsciously. We feel happy when such goals are achieved, and this acts as an emotional signal to continue with the task. If, however, we feel sad, this signals us to stop the task altogether, to search for a new plan, or to seek help. Anger signals that the task is being frustrated and that we should try harder; while fear signals us to stop the current task and attend vigilantly to the environment, to freeze or to escape. Oatley and Johnson-Laird explain other emotions, such as love, disgust and contempt, in a similar fashion. Our emotions, therefore, are critical to 'rational' thought; without them we would be entirely stymied in our interactions with the physical and social worlds.

It is the latter of these, the social world, that provides the greatest cognitive challenge to human beings. Indeed, it is coping with the demands of living in large social groups that provides the most likely explanation for the origin of human intelligence. And so it is not surprising that our more complex emotions relate directly to our social relationships. Without such emotions we would be socially inept; in fact, we would be unaware of the complexities and subtleties of the social world around us, and would fail entirely in our social relationships.

Robert Frank, an economist from Cornell University, made the most persuasive argument to this effect in his 1988 book, *Passions Without Reason*.[11] He argued that emotions such as guilt, envy and love predispose us to behave in ways that may be contrary to our immediate self-interest but that enable us to achieve social success in the longer term. Hence he describes emotions as having a 'strategic' role in our social lives. As important as having such emotions is the fact that others know that we possess them. Guilt is the classic example. If others know that you are liable to suffer from guilt, they will be more likely to trust you, because they will know that you are less likely to double-cross them; even though to do so may be in your own immediate interest, your anticipated feelings of guilt will prevent you from taking that course. Similarly, knowing that your partner 'loves' you enables you to enter into substantial commitments with them, such as sharing property and children, without the fear that they will leave you as soon as someone physically more attractive or wealthy comes along. If such emotions were absent and all decisions were made according to short-term calculations of 'rational' self-interest, the extent of cooperation within society would be minimal and would be restricted to those to whom one is closely biologically related. As Robert Frank made clear in his book, human society is simply not like that, for acts of cooperation and kindness from people who stand to gain no immediate benefit are pervasive.

It is for this reason that I strongly suspect that complex emotions were present in the societies of all large-brained hominids. Species such as Neanderthals lived in social groups much larger and more complex than those of non-human primates today; as I will make clear in chapters 14 and 15, cooperation was essential for their survival. The Neanderthals could only have survived for so long in the challenging ice-age conditions of Europe if they were not only capable of feeling happy, sad and angry, but also suffered the pains of guilt, shame and embarrassment, along with the elation of love.

Letting it be known

Robert Frank stresses the significance not only of experiencing emotions but also of displaying them to others via facial expressions, body language and vocal utterances. More than twenty years of study by the psychologist Paul Ekman have demonstrated that facial expressions for anger, fear, guilt, surprise, disgust, contempt, sadness, grief and happiness are universal – common to all people in all cultures.[12] Indeed, Charles Darwin himself, in *The Expression of Emotions in Man and Animals*, showed that the bases of such expressions are also found in apes. So we must expect that human ancestors were also expressive of their emotions, perhaps in a very similar fashion to modern humans.

The shoulder shrug with upturned palms and a raised brow, used to express a lack of knowledge or understanding, has been claimed as a universal gesture. Other elements of body language have not, to my knowledge, been studied on a cross-cultural basis, but their universality would not be surprising: arms folded across the chest as a defensive gesture; the hand-to-mouth gesture when telling a lie or covering embarrassment; rubbing the hands together to indicate a positive expectation; differing types of handshakes and kisses to indicate different attitudes towards the person being greeted.

Expressions and body language can be faked and hence one may seek to gain advantage by pretence. There are, however, limits imposed by the uncontrollable muscle movements and other physiological events that express one's true feelings. Ekman describes some facial muscles as 'reliable' in that they cannot be deliberately manipulated. Only 10 per cent of people, for instance, can intentionally pull the corner of their mouth down without moving their chin muscles. But almost everyone does so, without any intent, when they experience sadness or grief. People also constantly give away their true feelings by 'micro-expressions' – fleeting expressions or gestures, which might then be replaced by a quite different expression intended to reflect the emotion the individual wishes to project. Also, physiological changes, such as dilation of the pupils, blushing and perspiration, are very difficult – some say impossible – to fake and will give away feelings of high arousal. It is difficult, too, to prevent the pitch of one's voice increasing when one is emotionally upset.

We generally place considerable value on being able to 'read' the emotional state of someone from their expression, especially that immediately around the eyes. Simon Baron-Cohen, the Cambridge psychologist, claims that women are rather better at doing this than men and that this may reflect different types of cognitive skills.[13] Whether or not that is the case, it is evident from his work, along with that of Frank and Ekman, that humans have evolved not only emotions, in order to enable them to act intelligently in the world, but also the propensity to express them and the ability to interpret such expressions in others. One of the key means of such expression is the creation of music.

The language of music

We often make music, either by singing, playing instruments or listening to CDs, in order to express how we are feeling. My own house reverberates with Vivaldi's Gloria on sunny Sunday mornings, while the dirges of Billie Holiday seep from my study when I'm feeling down. But we also use music for something else: to induce an emotional mood within ourselves and within others. Hence when I am feeling 'old', I play Britney Spears, which is, as my children tell me, 'rather sad'; and when I need to stop them squabbling I fill the house with the music of Bach.

A mood is slightly different from an emotion; the former is a prolonged feeling that lasts over minutes, hours or even days, while the latter may be a very short-lived feeling. Dylan Evans, for instance, would describe the feeling of joy as an emotion, but happiness as a mood which one might enter into as a consequence of a joyful experience.

We all know that specific types of music can induce specific types of mood; we play soft, romantic music on a date to induce sexual love, uplifting, optimistic music at weddings, and dirges at funerals. Enlightened factory managers play music to improve employee morale when they have to undertake simple repetitive jobs, while dentists and surgeons use music to soothe and relax their patients, sometimes with such astonishing results that anaesthetics become redundant. Psychologists use music when they wish to induce particular moods in their subjects prior to setting them an experimental task to complete. When the psychologist Paula Niedenthal of Indiana University and her colleague Marc Setterlund needed subjects who were feeling happy they played them Vivaldi and Mozart; when they required sad subjects they used music by Mahler or Rachmaninov.[14]

In his 1959 book, *The Language of Music*, the musicologist Deryck Cooke attempted to identify which elements of music express and induce which emotions. He focused on the tonal music of Western Europe between 1400 and 1950, and argued that the source of its emotive power lies in the systems of relationship between different pitches – what he refers to as 'tonal tensions'. He began his work by supporting the commonly held idea that major scales express positive emotions such as joy, confidence, love, serenity and triumph, while the minor scales express negative emotions such as sorrow, fear, hate and despair. Such associations arose, Cooke argued, from the long-term history of Western music: having been used by some composers, the same associations were then used again and again until they became part of Western culture. He noted that other musical traditions, such as those of India, Africa or China, use quite different means to express the same emotions.

I am cautious about Cooke's reliance on a culturally inherited association between particular musical expressions and particular emotions, suspecting that there are associations universal to all humans that have yet to be fully explored. John Blacking was also cautious when he reflected on Cooke's ideas back in 1973:

> I find it hard to accept that there has been a continuous musical tradition between England in 1612 and Russia in 1893, in which certain musical figures

> have had corresponding emotional connotations. The only justification for such an argument would be that the emotional significance of certain intervals arises from fundamental features of human physiology and psychology. If this is so, some relationship between musical intervals and human feelings ought to be universal.[15]

Ongoing research is beginning to demonstrate that such universal relationships do indeed exist. In 2003, Hella Oelman and Bruno Lœng, psychologists from the University of Tromsø, demonstrated a significant degree of agreement between the emotional meanings that different individuals attributed to a particular musical interval. Their subjects were Norwegians who had been acculturated to a Western musical tradition. Oelman and Lœng noted that precisely the same associations between particular emotions and particular musical intervals had previously been found in connection with ancient Indian music – a radically different musical tradition. Their conclusion, that 'these meanings seem to be relevant for the emotional experience of musical intervals of humans across cultures and time, and thus might be universal',[16] confirms Blacking's intuition. Indeed, Cooke himself often appears to be sympathetic to this view, for he frequently refers to the 'naturalness' of the emotions arising from a particular tonal tension.

Cooke systematically worked his way through the relationships of the twelve-note tonal scale, drawing on musical examples from plainsong to Stravinsky to illustrate how particular relationships express particular emotions. The major third is, according to Cooke, an expression of pleasure or happiness; for this reason it forms the basis for Beethoven's 'Ode to Joy' in the Choral Symphony, and is used by Verdi in *La Traviata*, Wagner in *Das Rheingold*, Stravinsky in his *Symphony of Psalms*, and Shostakovich in his Fifth Symphony. Moreover, wishing to avoid the claim of elitism, Cooke explained that 'Polly-Wolly-Doodle' shows the same use of the major third at a popular level. Conversely, when such composers wished to express painful emotions they bring the minor third into prominence, as in the 'doom-laden' first movements of Beethoven's, Tchaikovsky's and Mahler's Fifth Symphonies. Overall, Cooke claimed, 'the strong contrast between the "natural" pleasurable major third and the "unnatural" painful third has been exploited throughout musical history'.[17]

Cooke continued his analysis with each of the other relationships in the twelve-note scale, making arguments such as that the minor seventh expresses mournfulness while the major seventh expresses violent longing and aspiration. Such tonal relationships, however, provide no more than a basis for the 'language' of music – Cooke described volume, rhythm, tempo and pitch as the 'vitalizing agents' of a piece of music.

Thus the louder the music is played, the more emphasis is given to whatever emotion is being expressed; conversely, the softer, the less emphasis. Rhythm and tempo are used to throw emphasis onto particular notes within a tonal sequence and hence to qualify the particular emotion being expressed. Cooke argued that the joy expressed by a certain progression of tonal tensions may be tumultuous if the tempo is allegro, easy-going if moderato and serene if adagio; the despair expressed by another progression may be hysterical if the tempo is presto or resigned if lento.

While pitch is felt by everyone to be the 'up-and-down' dimension of music, Cooke argued that we should also think of it as the 'out-and-in' and the 'away-and-back' of music. By doing so, we can more readily appreciate the significance of rising or falling pitch for the emotional content of music: 'The expressive quality of rising pitch', Cooke explained, 'is above all an "outgoing" of emotion: depending on the tonal, rhythmic and dynamic context, its effect can be active, assertive, affirmative, aggressive, striving, protesting, or aspiring. The expressive quality of falling pitch is of an "incoming" of emotion: depending on context, it can be relaxed, yielding, passive, assenting, welcoming, accepting, or enduring.'[18]

So, to rise in pitch in a major scale is normally to express an outgoing feeling of pleasure: an excited affirmation of joy, as in the fast, loud climax of Beethoven's Overture *Leonora* No. 3; a calm, emphatic affirmation of joy, as in the slow and loud 'Gratias' in Bach's B Minor Mass; or a calm, quiet, joyful aspiration, as in the slow and soft ending of Vaughan Williams's Fifth Symphony. Conversely, to fall in pitch in a minor scale is to express an incoming feeling of pain, such as fierce despair in the slow and loud opening of the finale to Tchaikovsky's *Pathétique* Symphony. Although these examples drawn from Cooke's work are very specific, he argued that all music has a continuous fluctuation of pitch which accords with the ebb and flow of the emotions being expressed.

Does it work in practice?

I have provided no more than a brief summary of Cooke's key arguments about the 'language of music'. His book contains many more examples – addressing, for instance, the role of repetition and staccato in providing further range to emotional expression. With very few exceptions, it draws on the works of great composers, which traditionally have been, and still are, played by great orchestras to the relatively small section of society that has the opportunity and/or the desire to attend concerts. But Cooke's arguments are supposed to be general and therefore should apply to all forms of popular music and all contexts of performance. Moreover, our concern is with the intuitive understanding that people have of how music can express emotion. So here we should turn to the work of Patrik Juslin, a psychologist at Uppsala University, and his careful experimentation and detailed statistical analyses investigating how variations in tempo, loudness, rhythm, pitch and so forth induce different types of emotions.[19]

In 1997, Juslin published the results of two related sets of experiments, in which the music was generated by electric guitars. First, he wanted to identify how musicians themselves sought to express specific emotions through their instruments. He gave three professional musicians five different melodies to play, chosen to cover somewhat varying emotional characters in their melodic structures: 'Greensleeves', 'Nobody Knows', 'Let It Be', 'When the Saints Go Marching In', and 'What Shall We Do with the Drunken Sailor'. They were asked to play each piece in whatever manner they thought appropriate so that they would sound happy, then sad, angry, fearful, and finally emotionally neutral. The musicians were not allowed to make any changes to the pitches, melodies or guitar sound. Under these constraints, they could play the pieces as they wished, but they would do so with no knowledge of how the others chose to express the specified emotions.

The results showed considerable concordance among the musicians, indicating that distinct styles of playing were associated with each of the intended emotional expressions. When seeking to express anger, the music was played loudly, with a fast tempo and legato articulation; for sadness, a slow tempo, legato articulation and low sound levels were chosen; happiness was expressed with a fast tempo, high sound level and staccato articulation; fear was associated with a low sound level, staccato articulation and a slow tempo.

However, identifying the manner in which particular emotions were musically expressed was just the first part of Juslin's experiment. The second was to explore whether listeners were able correctly to identify which emotion the musicians were attempting to express. For this, Juslin used twenty-four students. Half of these were musically trained; half had no close involvement with music. They were asked to listen to fifteen performances of 'When the Saints Go Marching In' and to rate each one as to whether it was happy, sad, angry or fearful.

Juslin found a very strong correlation between the emotion intended by the musicians and that which the listeners believed was being expressed. Happiness, sadness and anger were all relatively easy to identify; fear was a little more difficult but was successfully recognized in the great majority of 'fearful' performances. The listeners who were trained musicians had no greater success than those without expert musical knowledge, but women had slightly greater success than men. Juslin had expected this gender difference – research, as mentioned above, has shown that women are better at recognizing emotional states from facial expressions and body language.[20]

Using the voice

While Juslin's work has shown that one requires neither an orchestra nor a score by Beethoven, Vivaldi or Tchaikovsky in order to express emotion through music and for listeners to be able to identify that emotion, the music from electric guitars is rather distant from that which our human ancestors would have made. But similar experiments exploring how emotion is expressed by the voice alone have produced similar results. Klaus Scherer, a psychologist from the University of Geneva, has specialized in this area.[21] Once again, it is intuitively obvious to all of us that this is a routine and key aspect of human speech; orators throughout history, from Cicero to Churchill, have used the voice to induce emotions and stir people to action. Yet our scientific understanding of how vocal acoustics express emotion remains limited.

Scherer's experiments required that any semantic meaning be removed from the words so that the emotions were expressed via acoustic properties alone. Like Juslin, he tested whether listeners could correctly identify the intended emotion. In one experiment, the voice was employed to list the alphabet or a numerical sequence, while listeners attempted to recognize what emotion it was endeavouring to express. His findings confirmed those of numerous other investigators that a success rate of at least 60 per cent is usual, even when the complex emotions of jealousy, love and pride were involved in addition to the basic emotions of anger, sadness, happiness and fear. Of these, sadness and anger appear to be the easiest to recognize.

Some of Scherer's other experiments have attempted to identify the relationship between specific acoustic properties, such as pitch level, tempo, tonality and rhythm, and the specific emotion expressed – just as Juslin did with his guitar players. The results were very similar. Thus vocal expression that was slow and at a low pitch was recognized as conveying sadness, while large pitch variation and a fast tempo were identified with happiness.

The analysis by Cooke and the experiments of Juslin and Scherer place our intuitive knowledge and common experience of how emotion is expressed by music on a scientific and objective footing – whether the music is generated by the voice alone, by an instrument, or by a complete orchestra.

Music can induce emotional states

Music induces emotional states both in those who perform and in those who simply listen. Although this is a widely held assumption, it is difficult to test formally. One can, of course, simply ask people how they feel after listening to a piece of music. But when making such subjective reports, listeners are liable to confuse the emotions they recognize in the music with those that they feel themselves.[22]

The only way of avoiding this is to rely upon physiological correlatives of emotional states, which can be objectively measured. Carol Krumhansl of Cornell University presented thirty-eight students with six excerpts of classical music that had been independently assessed as expressing emotions of sadness, fear, happiness and tension.[23] She wired the students so that their respiratory and cardiovascular systems could be monitored while they listened to the music and immediately afterwards. Significant physiological changes were found to have occurred, and these differed according to the piece of music, and hence emotion, to which the subjects had been exposed.

Music that was expected to induce sadness produced large changes in heart rate, blood pressure, skin conductance and temperature; 'fearful' music produced large changes in pulse rate and amplitude; 'happy' music produced large changes in respiratory patterns. Such results have been replicated and developed by further experiments, but their utility has been questioned because of their reliance on the unfolding of a piece of music over time rather than the identification of the specific musical structures that induce specific emotions.[24]

The possible mechanisms by which music induces emotional states have been considered by Klaus Scherer and his University of Geneva colleague Marcel Zentner.[25] A piece of music might be subconsciously appraised in the same way as visual stimuli – a snake or a spider, for example – are automatically appraised and provoke an emotional response. Alternatively – or, more likely, in addition – a piece of music may cue memories of a past emotional experience, or else facilitate empathy for the performer. The last case might arise from mimicking the same movements as the performer, resulting in the achievement of a similar physiological state.

Scherer and Zentner argued that four factors influence the extent to which an emotional state will be induced by a piece of music: (1) the acoustic qualities of the music itself – its melody, rhythm, tempo, loudness and other features; (2) the manner in which it is performed – as was evident from Juslin's guitar-playing experiments; (3) the state of the listener, in terms of their musical expertise, general disposition and current mood;[26] (4) the context in which the music is performed and heard – whether it is a formal or informal occasion, whether it occurs without interruption, and any other factors that might influence the acoustics and ambience of the listening experience.

Music as medicine and therapy

In some cases, musicians deliberately create the highest possible emotional impact by carefully manipulating all of the factors just listed. A key example of this is in music therapy,[27] which is used to support a wide range of adults and children who have a variety of healthcare needs. The success of music therapy further demonstrates how music can be used both to express and to arouse a wide range of emotions, and also to lead to substantial improvements in mental and physical health. There is perhaps no better rebuttal of Steven Pinker's claim that music is biologically useless than the achievements of music therapy.

Music therapy as an explicit set of practices first developed in the West during the twentieth century – especially during the First World War, when doctors and nurses witnessed the effect that music had on the psychological, physiological, cognitive and emotional states of the wounded. The first major academic study of music's medicinal properties was published in 1948, partly as a response to the continued use of music therapy in military hospitals and in factories during the Second World War.[28] Music therapy is now widely used for those with mental and/or physical disabilities or illnesses. In chapter 10 I will describe some striking results of how music therapy has enabled those suffering from brain degeneration to regain physical movement. My concern here, however, is with the emotional impact of music therapy.

One of its most significant functions is to relax patients who are preparing for, undergoing or recovering from surgery, notably dental, burns and coronary treatments. It is now well attested that music with slow, steady tempos, legato passages, gentle rhythms, predictable change, simple sustained melodies and narrow pitch ranges is conducive to relaxation and can lead to a significant decrease in anxiety. By the use of such 'sedative music' doctors have been able to reduce doses of anaesthetics and other pain-relief medication, while patients have benefited from shorter recovery periods and have expressed higher degrees of satisfaction.[29] To quote one recent report on the use of music at the Spingte Clinic in Germany:

> Fifteen minutes of soothing music lulls the patient into such a state of well-being that only 50 per cent of recommended doses of sedatives and anaesthetic drugs are needed to perform otherwise very painful operations. Indeed, some procedures are now undertaken without any anaesthetic ... More invigorating music then alerts the patients' systems so they can actively respond to the surgeon. Once this is complete the music then takes the patient back into a relaxed state of recovery.[30]

During the 1990s, the psychologist Susan Mandel worked as a music therapist at the Lake Hospital in Ohio.[31] While her reports provide further evidence for the healing power of music, they also demonstrate some of the difficulties in evaluating the effectiveness of such therapy. Her particular concern was the use of music to reduce stress in those who had suffered heart attacks, as stress increases the likelihood of a second attack. Mandel introduced music therapy sessions for outpatients at Lake Hospital, 60 per cent of whom were judged to be at risk from stress. The sessions were for small groups or individuals, and used either live or taped music in conjunction with verbal discussion to encourage the expression of feelings and to reduce anxiety. Mandel judged the scheme to have been a considerable success. But, as with the majority of music therapy schemes, any formal measurement of success is difficult in the absence of a control sample. Mandel's case is also complicated because her music sessions involved lyrics, and it could just as easily have been the words rather than their musical expression that reduced stress and anxiety. However, judging from the high numbers of patients who chose to return for more than one music therapy session, Mandel appears to be justified in arguing that music was having some level of positive effect.[32]

In other applications, music therapy is used to stimulate rather than to sedate the participants – to enhance self-esteem or to facilitate personal relationships. It is increasingly used with those suffering from mental disorders such as autism, obsessive–compulsive disorder and attention-deficit disorder.[33] In these cases it has been shown to foster cooperation, promote body awareness and self-awareness, facilitate self-expression, and reinforce or structure learning. But perhaps the most valuable aspect of music therapy is that it can be a self-help system – one does not need a therapist at all, just a piano to play, a bath to sing in, or a collection of CDs with which to manipulate one's own mood. Overall, Mandel appears to be quite right to argue that 'music is a potent force for wellness and sound health'.[34]

This would not be a surprise to a social anthropologist, as the role of music as a force of healing is the theme of several anthropological studies; one of the unfortunate features of music therapy literature is that it ignores the anthropological literature, and vice versa.[35] Evans-Pritchard's classic 1937 account of Azande magic was one of the first to describe witch doctors' séances during which divinatory medicines made of magical trees and herbs were activated by drumming, singing and dancing.[36] In the 1990s, three important anthropological studies were published concerning the use of music for healing by traditional societies in the Malaysian rainforest and in central and southern Africa.[37]

Neither would Mandel's statement be news to historians, as music therapy has in fact been used since antiquity. Peregrin Horden in 2000 edited a fascinating collection of essays that describe the use of music and medicine in the classical, medieval, Renaissance and early modern periods of Europe, as well as in Jewish, Muslim and Indian traditions. Another important collection appeared in the same year, edited by Penelope Gouk and entitled *Music Healing in Cultural Contexts*. I believe that this was the first study to draw together the work of music therapists, anthropologists, historians and others concerned with the healing powers of music into a single volume – and I very much wish I had been at the 1997 conference from which the volume arose.[38]

The consequences of being happy

We have so far established that: (1) emotions and their expression are at the very centre of human life and thought; and (2) music can not only express emotional states but induce them in oneself and in other individuals. Such findings have important implications for how the capacity for music evolved: they suggest that those individuals in the past who had, by chance genetic mutation, enhanced musical abilities may have gained a reproductive advantage. To further this argument, we must appreciate the advantage of being able to manipulate the emotional states of other individuals. We can begin by examining the consequences of being happy.

In the 1970s, the psychologist Alice Isen started to explore this idea through a series of simple yet ingenious experiments.[39] In one of these she gave her experimental subjects a perceptual–motor skills test and then told a random sample that they had passed – inducing a mildly happy state – and the remainder that they had failed. Each person was then presented with a situation that, they thought, was quite unrelated to the test: the sight of a stranger who had dropped a pile of books. Those who had been made happy were more inclined to help gather up the books than those who had been told of their failure.

Isen followed this up over the next two decades with a suite of experiments that involved various methods of inducing happiness. In all cases, she compared the behaviour of her subjects with that of others in a neutral or a negative mood. Many of the experiments related to quite simple situations such as might arise in the course of one's daily business. She found, for example, that individuals who were given free biscuits became more helpful and more willing to keep quiet when asked to do so in a library; those who had watched comedy films or had been given sweets were more responsive when asked to solve problems; and when children were rewarded with stickers they advanced more rapidly in their studies.

One of her experiments shows how happiness enables people to think more creatively.[40] Subjects were given either a funny film to watch, to induce happiness, or an emotionally neutral film. They were then given a problem to solve: how to fix a candle to a cork board attached to a wall, using only the candle, a box of tacks, and the matches to light the candle. The only way to achieve this was to empty out the tacks, pin the box to the wall, and stand the candle in the box. A significantly higher proportion of the 'happy' subjects than of those whose mood was neutral was able to solve the problem within ten minutes – 75 per cent in one experiment and 58 per cent in another, as opposed to 20 per cent.

Mood has also been shown to influence our perceptions of others and the judgements we make about them. One experiment undertaken by the psychologist Robert Baron involved inducing states of either happiness or sadness in subjects who were about to conduct an interview with a job applicant.[41] The applicant was an accomplice of the experimenter and always gave the same answers to the interviewer, which involved both negative and positive comments about herself. It was found that 'happy' interviewers rated the applicant far more positively than the despondent interviewers, both on her potential ability to undertake the job and on her personal qualities. They were also more inclined to recall the positive things that the applicant had said about herself than the negative things.

Self-evaluation is also influenced by one's mood. Joseph Forgas and Stephanie Moylan interviewed nearly a thousand people just after they had left the cinema, having watched films that had been classified as happy, sad or aggressive.[42] Questions were asked about many topics, including political figures, future events and their own life-satisfaction. Those who had seen a 'happy' film made judgements that were more positive, lenient or optimistic than those who had watched the sad or aggressive films. Such mood biases were universal, irrespective of the demographic background of the subjects.

Many more experiments such as these could be described. The point is quite simple but highly significant: one's emotional state influences the way one thinks and behaves. People who are happy tend to be more helpful and cooperative; they evaluate themselves and others more positively, and think more creatively. One is immediately struck by the thought of how good it would be if one were always surrounded by happy people – and if they were not happy on their own account, how nice it would be to induce a little happiness into their lives, perhaps by singing them a happy song.

Music can facilitate helpfulness and learning

That music can increase cooperation and helpfulness by inducing good moods has, in fact, been demonstrated experimentally. Having been inspired by Isen's work, Rona Fried and Leonard Berkowitz undertook a study with their students at the University of Wisconsin.[43] They divided them into four groups and induced different moods in three of them by playing them different pieces of music. Two selections from Mendelssohn's 'Songs Without Words' were chosen to instil a soothing mood in one group; Duke Ellington's 'One O'Clock Jump' was played to create feelings of excitement in another; and John Coltrane's 'Meditations' was used to instil negative emotions, of sadness and despondency, in the third group. The fourth, control group simply sat in silence for the seven-minute duration of the musical recordings.[44] The students had to complete a mood questionnaire both before and after listening to the music, and this confirmed that the music had made a significance difference to their feelings.

Just before they were dismissed, the experimenter asked for volunteers to help with another, quite unrelated experiment which would require anywhere between fifteen minutes and two hours of their time. They were requested to complete a form to specify whether they were prepared to help, and if so for what amount of time. This, of course, was the test of helpfulness – the experimenter wanted to discover whether the four groups varied in their willingness to help according to the type of music to which they had been listening.

This proved to be the case. Those who had listened to the Mendelssohn pieces turned out to be the most helpful, as measured by their willingness to help with the second experiment and the length of time they were prepared to offer. On both measures, the students who had listened to Coltrane's music, leading to adverse moods, were the least willing to be helpful.

The cognitive impact of music, achieved by inducing particular emotional moods, is the likely explanation of the so-called 'Mozart effect'. The belief that listening to Mozart can make children, babies and even the unborn smarter, has been particularly espoused in the USA. Listening to Mozart has indeed been shown to improve performance in short-term reasoning tasks, but the claims for general enhancements of intelligence are quite unfounded. The short-term impact should not be at all surprising. Mozart's music is often selected to induce calm and happy moods, and Isen's and others' experiments have shown that such moods can improve creative thinking. There is a Mozart effect, one that works via the manipulation of mood, and it is likely that it has been exploited by manipulative musicians throughout our evolutionary history.

Summary

We all intuitively appreciate that music is, in some form, the 'language of emotion'. This chapter has attempted to elucidate what that might mean as a basis for the evolutionary history of music and language that will follow in the remainder of this book. It began by explaining that emotions are fundamentally important to thought and behaviour – they provide a guide to action – and the fact that music is so intimately related to our emotional lives counters the argument that it might be some recent invention or a spin-off from some other evolved capacity, such as language. I summarized some of the arguments made by Deryck Cooke in *The Language of Music*, which attempted to identify the specific emotions expressed by tonal tensions within a piece of music and how such emotions are moderated by volume, rhythm, tempo and pitch. This led us to consider some more recent and scientifically based work by Patrik Juslin and Klaus Scherer investigating how music both expresses and induces emotional states, some of which can be detected by measures of physiological arousal. Emotional states have direct consequences for behaviour, and hence a person's behaviour itself can be influenced by the music they hear, as was illustrated in the studies by Fried and Berkowitz.

In summary, music can be used to express our emotions, and to manipulate the emotions and behaviour of others. In modern Western society, and probably in those of all modern humans, music is rarely used in this manner other than for entertainment, because we have a far more powerful means of telling someone what we are feeling: language. But there would have been a time when our ancestors lacked language even though they had a complex range of emotions and the need at times to influence the behaviour of other individuals. This is, in fact, the situation for our close relatives who are alive today, the great apes. And it is to these that we must now turn as we begin to examine how music and language evolved.

Part Two
The Past

8 Grunts, barks and gestures

Communication by monkeys and apes

My initial chapters have provided the pertinent 'facts' about music and language today – facts to which I will return in the final chapter of my study. This chapter begins the evolutionary history that will provide the required explanation, although it also remains with the present-day world. It concerns the natural communication systems of living non-human primates, principally the African apes since these are likely to be similar to the communication systems of our earliest human ancestors.

By 'natural', I mean those communication systems used in the wild, as opposed to those that have been acquired by chimpanzees and other primates in laboratory settings under human instruction and/or support. Since the 1950s, psychologists have explored the extent to which non-human primates can learn to communicate by symbols, most notably through the use of simplified computer keyboards, in an attempt to discover whether the cognitive foundations for language are present in these species. At the risk of making a sweeping generalization, such work has found that chimpanzees and bonobos can indeed learn to use symbols to communicate. But few have learnt more than 250 symbols,[1] and they appear to possess minimal ability, if any at all, to combine symbols to make complex phrases.[2]

It is not certain whether the use of symbols by apes has any relevance to the understanding of the cognitive foundations of language; I have severe doubts and believe that these experiments tell us no more than that primates are quite 'clever', in the sense of being able to learn and to use associations between symbols and their referents. In other words, it is quite easy to generate a limited number of additional neural circuits in their brains by providing the right stimuli – in this case, interaction with language-using humans. Consequently, my concern is principally with the communication systems of non-human primates in the wild, because it is within these that the roots of language and music are to be found.[3]

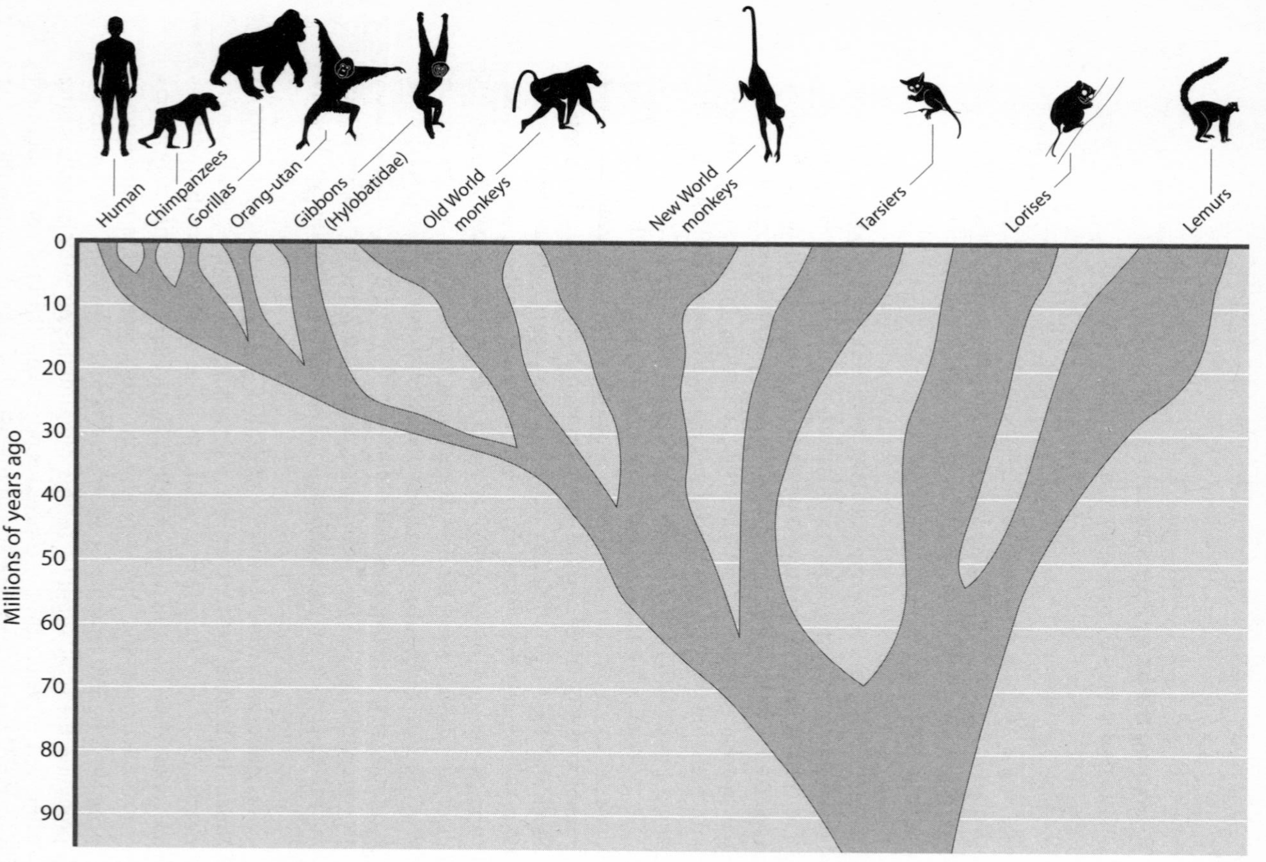

Figure 6 Evolutionary relationships between living primates, as established by molecular genetics.

The African apes

There are three species of African apes: the gorilla, the bonobo, and the chimpanzee. We most likely shared a common ancestor with all of these apes at some time between 8 and 7 million years ago. After that date there would have been two separate lineages, one of which led to the modern-day gorilla. The other line saw a further divergence at between 6 and 5 million years ago, with one branch leading to *Homo* and the other splitting at between 2 and 1 million years ago, leading to the chimpanzee on the one hand and the bonobo on the other.[4]

In view of their close evolutionary relatedness, it is not surprising that the African apes share gross similarities in the acoustic structures of their calls, these being variations on grunts, barks, screams and hoots.[5] Indeed, some of the variation reflects no more than size; bonobos have higher-pitched calls than chimpanzees because their vocal tracts are smaller.[6] It is reasonable to assume, therefore, that the ancestors we shared with these apes at between 8 and 5 million years ago also had a similar repertoire of calls, and that these provided the evolutionary precursor to modern human language and music.

The great ape communication system is not particularly distinguished; indeed, the African apes have no special vocal skills at all, so they seem to provide a most unpromising start for my evolutionary history of language and music. Surprisingly, the most language-like and music-like calls are found in primates that are far more distantly related to humans. Vervet monkeys – with whom we shared an ancestor at around 30 million years ago – have specific calls for specific predators, which have been claimed to be something close to words; gelada baboons chatter to each other in what sounds uncannily like human conversation, while male and female gibbons sing duets.[7]

So, before we turn to the African apes, we need to pay some attention to these other non-human primates, since their communication systems may help us to understand those of the earliest hominids and later members of the *Homo* genus. Unlike the ape calls, they are unlikely to provide direct evolutionary precursors to human language and music. They may, however, provide a strong analogy, because the communications systems that evolved independently among the early hominids did so in order to solve problems similar to those faced by non-human primates today.

Vervet alarm calls: are they using 'words'?

In 1967 the ethologist Thomas Struhsaker made some intriguing observations regarding the calls of vervet monkeys in the Amboseli National Park in Tanzania.[8] Vervet monkeys are about the size of a domestic cat and live in troops in savannah woodland, spending approximately equal amounts of time on the ground and in the trees. Struhsaker found that vervets use acoustically distinctive alarm calls for different types of predators. At the sight of a leopard, a monkey barks loudly while running into the trees, encouraging the other vervets to escape the potential danger by doing the same. If the predator is an eagle, the alarm call will be a short double-syllable cough. On hearing this, the monkeys will look up and, if necessary, run for cover. The sight of a snake leads to a 'chutter'. The monkeys will then stand on their hind legs and peer into the grass to find the snake, which might then be mobbed in order to shoo it away. Each of these three acoustically distinct alarm calls relates to a different type of predator and each results in a particular type of behaviour that is appropriate for the predator concerned.

A decade after Struhsaker's observations, Dorothy Cheney, Robert Seyfarth and Peter Marler began an innovative set of studies to discover more about the subject. The key question they addressed was whether the monkeys were referring to predators in the same manner as we do when we use words to refer to objects and events in the world. The alternative possibility was that the monkeys were simply expressing different levels of fearful and aggressive emotions when faced with different types of predators.

Cheney and her colleagues approached their task from a number of angles: they tape-recorded the alarm calls and then analysed their acoustic properties; they used playback experiments to investigate systematically how the monkeys behaved in response to the predator alarm calls in different situations; they explored how infant and juvenile monkeys learned which calls to make – if, indeed, there is any learning involved; and they analysed the 'grunts' that the monkeys make to each other that superficially sound like human conversation.

Their results and conclusions were published in Cheney and Seyfarth's seminal 1990 volume, *How Monkeys See the World*. This was one of the first books to make an explicit attempt to establish what, if anything, a non-human primate is thinking when making vocalizations or, indeed, engaging in any particular type of behaviour. The original observations of Thomas Struhsaker were confirmed, but it proved impossible to establish whether the alarm calls really were functioning as words in the minds of the monkeys in the same way as the word 'leopard' or 'eagle' would do for us.

The evidence against this interpretation was the limited use of the alarm calls. They were very rarely used in contexts other than when the predators were actually present.[9] And they were never combined with any other vocalization to provide additional information, in the manner in which we combine words into sentences. Moreover, the actual number of distinctive calls was extremely limited.[10] Cheney and Seyfarth could easily think of many other calls that would be adaptively useful for the monkeys to use, such as one from a mother to her infant when she is moving on and risks leaving the infant exposed. Cheney and Seyfarth had to conclude that there is no evidence that vervet monkeys can understand 'referential relations' – the fact that a call can actually refer to an entity in the outside world.

Other than Cheney and Seyfarth's own analysis and discussion, the most perceptive interpretation of the vervet vocalizations has come from Alison Wray, the linguist at Cardiff University whose notion of holistic proto-language I introduced in my initial chapter.[11] She has argued that the vervet alarm calls should be compared with complete messages rather than with individual words in human language. So the 'chutter' that arises from the sight of a snake should not be thought of as being the monkey word for 'snake', but as something like 'beware of the snake'; similarly, the eagle alarm cough should be interpreted as a 'beware of the eagle' type message, rather than as a word for eagle, or even as a 'look up at the sky and run for cover' message. Wray describes the alarm calls as 'holistic' because they have no internal structure and were never combined with any other vocalization to form a multi-component message. She also suggests that the alarm calls should be considered as 'manipulative' rather than 'referential'. This is a key distinction: although such calls may be 'functionally referential', monkeys are not trying to tell their companions about something in the world, they are simply attempting to manipulate their behaviour.[12]

Diana monkeys and Campbell monkeys have similar alarm call systems to those of vervets, while Diana monkeys also appear to 'understand' the alarm calls of guinea fowl and the different kinds of chimpanzee screams.[13] Experimental studies with rhesus monkeys have shown relationships between their calls and food availability. The rate at which they call appears to be related to how hungry they feel, while the acoustic structure of their call relates to food type and quantity.[14] Rhesus monkeys have also shown some degree of self-control over whether or not to call. Those individuals who find food but do not call are prone to aggression from other individuals, seemingly as a form of 'punishment'.

Chattering geladas

While the studies of vervets have taught us how monkeys use functionally referential calls to warn about potential predators, studies of gelada monkeys have been important in understanding how acoustically variable calls, many of which sound distinctly musical, mediate social interactions. Geladas are larger than vervets. Their natural habitat is the Ethiopian mountains, where they live in large groups feeding on roots, leaves and fruit. The range and frequency of the vocalizations made by geladas are more extensive than in the case of vervets, although there are no records of predator alarm calls or other associations between particular calls and particular entities. The gelada vocalizations are used in social interaction: 'As they approach one another, walk past one another or take leave of one another, as they start or stop social grooming, as they threaten someone because he is too close to a partner, solicit someone's support or reassurance, in fact, as they do the infinite variety of different social actions that make up the minute-to-minute substance of their social lives they always accompany these actions with vocalizing.'[15]

That description was provided by Bruce Richman, who spent eight years recording and analyzing the vocalizations of captive, but free-ranging, gelada monkeys. The sounds they made were diverse: nasalized grunts, high- and low-pitched calls, fricatives, glides, 'tight' and 'muffled' voices. Although any 'meaning' has proved impossible to tease out, these vocalizations evidently play a key role in acknowledging and maintaining social bonds between individuals and the group as a whole. Geladas follow the vocalizations of other individuals in detail. They often synchronize their own vocalizations with those of others, alternate with them, or even complete unusual patterns of sounds.

The acoustic feature that most interested Richman was the great variety of rhythms and melodies that the geladas use: 'Fast rhythms, slow rhythms, staccato rhythms, glissando rhythms; first-beat accented rhythms, end-accented rhythms; melodies that have evenly spaced musical intervals covering a range of two or three octaves; melodies that repeat exactly, previously produced, rising or falling musical intervals; and on and on: geladas vocalize a profusion of rhythmic and melodic forms.'[16] Richman asked what function was served by the rhythm and melody that he found so abundant in the gelada vocalizations. After making a detailed description of their use and exploring the contexts in which they arose, he concluded that they performed much the same function as the rhythm and melody that is found in human speech and singing. In essence, the geladas used changes in rhythm and melody to designate the start and end of an utterance; to parse an utterance, so allowing others to follow along; to enable others to appreciate that the utterance was being addressed to them; and to enable others to make their own contribution at the most appropriate moment. In fact, Richman's interpretation of how geladas use rhythm and melody appears strongly analogous to its use in the early and non-linguistic stages of infant-directed speech (IDS), as described in chapter 6.

A further function of gelada rhythm and melody identified by Richman is the resolution of the frequent emotional conflicts arising from the complex and ever-changing social situations of monkey life. He describes one instance when an adult male was walking slowly back to his group after engaging in a bout of threats with some bachelor males. As he approached the females, he produced what Richman described as a 'long series' – a fast, long sequence of sounds with a steady rhythm and melody, which was clearly a sign of a friendly approach and positive emotion. But the gelada also added what Richman described as 'tight features' – an exaggeration of the higher frequencies, caused by a tongue constriction at the back of the mouth. This is typically used to express anger and an aggressive intent. Richman explained that by expressing the 'long series' with 'tight features', the male gelada found a means to approach the females without making them fearful while also expressing to them his angry state. Thus he both resolved his own emotional conflict and provided as much information as possible to the females about his current state.

Continuity with *Homo*?

The chattering of monkeys often sounds very much like human conversation, even though it is quite evident that words are not being used. This may derive from similarities in the variations of pitch used by monkeys and by humans to express different types of emotion in different types of social context. Lea Leinonen, a psychologist from Helsinki University, and her colleagues tested whether adults and children were able to recognize the emotional content of monkey calls, using those of the macaque (*Macaca arctoides*).[17] Leinonen recorded a large number of macaque vocalizations and drew on their social contexts in order to categorize them as expressions of contentment, pleading, dominance, anger or fear. These were then played to the human subjects, who had a high degree of success in interpreting the emotional content of each call. She found that children gradually improved in their ability to interpret the calls until, at the age of nine or ten, they reached the level achieved by adults. Leinonen argues that such results indicate that monkeys and *Homo* share the same vocal cues in emotional communication.

This idea was further examined in a study that made an explicit comparison between the waveforms of each category of macaque vocalization and those of words used by speakers of Finnish and of English in similar social situations.[18] For the test words Leinonen chose the forenames 'Saara' (Finnish) and 'Sarah' (English), as these are emotionally neutral and have long vowels that can be expressed in various ways. The human subjects then had to use the test word in the course of a story that had been designed with the social contexts of the monkey vocalizations in mind.

The similarities in the waveforms of the monkey and human utterances were striking: the same types of pitch change were used to express each of the emotional states. Leinonen and her colleagues argue that this reflects a strong 'primate inheritance' in human emotionally laden speech. This derives, they suggest, from the need of both monkey and human mothers to communicate different emotions to their newborn infants with utterances that can be clearly distinguished, and their characteristics survive into human adulthood despite strong cultural influences on adult behaviour.

Duets in the forests of south-east Asia

If the vocalizations of geladas and macaques sound like conversation to the human ear, then those of gibbons certainly sound like music. There are twelve species of gibbon, all of which live in the rainforests of south-east Asia, and which shared a common ancestor with humans at around 18 million years ago. Gibbons form small family groups, typically consisting of a mated, monogamous pair with up to four offspring, who remain with their parents until they are eight or nine years old.[19] Males and females of all species 'sing' alone both before and after mating, and the mated pairs of all but two species also 'sing' duets. Gibbon songs are species-specific and appear to be largely a product of biological inheritance rather than of learning.[20]

The female song is referred to as the 'great call'. It can last for anything between six and eighty minutes, and is typically a rhythmic series of long notes uttered with an increasing tempo and/or increasing peak frequency. Claims that the function of their songs is to attract mates have been tested and found wanting; rather, they are most likely to sing to defend their territory.[21]

Males produce one of several distinct types of phrases, which often become increasingly complex as the song bout proceeds.[22] When singing alone, their calls most likely function as advertisements of themselves to unmated females, much as in the case of male birds.[23] Although the female and male calls/songs sound as if they contain a series of distinct components, these do not appear to have any separate meanings that are being composed together to form a whole. As with the vervet and gelada vocalizations, those of the gibbons are holistic in character – they are equivalent to complete messages rather than strings of individual words.

Thomas Geissmann of the Institute of Zoology, Hanover, is one of the leading experts in the study of gibbon song and provides a concise description of how males and females combine their songs to form duets: 'Males usually stop vocalizing at the beginning of each great call and provide a special reply phrase (coda) to the great call before resuming their common short phrases. In addition, one or both partners often exhibit an acrobatic display at the climax of the great call, which may be accompanied by piloerection and branch shaking. The combination of the female great call and male coda is termed a call sequence, and it may be repeated many times during a single song bout.'[24]

Why should males and females sing together? The most attractive theory is that it strengthens the pair-bond, an idea that Geissmann tested on siamang gibbons, a species that performs some of the most complex vocal interactions.[25] He kept records on ten siamang groups in various zoos, measuring the extent of their duetting against three widely accepted measures of pair-bond strength: mutual grooming, spatial distance between mates, and behavioural synchronization. He found a strong correlation: the pairs that sung together most frequently also exhibited high levels of grooming, kept close spatial proximity to each other, and undertook similar behaviours at similar times. What remains unclear, however, is cause and effect: do siamang gibbons develop strong pair-bonds because the mated pair sing together, or do they sing together because they have a strong pair-bond?

Geissmann was careful not to generalize this finding to the other gibbon species whose duetting is less striking, suggesting that its role in strengthening the pair-bond may be specific to the siamang. Even in that species, duetting may have additional functions. The loudness of the siamang song might function as an advertisement of the pair-bond and a means of territorial defence.[26] The latter was shown to be the case for a population of Mueller's gibbons in the Kutai game reserve of eastern Kalimantan, Indonesia. John Mitani of the University of Michigan played recordings of singing by neighbouring groups at different locations within a group's territory and found that the intensity of the duet invoked varied markedly. It was greatest when the group felt a need to defend its territory.[27]

Whether gibbon duetting is about strengthening the pair-bond, advertising the pair to others, or territorial defence, it is undertaken by individuals who have entered a long-term, monogamous relationship and hence are dependent upon cooperation with one another for their own survival and reproductive success. Duetting is found in a few other types of primates and these, too, have monogamous lifestyles.

Gorilla, chimpanzee and bonobo vocalizations

After the predator alarm calls of the vervets, the use of rhythm and melody by geladas, and the duets of gibbons, the vocalizations of our closest living relatives do seem rather limited – particularly as they lack any distinctive musical qualities and have very few clear associations with events or objects. All such calls grade into each other, rather than falling into discrete categories comparable to human words.[28] And yet it is in this African ape repertoire, and not those of monkeys or gibbons, that the roots of human language and music must be found. The apparently greater complexity of monkey calls must be an illusion, one that simply reflects our limited understanding of ape calls.

While there are strong similarities in the call repertoires of the three types of African apes, important differences also exist. By understanding how these differences relate to the particular lifestyle of each species, we should be able to predict how the call repertoire of our hominid ancestors is likely to have evolved as their lifestyles diverged from that of the forest-living apes, as we will see in the following chapter.[29]

As regards similarities, the African apes effectively all have the same range of calls, reflecting their close evolutionary relatedness. The chimpanzees' so-called 'pant-hoot' is the same as the gorillas' 'hoot-series', and the chimpanzees' drumming is the same as the gorillas' chest-beat.[30] The adult males are always the most vocal and their calls are used in the same range of situations. Prominent among these are screams by subordinate individuals who are the object of aggression from dominant individuals, and 'laughter' by juveniles when playing.

The most frequent call made by chimpanzees is the pant-hoot. These vary between two and twenty-three seconds in length. They often involve a brief introduction consisting of unmodulated elements, which then become increasingly shorter and louder. These build to a climax – one or several long, frequency-modulated elements which resemble a scream. The pant-hoot then concludes in a similar fashion to its build-up.

Pant-hoots are made in several situations: when an individual rejoins a group; when strangers are present; when arriving at sources of food; during agonistic displays; and when prey has been captured. They can, however, be divided into two calls: the long-distance calls, which are made to regulate spacing between individuals; and shorter-range pant-grunts, which seem to be about maintaining dominance in relationships. The long-distance calls show greater degrees of variation between individuals, presumably as a means of self-identification which is unnecessary in the short-range pant-grunts because the caller would be within eyesight of the recipients.[31]

Studies have failed to identify differences between the pant-hoots used in feeding and non-feeding situations. Hence it appears unlikely that chimpanzees use such calls to transmit to other individuals information about the quantities or types of available foodstuffs. Although each individual has a distinctive pant-hoot when calling by itself, when males 'chorus' together they appear to match one another's calls.

Differences in call repertoires have also been found between different chimpanzee groups. The males from the two most intensively studied populations of chimpanzees, those from Gombe and Mahale, have acoustically distinctive calls, which are most likely the product of vocal learning. This may reflect the need for chimpanzees to identify 'friend' or 'foe' at a distance, as neighbouring chimpanzee groups are often aggressive towards each other.[32]

A recent study by Catherine Crockford and her colleagues from the Max Planck Institute for Evolutionary Anthropology has found that each of four communities of chimpanzees in the Taï Forest of West Africa have community-specific pant-hoots.[33] They explained this by attributing to chimpanzees a propensity actively to modify their pant-hoots so that they are different from those of neighbouring communities.

The most frequent call made by gorillas is a 'double grunt', produced with the mouth kept shut.[34] These are made in a variety of situations: when feeding, resting, playing, or moving from one location to another. Double grunts are made both when in full view of others and when partially or wholly obscured by vegetation. Detailed studies of their acoustics have shown that double grunts are individually distinctive, and that they fall into two classes. The 'spontaneous' double grunt is relatively high-pitched and is made after a period of silence; the 'answer' is lower in pitch and is given by another animal within five seconds of the first call.

As with chimpanzees, the function of these calls appears to be primarily social. Some mediate competition between individuals: if a gorilla is feeding and is approached by a higher-ranked individual, an exchange of grunts often leads to the subordinate animal leaving the food source. In other circumstances, double grunts appear to be employed in order to control the spacing between individuals while feeding. In these situations the calls are evidently manipulative rather than referential – they are simply being used to direct another individual's behaviour. Indeed, this interpretation most probably applies to all African ape vocalizations.

Gorillas are more vocal than chimpanzees – a typical individual will call about eight times every hour rather than just twice. This is most probably a consequence of gorillas living in groups that are more stable in size and composition than the loosely organized communities of chimpanzees. Gorillas rest together after every long feeding bout, socializing by physical contact, especially grooming.

Why are the call repertoires so limited?

If the African apes are closely related to language-using humans and have far bigger brains than monkeys,[35] why should their vocal behaviour appear so much simpler than that of the vervets and geladas? This might be an illusion, no more than a reflection of our limited understanding of what pant-hoots, grunts and screams actually mean, combined with our failure to detect the subtle acoustic variability that may be perceived by the apes themselves.[36] But if, as seems likely, it is indeed the case that African apes really do have limited call repertoires, we must ask why this is the case.

There are several possibilities, all of which may be contributing factors.

One is neural inflexibility – apes are simply unable to learn new calls in the wild and so they maintain a very limited repertoire. There is, however, some evidence for vocal learning by non-human primates in the wild, such as the distinctive, community-specific pant-hoots of the chimpanzees in the Taï forest mentioned above. If vocal learning does exist, however, it is relatively insignificant compared with that found in human speech and birdsong.[37]

A second reason for such limited vocal diversity, especially when compared with modern humans, is a constraint on what sounds can be produced imposed by the anatomy of the ape vocal tract.[38] The larynx is the valve that prevents food and water from entering the lungs, with two surrounding flaps of flesh that form the vocal cords. Humans can make such a rich mix of tones because the sound waves that arise from air passing between the vocal cords can reverberate within the throat. These sound waves have the opportunity to interfere with other waves and then become further shaped by the mouth. By changing the position of the tongue and lips, the size and shape of the chamber within the mouth can be altered, which enables a wide variety of sounds to be generated. The great apes, however, have far less throat to play with, owing to the relatively high position of their larynx. In addition, they have relatively inflexible mouth chambers, because of the substantial size of their teeth and relatively long, shallow tongues. With this morphology, the production of vowel sounds is practically impossible.[39]

Even so, one might reasonably expect African apes to have a larger call repertoire than they do. One might argue that there is simply no need for a greater number and diversity of vocalizations – the African apes are extremely well adapted to their environments, with no selective pressure to develop their call repertoires. But such arguments are not persuasive. Consider chimpanzee hunting, for instance.[40] This appears to involve substantial coordination between different individuals, who adopt specific roles as they pursue monkeys in the treetops. It is, however, undertaken with limited, if any, vocal communication, and certainly with no signs of prior planning. The frequency of success is often rather low, and one can readily imagine that it could be enhanced by greater communication between the participants in the hunt. Similarly, the rate at which juveniles learn tasks such as the cracking of nuts would surely be enhanced if their mothers could 'tell' them how to strike the nuts with hammer-stones.[41]

The limited use of vocal communication, especially the apparent absence of calls with direct reference to objects or actions, is particularly surprising in the light of the ability of apes to learn to use symbols in laboratory studies. There does not appear to be any a priori cognitive constraint on the use of symbols, and yet this is something that does not occur in the wild, or in captivity without human influence.

Some argue that the key constraint on African apes is their inability to recognize that another individual's knowledge of the world is different from their own.[42] While a mother ape 'knows' how to crack nuts open with hammer-stones, she cannot appreciate that her infant lacks that knowledge. So she has no incentive either to 'explain', by gestures or calls, how it is done or to manipulate her infant to do it. If one assumes that another individual has the same knowledge and intentions as one's own, there is no need to communicate that knowledge or to manipulate their behaviour.

An appreciation that another individual's knowledge, belief and desires are different from one's own requires a 'mind-reading' ability, sometimes referred to as 'theory of mind', which is central to the social lives of modern humans. It may be that all non-human primates lack such mind-reading abilities. Following the publication of Cheney and Seyfarth's *How Monkeys See the World*, there has been widespread agreement among academics that monkeys lack a theory of mind, but there is still considerable disagreement as regards chimpanzees.[43] Some believe that occasional observations of deceptive behaviour suggest that chimpanzees can appreciate the difference between their own thoughts and those of another individual, since deceiving involves providing another with false beliefs.

The psychologists Richard Byrne and Andy Whiten championed this view in the early 1990s, bringing together a great many isolated observations of apparently deceptive behaviour to produce a strong case that chimpanzees have a theory of mind. Because many of their observations were 'anecdotal' reports, questions remained regarding the validity of their conclusion, and some (predominantly laboratory-based) psychologists refused to accept it. However, recent laboratory experiments have demonstrated that chimpanzees can indeed understand psychological states, vindicating the inferences that Byrne and Whiten made from observations of 'natural' behaviour.[44]

This certainly does not mean that the chimpanzee theory of mind is the same as that of humans. The term 'theory of mind' must encompass a range of cognitive skills. There is, for instance, the ability to appreciate that another individual has a particular desire – for food, for example – and this may be different from appreciating that an individual has a particular belief. Another distinction concerns what have become known as 'orders of intentionality'. If I know what I think, then I am termed as having a single order of intentionality; if I know what someone else thinks, then I have two orders of intentionality; if I know what someone else thinks that a third party thinks, then I have three orders of intentionality – and so forth. Whereas humans routinely use three or four orders of intentionality in their social lives, apes might be limited to two orders at the most.[45] Hence, although apes may have a theory of mind, it may be of a quite different nature from that possessed by humans. In consequence, they may have a relatively limited understanding of what other individuals are thinking, and this may be a key factor in explaining their very limited call repertoire.

In summary, my overview of the call repertoires of African apes has described how gorillas, chimpanzees and bonobos share the same narrow range of calls, which reflects their close evolutionary relationships. The differences that exist can be explained by their physiology and lifestyles, especially their patterns of social organization. The pant-hoots of chimpanzees and the double grunts of gorillas appear to be individually distinctive and serve to mediate social relationships. They are quite unlike the words of human language, and can be characterized as being holistic and manipulative in character.

Gestural communication in apes

Gorillas, bonobos and chimpanzees communicate by gesture as well as by vocalizing. Unfortunately, the number of studies of this, especially those dealing with wild populations, is limited, and ape gestures are often subtle; unless they are being studied in their own right they may go unnoticed when an observer is concentrating on some other form of activity. Consequently, the key evidence we have comes from captive apes, albeit ones that live in large, naturalistic enclosures and have not received any instruction in human-like communication. Although this evidence is sparse, it indicates that even among apes communication is multi-modal, involving a synchronized use of vocalizations and gestures.

Joanne Tanner and Richard Byrne of the University of St Andrews studied gestures used by gorillas in San Francisco Zoo, focusing on those made by Kubie, a thirteen-year-old male, to Zura, a seven-year-old female.[46] More than thirty different types of gestures were recorded within the gorilla group as a whole, but their study concentrated on nine gestures that dominated Kubie's repertoire. Some of these were tactile: Kubie would touch Zura's body and move his hand in the direction in which he wished her to move, but without forcing her to do so. Other gestures were purely visual: when he had her attention, Kubie would move his hand or arm in space to indicate the direction in which he wished Zura to move.

The most frequent gesture used by Kubie was a 'head nod'. This was used when Kubie's hands were either not available, because he was knuckle-walking or holding something, or when Zura was too far away to touch. The head nod served to capture Zura's attention, and would often draw her attention to Kubie's body so that he could make further gestures. The tactile gestures achieved a high success rate in encouraging Zura to move her body in the direction that Kubie desired. They included his moving his hand down Zura's body, patting her head, back or bottom, pushing her head gently downwards and light tapping. When the two gorillas were a short distance apart, Kubie would simply give Zura a body tap. This frequently indicated that Kubie wanted to play, as it was accompanied by his play face and was sometimes followed by his 'armswing under' gesture, which drew Zura's attention to his genital area.

When considered as a whole, there are three key features of the gorillas' gestures that we must note. First, they are 'iconic' in nature: the path of the gesture matches the desired path of body movement. This contrasts with the symbolic gestures used in modern human sign language, where there is an arbitrary relationship between the shape or movement of the gesture and its meaning – although it should be noted that the majority of human gestures, and many of those in sign language also have an iconic element. Secondly, the gorilla gestures are complete acts in themselves: they are not composed into sequences involving multiple gestures that have a meaning beyond those of the constituent parts. They are holistic, in the same way as the monkey and ape vocalizations described above are holistic utterances. The third key feature of gorilla gestures is also shared with monkey and ape vocalizations: they are manipulative rather than referential.[47] Kubie and the other gorillas use gestures to manipulate the movement and actions of another individual, rather than to inform them about the world.

None of the highly visual gorilla gestures that Tanner and Byrne observed in the San Francisco Zoo have been reported for gorillas in other zoos or in the wild. This is unlikely to be explained by a failure to record them, as Richard Byrne himself has made extensive observations of gorillas in Rwanda. Tanner and Byrne suspect that Kubie adopted a more extensive gestural range than is usual for gorillas because he had to share the zoo compound with two mature silverback males. Hence Kubie needed cooperation from the females in order to interact with them and prevent the silverbacks from interfering. By using gestures, Kubie could communicate with the females in silence and thereby avoid the attention of the older males.

The fact that Kubie was able spontaneously to develop gestural communication indicates that the mental capacity to do so is present in the gorilla today, and was most likely present in the common ancestor of gorillas and modern humans which lived between 8 and 7 million years ago. Prior to Tanner and Byrne's work this had already been established for the common ancestor of chimpanzees and humans from observations of the gestures of modern-day bonobos and chimpanzees.

Tanner and Byrne cited the description provided by Wolfgang Kohler in 1925: '... one chimpanzee which desires to be accompanied by another, gives the latter a nudge, or pulls his hand, looking at him and making the movements of "walking" in the direction desired. One who wishes to receive bananas from another, imitates the movement of snatching or grasping, accompanied by intensely pleading glances and pouts. The summoning of another animal from a considerable distance is often accompanied by a beckoning very human in character.'[48] Iconic gestures were documented for bonobos in a study undertaken during the 1970s by Sue Savage-Rumbaugh and her colleagues. The results were essentially the same as those of Tanner and Byrne for gorillas: bonobo gestures depict the action desired from the recipient of the gesture.[49]

Summary

The communication systems of monkeys and apes remain little understood. It was once believed that their vocalizations were entirely involuntary and occurred only in highly emotional contexts. Their calls were believed to represent the individual's emotional state or their imminent behaviour.[50] The studies I have described in this chapter have contributed to a quite different view, one that recognizes that monkey and ape vocalizations are often deliberate and play a key role in social life.

My review of selected aspects of vervet, gelada, gibbon, chimpanzee, bonobo and gorilla communications has acknowledged considerable inter-species variability, reflecting the contrasting physiology and adaptive strategies that these primates have evolved. But there are some common features. First, none of the vocalizations or gestures are equivalent to human words. They lack consistent and arbitrary meanings, and are not composed into utterances by a grammar that provides an additional level of meaning. The term that Alison Wray uses for the vervet alarm calls is generally applicable to non-human primate vocalizations and gestures: they are holistic.

Secondly, the term 'manipulative' is also generally applicable. The vocalizations and gestures do not appear to be telling another individual about the world in the same way as we refer to objects, events and ideas when talking to another individual. Monkeys and apes probably simply do not appreciate that other individuals lack the knowledge and intentions that they themselves possess. Rather than being referential, their calls and gestures are manipulative: they are trying to generate some form of desired behaviour in another individual, whether that is Kubie 'wishing' Zura to engage in play, or a vervet monkey 'wishing' the other monkeys to escape from a predator.[51]

A third feature may be applicable to the African apes alone: their communication systems are multi-modal, in the sense that they use gesture as well as vocalization. In this regard, they are similar to human language. Whether monkeys also use gesture depends upon how the term is defined.

If a vervet sees another standing upright and looking around, then it may do the same because that stance communicates that a snake has been seen. But this is a rather generous use of the term 'gesture'. Monkeys are unlikely to have the motor control or neural flexibility to acquire gestures; zoo populations of monkeys have never been observed using idiosyncratic gestures equivalent to those seen among zoo populations of gorillas and chimpanzees.[52] Hence, apes may be the only non-human primates that employ multi-modal communication.

Finally, a key feature of the gelada and gibbon communication systems is that they are musical in nature, in the sense that they make substantial use of rhythm and melody, and involve synchronization and turn-taking. Again, depending upon how one would wish to define 'musical', this term could be applied to non-human primate communication systems as a whole.

The holistic, manipulative, multi-modal, and musical characteristics of ape communication systems provide the ingredients for that of the earliest human ancestors, living in Africa 6 million years ago, from which human language and music ultimately evolved. We can now begin to trace that evolutionary history by examining the fossil and archaeological records.

9 Songs on the savannah

The origin of 'Hmmmm' communication

Prelude in C Major by J. S. Bach: australopithecines waking up in their treetop nests

'Hominids' refers to primates that are classified with modern humans in the family Hominidae; other than *Homo sapiens*, all such species are extinct. The first to exist was the common ancestor to humans and chimpanzees, which lived between 6 and 5 million years ago but which is 'known' only from the extent of the difference between human and chimpanzee DNA; there are no fossil specimens. Some palaeoanthropologists thought that one such fossil might have been discovered in 2002, when a well-preserved ape-like skull was found in the Sahel region of Africa. Later reflection suggested that the hominid status of this specimen was questionable, leaving the earliest incontrovertible hominid fossils dating from 4.5 million years ago and belonging to a species designated as *Ardipithecus ramidus*.[1]

The australopithecines and early *Homo*

The fossilized fragments of *Ardipithecus ramidus* were discovered in 1992 and 1994 in the Middle Awash region of Ethiopia.[2] A full description and interpretation of the finds has not yet been published. However, from preliminary reports it appears to have been about the size of a chimpanzee, and its teeth look to have been more similar to living African apes than to other hominids, while it may have spent more time standing or even walking on two legs than do modern chimpanzees or gorillas. This is because its foramen magnum, the opening in the skull through which the spinal cord passes, is located underneath the skull, as in other hominids, rather than to its rear, as in living African apes. This suggests an upright posture, which is normally associated with bipedalism.

Geological sediments that date to between 4.5 and 1.8 million years ago have provided fossils from what are believed to be seven other species of hominid, although estimates for the specific number varies between researchers. The majority of fossils are classified in the genus *Australopithecus*. They are a very mixed group, some having a relatively slight build, referred to as 'gracile', and appearing to be generalist foragers, while others have massive jaws and teeth which reflect a specialized adaptation to feeding on coarse plant material. With the exception of a specimen from Chad, all the fossils have come from either southern or eastern Africa. This most likely reflects no more than the distribution of geological exposures of the appropriate date; it seems probable that the australopithecines lived throughout much of the African continent.

As a whole, the genus shares three key features. First, as far as we can tell, they were all partially bipedal. Secondly, their brain sizes were between 400 and 500 cubic centimetres, and hence equivalent to those of modern-day African apes and about one third of the size of a modern human brain. Thirdly – and another important similarity with chimpanzees and gorillas – the australopithecines were sexually dimorphic, the males being considerably larger than the females. A further similarity that is often assumed is that the australopithecines had similar amounts of body hair to modern apes. This assumption is based on ideas about why body hair became reduced during a later period of human evolution, which I will consider in a later chapter.

The best-known of these hominid species is *Australopithecus afarensis*, of which the most complete specimen is A.L. 288-1, popularly known as 'Lucy'. This specimen consists of forty-seven bones, including those of the upper and lower limbs, the backbone, ribs and pelvis – a remarkable number to have survived from a single individual. Fragments from several other members of this species have also been discovered, including a cluster of more than two hundred specimens from thirteen individuals, dubbed the 'first family'. Some palaeoanthropologists give *Australopithecus afarensis* a special place in human evolution as the species from which the first member of the bigger-brained *Homo* genus evolved.

From the shape of Lucy's pelvis, knee and foot bones, we know that *Australopithecus afarensis* habitually spent time on two legs and stood at just over one metre tall – about the same size as a chimpanzee today. But Lucy also had anatomical adaptations for tree-climbing, or perhaps hanging from branches: a relatively short femur, curved finger and foot bones, a highly mobile wrist, and powerful arms. These tree-climbing features may represent no more than the retention of ancestral anatomical features, which were no longer being used. The same explanation has been offered for wrist bones that suggest knuckle-walking.[3] It seems most likely, however, that, although being able to walk on two legs, Lucy also spent a considerable amount of time in the trees, just as the great apes do today. This would have been for finding food – leaves, fruits, insects, eggs – and for making leafy nests in which to sleep at night.[4]

Relatively small teeth and large brains distinguish two of the hominid

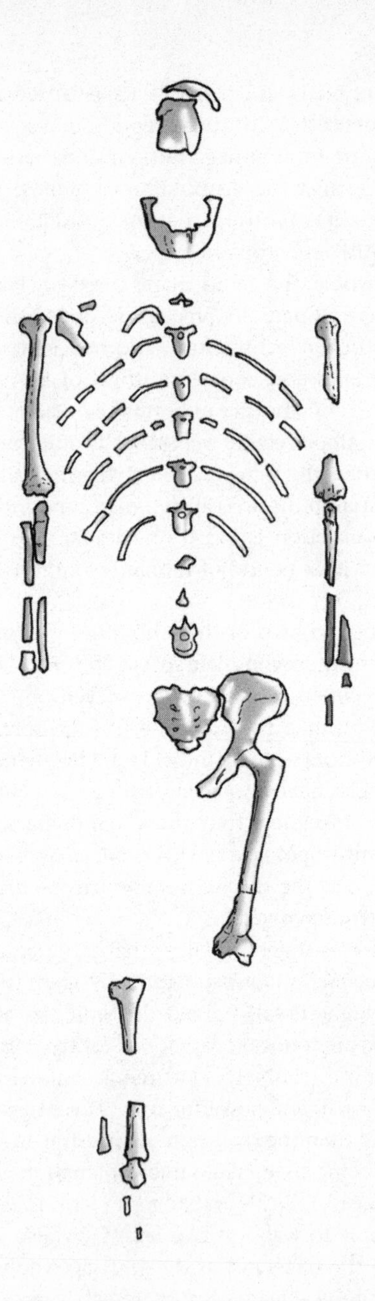

Figure 7 Skeleton of *Australopithecus afarensis* from Hadar, Ethiopia. Specimen A.L. 288-1, otherwise known as Lucy and dating to around 3.5 million years ago. Lucy would have stood just over 1 metre tall.

species that date to between 2.5 and 1.8 million years ago. The significance of these features remains debatable; the bigger brain might be no more than a reflection of a bigger body. This is difficult to assess because very few skulls are found with the corresponding skeletal remains intact. However, these differences from the australopithecines have been thought to be sufficiently distinct and important to permit the definition of a new genus of hominids, *Homo*.[5]

As with the australopithecines, there appears to be considerable variation in size, dental patterns and body design among these specimens, with some looking more human-like and some more ape-like. Several of the fossils are classified as *Homo habilis*, the key specimen being a skull from the Olduvai Gorge known as OH7, dating to 1.75 million years ago, which has a brain size of 674 cubic centimetres. Another key fossil comes from Koobi Fora. This is known as KNM-ER 1470, which has a larger brain size, at 775 cubic centimetres, and is classified as *Homo rudolfensis*.

All in all, the period between 4.5 and 1.8 million years ago was one in which several 'bipedal apes' lived in Africa, each species having a slightly different body shape and dentition, reflecting differences in behaviour and diet.

Hominid lifestyles

At least one of these bipedal apes, whether australopithecine or *Homo*, was routinely making stone artefacts from nodules of basalt, chert and occasionally other materials. Archaeologists refer to these artefacts as constituting the 'Oldowan industry'. This is characterized both by the sharp flakes that were detached from the nodules of stone and by the minimally shaped 'chopping tools' the process left behind. It appears unlikely that the hominids had a particular shape of tool 'in mind' when they were striking one stone against another: they simply needed both sharp-edged flakes and more robust stone implements. Such tools were used for cutting through the tendons and flesh of animal carcasses; we have the cut marks on the bones as evidence. This resulted in accumulations of fragmented animal bones and discarded stone tools, which were eventually discovered as archaeological sites, such as those of Olduvai Gorge and Koobi Fora.[6]

The stone tools may well also have been used for other tasks, such as cutting plant material and sharpening sticks. In addition, the early hominids most probably had their equivalents of chimpanzee termite sticks, ant probes, leaf sponges and so forth, none of which have survived in the archaeological record.

Since the 1970s, archaeologists have meticulously excavated and analysed many clusters of stone artefacts and fragmented bones in an effort to reconstruct the behaviour and environmental setting of the early hominids. Considerable debate continues as to whether they were hunters or merely scavengers of carnivore kills, whether they lived in 'home bases' that were loci for food sharing and childcare, and whether their stone tools are significantly more complex than the tools made by chimpanzees today.[7]

Some aspects of their lifestyle are beyond dispute, and can be used to propose how the typical ape-like call repertoire that we must attribute to the 6-million-year-old common ancestor may have developed. As regards their environment, it is clear that by 2 million years ago australopithecines and early *Homo* were adapted for living in relatively open landscapes, in contrast to the woodlands inhabited by the African apes today. This is evident from the animals with which the fossil finds are associated – antelopes, bovids, zebras, and other species adapted to open grasslands and scrubby savannahs. A key consequence of this is that the hominids are likely to have chosen to live in much larger groups than their forest-dwelling relatives, such as the 4.5-million-year-old *Ardipithecus ramidus* and the elusive common ancestor of humans and chimpanzees of 6 million years ago.

We can be confident about this because a strong relationship exists among modern-day primates between group size and habitat; primates living among trees live in smaller groups. The reason for this is largely to do with predator risk; trees provide a ready means of escape from terrestrial carnivores and make it difficult for birds to spot their prey. Away from the cover of trees, safety can only be found in numbers, which provide more eyes to notice predators and lessen the chances that any particular individual will be attacked. There is, however, a cost: social tensions leading to conflict can easily arise when large numbers have to live continually in close proximity to one another.[8]

Although living in larger groups than their immediate ancestors and their modern-day ape relatives, australopithecine and early *Homo* social groups would have been very intimate communities, with each individual having considerable knowledge about all the other members of the group. They would have shared a great deal of knowledge about existing and past social relationships and about their local environment.

Meat most likely played a more prominent part in the diets of the gracile australopithecines, *Homo habilis* and *Homo rudolfensis* than it does in the diet of modern-day apes. There are several indications of this from the archaeological sites: not only the large collections of fragmented and cut-marked animal bones but also, in some cases, a predominance of those bones that carry the most meat. Although interpretations vary as to how much meat, marrow, fat and organs were eaten, and how these were acquired, the most reasonable interpretation is that the hominids were opportunists, sometimes picking the leftover morsels from carnivore kills and sometimes killing an animal themselves. As the hominids' hunting weapons appear to have been restricted to stone nodules and, possibly, sharpened sticks, they would probably only have killed injured or sickly animals.

Further aspects of early hominid behaviour remain more contentious. I favour the idea that as well as living in larger groups, hominids were foraging over much larger areas than are covered by the modern apes. Each day, different individuals or small parties are likely to have worked quite separately, and then to have shared the food they collected at the end of the day. This is the 'home base/food sharing' hypothesis, originally proposed in the 1970s to explain why the excavated bone fragments at archaeological sites often include such a diverse range of animal species.[9] Animals from different habitats are frequently found together, suggesting that parts of carcasses were transported around the landscape. Nodules of stone were certainly being carried by the hominids because they are found far from their sources, as are the stone flakes that were detached from them. It seems reasonable to suppose that plant foods were also carried.

The anatomy of early hominids supports these inferences about behaviour in the savannah. Studies of living primate species have established that those with larger brains tend to live in larger groups. And so *Homo habilis* and *Homo rudolfensis*, with brains up to 50 per cent larger than those of modern chimpanzees, are likely to have lived in proportionally larger groups, perhaps reaching eighty or ninety individuals.[10] Meat-eating is suggested by the reduction in tooth size and also, perhaps, by the larger brain, because meat may have been necessary to 'fuel' that metabolically expensive organ.[11] A greater degree of movement around the landscape is suggested by the trend to bipedalism, as will be fully explored in the next chapter.

Intellect and emotion

Whether or not the striking of stone nodules to make sharp-edged flakes and choppers indicates greater dexterity and/or cognitive skill than is found in modern apes remains widely debated.[12] The complete absence of deliberate stone-flaking among wild chimpanzees might reflect a lack of the need for such tools rather than an inability to make them. Just one systematic laboratory-based experiment has been undertaken to test whether an African ape can make Oldowan-type stone artefacts. The ape in question was Kanzi, a bonobo that had already shown a notable ability for using symbols.[13] He failed to master the flaking of stone nodules, but this result is hardly conclusive – a younger animal may have fared much better.

However, we can be confident that the hominids would have been emotional creatures, feeling and expressing happiness, sadness, anger and fear.

Moreover, in the light of their social complexity, I suspect that feelings of guilt, embarrassment, contempt and shame would also have been experienced. As I explained in chapter 7, all such emotions would have been guides to action in their challenging natural and social environments. Individuals with social skills would have been at an advantage; they were the ones who would know when and how to elicit cooperation, who to fight, to trust, to avoid. An emotional intelligence would have been just as important to survival as the capacity for rationally weighing up the costs, benefits and risks of alternative strategies.

Just as with modern apes and humans, the facial expressions, body language, actions and vocalizations of early hominids would have communicated feelings and led to appropriate social interactions. Those who displayed signs of anger would be avoided; those who appeared calm and happy would be approached for grooming, shared feeding and games. Being able to express one's emotions vocally would have been essential to social success.

One cognitive capability that may have been significantly different from that of the African apes today is the hominids' theory of mind. In the previous chapter I explained that this is the ability to imagine that the beliefs, desires, intentions and emotional states of another individual might be different from one's own. The complete absence of or, at most, a very weak theory of mind capability among monkeys and apes was the explanation offered for their small range of manipulative calls.

The relatively large brains of *Homo habilis* and *Homo rudolfensis*, when compared with the australopithecines and modern-day apes, might reflect an enhanced theory of mind capability.[14] The selective pressure for this leap in understanding would have arisen from living in larger groups and the consequent increase in the complexity of social interactions. Those individuals who were more able to predict the behaviour of others because they could 'read their minds' would have had a competitive advantage within the group.

Anatomy and the hominid call repertoire

If our woodland-living, 6-million-year-old ancestor had the same call repertoire as modern African apes – one that I characterized in the previous chapter as holistic, manipulative and multi-modal – how would this have evolved by 1.8 million years ago among the hominids that lived on the open savannah?

We should first note that the anatomical differences between the early hominids, especially *Homo*, and the modern-day apes would have provided the potential for a more diverse range of vocal sounds. The key difference is the reduction in the size of the teeth and jaws because of the dietary trend towards meat-eating. This would have changed the shape and volume of the final section of the vocal tract. The more upright stance deriving from partial bipedalism would also have changed the vocal tract – a development I will fully explore in the next chapter.

The changes to the teeth and jaws, and hence the potential movement of the tongue and lips, are important because we can think of sounds emitted from the mouth as deriving from 'gestures', each created by a particular position of the so-called articulatory machinery – the muscles of the tongue, lips, jaws and velum (soft palate).[15] When we say the word 'bad', for instance, we begin with a gesture of the lips pursed together, whereas the word 'dad' begins with a gesture involving the tip of the tongue and the hard palate. So each of our syllables relates to a particular oral gesture.

The psychologist Michael Studdert-Kennedy argues that such gestures provide the fundamental units of speech, just as they form the units of ape vocalizations today and hominid vocalizations in the past. As motor actions, such gestures ultimately derive from ancient mammalian capacities for sucking, licking, swallowing and chewing. These began the neuroanatomical differentiation of the tongue that has enabled the tongue tip, tongue body and tongue root to be used independently from each other in order to create particular gestures, which in turn create particular sounds, some of which involve a combination of gestures. Consequently, even though we should think of the hominid vocalizations as holistic in character, they must have been constituted by a series of syllables deriving from oral gestures. These, therefore, had the potential ultimately to be identified as discrete units in themselves, or in combination with one another, which could be used in a compositional language.

As the size of the dentition and jaws in the early *Homo* species became reduced, a different range and a greater diversity of oral gestures would have become possible, compared with those available to their australopithecine ancestors. Although we do not know exactly how the potential range of vocalizations would have varied between the australopithecines, early *Homo* and the modern African apes, one thing is certain: hominids would have been more sensitive to high-frequency sounds than are modern humans. When cleaning the exposed middle-ear cavity of the fossil specimen known as Stw 151 from Sterkfontein in South Africa, Jacopo Moggi-Cecchi, a palaeoanthropologist from the University of Florence, found the fossilized stapes of that individual. This is one of the three bones of the middle ear (the other two being the malleus and incus) and the only early hominid ear bone ever found. A detailed study of its size and shape, particularly the part of the bone known as the 'footplate', showed it to be more similar to that of modern apes than of humans. There is a known correlation in mammals between the size of the footplate and the range of audible frequencies; the footplate of Stw 151 was smaller than that of modern humans, and this indicates a greater sensitivity to high-frequency sounds.[16]

Broca's area and mirror neurons

From the evidence of the surviving crania, it appears that the brains of early *Homo* were larger than those of the australopithecines. Brain size alone, however, does not tell us how the brain was organized – what type of neural networks were present. The only possible indicator for this is the impression left on the inside of the cranium, which might reflect the specific shape of the brain – but even this might provide an entirely false hope for exploring the structure of hominid brains.

Some fossil crania have been found full of sediment; when the fossil bones were removed, a natural cast of the inside of the cranium, formed by the consolidated sediment, was left behind – referred to as an 'endocast'. In other cases, artificial endocasts have been made using rubber solution. When the distinguished South African palaeoanthropologist Phillip Tobias examined endocasts of early *Homo* in the late 1970s, especially the specimen known as KNM-ER 1470, he thought he could detect the impression of the specific sulcal patterns that mark the presence of Broca's area in the modern brain. A study by Dean Falk from New York University in 1983 confirmed these findings, and she also argued that such sulcal patterns are absent from australopithecine endocasts.[17]

I referred to Broca's area – an area of the brain that was once thought to be dedicated to and essential for speech – in chapter 3. It is now known to be important for motor activity in general, while the neural networks for language are recognized as being more widely distributed throughout the brain than was previously thought. Nevertheless, Broca's area remains important and it has recently become the subject of renewed interest owing to the discovery of so-called 'mirror neurons'.

The brain scientists Giacomo Rizzolatti and Michael Arbib, based at the universities of Parma and Southern California, discovered mirror neurons in the 1990s when studying the area of the monkey brain known as F5.[18] They had already found that some neurons in this area of the monkey's brain become active when the animal undertakes an action such as grasping, holding or tearing. They then discovered that the same neurons become active when the monkey sees another monkey (or a human) undertake a similar action. Rizzolatti and Arbib dubbed these 'mirror neurons' and argued that they provide the basis for imitative behaviour in humans, especially in relation to speech. A key aspect of their argument is that Broca's area in the modern human brain is homologous with area F5 in the monkey brain; in other words, F5 is the direct evolutionary precursor of Broca's area.

Rizzolatti and Arbib describe the activity of mirror neurons as 'representing action': 'This representation can be used for imitating actions and for understanding them. By "understanding" we mean the capacity that individuals have to recognize that another individual is performing an action, to differentiate the observed action from other actions, and to use this information to act appropriately. According to this view, mirror neurons represent the link between sender and receiver that . . . [is a] necessary prerequisite for any type of communication.'[19]

The presence of mirror neurons within Broca's area would help explain its significance for speech, and especially the acquisition of the oral gestures that Michael Studdert-Kennedy describes as the fundamental units of human speech. A great deal of language acquisition by children is achieved by vocal imitation; children copy their parents' motor actions and produce similar sounds, but with no notion of where a particular word begins or ends.[20] Such vocal imitation is a remarkable achievement. As Studdert-Kennedy describes it: 'To imitate a spoken utterance, imitators must first find in the acoustic signal information specifying which organs moved, as well as where, when, and how they moved with respect to each other, and must then engage their own articulatory systems in a corresponding pattern of action.'[21]

Vocal imitation by children must be an extension of the facial imitation that infants, and even one-hour-old neonates, find irresistible; stick your tongue out at an infant and it will stick its tongue out. Studdert-Kennedy stresses that initially infants simply move the same part of the face rather than imitating the action precisely – for example, the tongue might protrude from the side of the mouth rather than straight out.

The mirror neurons described by Rizzolatti and Arbib might play a key role in the imitation of oral gestures and hence in enabling children to acquire language. Whether they play a more fundamental role in the evolution of language, as Rizzolatti and Arbib have argued,[22] is rather more contentious. But they are likely to be correct that 'the human capacity to communicate beyond . . . other primates depended on the progressive evolution of the mirror system.'[23]

It is, therefore, of some interest that the brains of early *Homo* may show the first evidence for an evolved Broca's area – an enlargement of the F5 area suggesting an enhanced number and use of mirror neurons. If this were the case, then *Homo* may have been using mirror neurons not only for the imitation of hand, arm and body gestures – as might be found in monkeys and is clearly indicated by the development of Oldowan technology in *Homo* – but also for the imitation of oral gestures that now constitute the fundamental units of speech. This may have enabled a greater range and diversity of oral gestures to be used, as compared with those of the australopithecines and with apes today, which then came to be culturally as well as genetically transmitted across the generations.

Foraging and extensions of the call repertoire

The changes in facial anatomy and the evolution of Broca's area most likely occurred first for reasons quite unrelated to vocalization, but they had the consequence of causing the hominid call repertoire to diverge from that of the common ancestor to apes and humans. Changes in lifestyles would have had a similar impact. The key developments by 2 million years ago were an increased consumption of meat and more extensive daily movement through an open landscape. This lifestyle is likely to have caused relatively greater exposure of the hominids to predators than is experienced by the African apes today; we know from carnivore-gnaw and eagle-talon marks on hominid bones that hominids were indeed taken as prey.[24]

One such specimen was the very first australopithecine discovered, the so-called Taung child, found in South Africa in 1924. This skull was of an infant of three to four years old and came from a collection of animal bones that included bones from small baboons and hares along with tortoise carapaces and crab shells. The bones of medium-sized mammals, which have regularly been found alongside hominid fossils discovered later in the century, were notably absent. A study of the Taung bones undertaken in 1994 found telltale signs that they had in fact originally been collected by a large bird of prey. The baboon skulls had been broken into from behind, in order to gain access to the brain, and carried distinctive talon and beak marks, a few of which were also found on the Taung child itself. The species represented in the collection are typical of those preyed upon by eagles, the bones of which accumulate within and immediately below their nests. And so it appears that, at around 2.3 million years ago, an eagle swept down to snatch an australopithecine infant – no doubt to the evident distress of its mother.[25]

Eagles and carnivores would have been a persistent threat to the hominids when foraging in environments with far less tree cover than those inhabited by the African apes today. Consequently, long-distance calls given by lone hominids or small groups are unlikely to have been as prominent as the pant-hoots of modern-day chimpanzees, for such calls would have further increased vulnerability to predators. The quietness that is characteristic of chimpanzees when hunting or travelling at the border of their territory is likely to have been pervasive among early hominids, especially when operating outside the security of a large group. One consequence of this may have been the development of greater control over the loudness of vocalization; the most successful hominids are likely to have been those who could, when occasion required, speak quietly as a means of meeting the need for greater communication while avoiding attracting attention to themselves or the group.

A second consequence is that gesture may have become more important than it is among apes today. In the previous chapter we noted that Kubie's use of gesture may have developed from the need to avoid attracting the attention of other male gorillas. It is reasonable to suppose that gestural communication during foraging may also have developed among early hominids in order to minimize the risk of attracting predators – this being made possible by the freeing of the hands from a locomotory role that occurred with the development of bipedalism.

Increased risk from predators may have placed selective pressures on the hominids to develop alarm calls similar to, and perhaps more extensive than, those used by vervet monkeys today. Eagle, leopard and snake alarm calls would have been of advantage to hominids on the savannah, especially when the attention of some individuals was absorbed in butchering a carcass or flaking nodules of stone.

The discovery of a carcass might itself have been an occasion for another type of manipulative call. If the carcass required butchery, to acquire joints of meat or marrow-rich bones, the hominid(s) might have experienced conflicting pressures: should a call be made to elicit help or should one remain quiet so as not to attract the attention of hyenas or lions? Speed in completing butchery and then finding safety is likely to have been imperative and was most probably a factor that encouraged the use of stone flakes for cutting hide and tendons. Hence we might expect the development of calls geared towards soliciting help from others for the processing of a carcass – perhaps similar to those used by rhesus monkeys that appear to encode the quantity of food, which we considered briefly in the previous chapter.

Richard Leakey discovered a site where such calls might have once been made, located fifteen miles east of Lake Turkana in East Africa. The late Glynn Isaac, who became professor of archaeology at Harvard before dying tragically young on a field expedition, excavated the site in the early 1970s. He called it the Hippopotamus Artefact Site, HAS for short, because it was dominated by the fossilized skeleton of a single hippopotamus, surrounded by a scatter of stone flakes and chopping tools that dated to 1.6 million years ago. The animal had originally been lying in a puddle or depression within a river delta. Isaac thought it unlikely that the hominids had killed the animal themselves but rather that they had found it dead and then scavenged upon a complete carcass.[26]

This must have been a bonanza: hippopotamus steaks to feed many individuals rather than the mere morsels of meat, fat and marrow typically found on carcasses from carnivore kills after they had been scavenged upon by hyenas and vultures. And so one might imagine food-related calls to elicit help from other hominids who could bring stone flakes for butchery – Isaac found that there were no suitable stones for making flakes in the immediate vicinity of the carcass. Quick work would have been essential, as the butchery of a carcass in a river delta would have exposed the hominids to predators and other scavengers searching for a free lunch.

It seems likely, therefore, that the lifestyle of hunting and scavenging for meat on the savannah would have placed pressure on hominids to expand their repertoire of holistic, manipulative gestures and calls. This extension might have taken place either through a process of learning or through one of biological evolution with those individuals that could make such calls gaining enhanced reproductive success. There is no reason, however, to think that vocalizations equivalent to human words were required, or that any grammatical rules evolved governing how utterances might be joined together to make phrases.

Social life and extensions of the call repertoire

The savannah-dwelling australopithecines and early *Homo* are likely to have lived in larger social groups than either the woodland-dwelling *Ardipithecus* or the common ancestor of chimpanzees and humans. Like the African apes today, they are likely to have lived in multi-male and multi-female groups, with dominance hierarchies. Sexual dimorphism suggests that dominant males had control of harems and that ape-like patterns of intra-group competition for mates and status were present.

Adult male hominids are likely to have used calls equivalent to the gorillas' double grunt to mediate their social interactions. As the number of competitive situations is likely to have been directly proportional to the size of the group, the frequency of such calls would have increased. Moreover, if such calls were to maintain their individual distinctiveness, as in the case of gorillas today, there would have been selective pressure on the hominids to generate a greater range of acoustic variation within this single class of call. One can also readily imagine that the 'spontaneous' and 'answer' grunts might have been joined by a second answer, one coming from either the original caller or a third party.

Larger group size and enhanced competition between males may also have created selective pressures for gestural communication. As we saw in the previous chapter, Kubie adopted gestures so that he would not attract the attention of higher-ranking males when interacting with a female. This arose in the unnatural context of an enclosure in San Francisco Zoo, but the same situation may have arisen in the natural context of early hominid groups. Indeed, competition between males for mates is likely to have been intense, with younger, smaller males having to engage in surreptitious mating opportunities, for which either silent gestural communication or quiet vocalizations may well have been essential.

Just as is found among ape communities today, there would have been a complex and ever-changing web of friendships and alliances between hominid individuals and small groups. Non-human primates today express their commitment to each other by grooming – the longer the time spent grooming, the stronger the relationship. As groups become larger, each individual has to invest more time in grooming others in order to service the increasing number of social relationships he or she must maintain. Grooming creates social bonding and cohesion – although exactly how and why remains unclear. It most probably releases chemicals known as opiates into the brain, which result in a feeling of contentment and pleasure. Social grooming provides the background to one of the most intriguing ideas about the evolution of language – the vocal grooming hypothesis.[27]

In 1993, the anthropologists Leslie Aiello and Robin Dunbar suggested that by the time of *Homo habilis* and *Homo rudolfensis* the size of early hominid groups had grown so large that grooming was no longer feasible as the only means to express one's social commitment to other members of the group. So long would have to be spent grooming that insufficient time would be left for activities such as finding food. They proposed that 'language' evolved as a form of vocal grooming: 'an expression of mutual interest and commitment that could be simultaneously shared with more than one individual'.[28] The advantage of 'language' would be twofold: one individual could vocally groom more than one other at a time, and this could be done while simultaneously pursuing other activities such as plant gathering.

Aiello and Dunbar suggested that a form of vocal grooming is precisely what one can observe among gelada monkeys, with their use of rhythm and melody, as described in the previous chapter. Their proposition is that among our hominid ancestors there would have been a gradual transition from social bonding based on physical contact to social bonding based on vocalization. Language evolved, they claim, as a means of 'gossip' – a way to talk about and express social relationships. And this remains, they contend, its principal function today.

Although the idea that language evolved in order to manage social relationships is compelling, there are problems with the 'gossip' hypothesis. One of these is the question of how early hominids could have expressed their social commitments, let alone talked about third-party social relationships, using no more than ape-like calls, even if these had gradually increased in number and acoustic diversity. And even if they were able to do this, how would the individual that was being vocally groomed be persuaded that this was a sign of social commitment? As the anthropologist Chris Knight has argued, words come cheap and are unreliable – and so would have been the holistic, manipulative calls made by early hominids. As such, they would be no substitute for a lengthy period of pleasure-inducing physical grooming.[29]

Yet while this is true for talking, it may not be the case for singing. If we conceive of early vocal grooming as relating to the origin of music rather than of language, then Aiello and Dunbar's arguments are more persuasive.[30] We know that singing together enables social bonding, and this may induce the same feelings of pleasure and contentment as being groomed, as I will explore in chapter 14. Aiello and Dunbar's characterization of vocal grooming in its earliest stages is in fact much closer to song than speech, for they prioritize its tone and emotional content over its information content.

We should, therefore, think of the early hominid social vocalizations as elaborations of the type of rhythmic and melodic utterances used by gelada monkeys. If such vocal behaviour were present, we can readily imagine pairs or small groups of hominids that would sound as if they were singing together, rather than gossiping, as a means of consolidating their social commitment to one another.

Expressing and inducing emotion, moreover, may have been just as important as demonstrating social commitment. Those individuals who were able to generate utterances with pitch sequences that induced emotions of happiness are likely to have had a selective advantage, because they would promote greater degrees of cooperative behaviour (see chapter 7). Cooperation would have been essential for hominid life, whether in terms of foraging behaviour, food sharing or social interaction. Social tension would have been considerable within the hominid groups, and hence communal song, inducing emotions of contentment while diffusing those of anger, may have been common. More generally, hominids would have begun to express their own emotions and to attempt to induce particular emotions in others, by means of vocalizations having greater acoustic variability than is found among the African apes. It is even conceivable that they utilized some of the specific pitch patterns that appear to induce particular emotions in modern humans, as described in chapter 7.

The theory of mind capability would have greatly enhanced this development. Possessed of a theory of mind, hominids would have had a greater appreciation of the emotions, beliefs, desires and intentions of other individuals, and would have been able to predict their behaviour in a manner impossible for modern monkeys and apes. Hence it is probable that the range of holistic and manipulative calls would have increased so as to maximize the likelihood that others would behave in the way most desirable to oneself. There may, for instance, have been a greater number of calls specifically directed from mothers to their infants once they became aware of their infants' lack of knowledge. Indeed, a theory of mind and the development of mother–infant communication were probably essential for the development of Oldowan technology, enabling the cultural transmission from one generation to the next of the technical skills for flaking stone nodules.[31]

At FxJj50

We can identify some of the sites in Africa where such vocalizations would once have been heard. These are the sites that Glynn Isaac interpreted as home bases for groups of hominids that had spent the day foraging individually or in small parties and had brought back pieces of animal carcasses and collections of plants to share out among the social group. While he believed such gatherings were primarily for food sharing, anthropologists have stressed their value for defence against predators, and they are therefore likely to have been sited among trees. A key example is FxJj50.[32] This is in the Koobi Fora region of East Africa and was excavated by Glynn Isaac and his colleagues in the late 1970s.

When occupied at 1.6 million years ago, FxJj50 was located next to a watercourse, close to groves of shady trees and clumps of fruiting bushes, with access to stones for flaking. Almost fifteen hundred pieces of chipped stone were excavated, along with more than two thousand fragments of animal bones. Several of these had cut-marks from stone artefacts and could be joined back together so that the specific patterns of butchery could be reconstructed. They had come from at least twenty different species of animals, including a variety of antelopes, giraffe, baboons, tortoises and birds. Isaac believed that such diversity was key evidence for food sharing – why else would such a range of animal types, coming from different local environments, have been brought to one single locality?

So at FxJj50 one should imagine thirty or more hominids gathered together: males, females, and infants; those of high status and those of low status; individuals with different personalities and emotions; those with resources to share and those wishing to consume some food. Emanating from the site would have been a variety of calls, reflecting the diversity of activities, how these changed through the day, and the varying emotional states of individuals and the group as a whole. One might have heard predator alarm calls; calls relating to food availability and requests for help with butchery; mother–infant communications; the sounds of pairs and small groups maintaining their social bonds by communicating with melodic calls; and the vocalizations of individuals expressing particular emotions and seeking to induce them in others. Finally, at dusk, one should perhaps imagine synchronized vocalizations – a communal song – that induced calm emotions in all individuals and faded away into silence as night fell and the hominids went to sleep in the trees.

Summary: the origin of 'Hmmmm' communication

Changes in anatomy, foraging behaviour and social life between the time of the common ancestor at 6 million years ago and the earliest *Homo* at 2 million years ago would have had a major impact on the hominid communication system. The key question to ask is whether we are dealing simply with an increase in diversity and quantity of vocalizations/gestures of essentially the same kind as are found among modern-day apes and monkeys, or whether a change in kind occurred. In the previous chapter I characterized the ape and monkey vocalizations/gestures as being holistic and manipulative, rather than compositional and referential. I see no reason why this should have been different for the early hominids. Although they had some degree of bipedalism, and some species had larger brains and made basic stone tools, up until 1.8 million years ago the hominids remained very 'ape-like' in their anatomy and behaviour. Hence my belief is that their vocal and gestural utterances remained holistic, in the sense that they were complete messages rather than words to be combined, and were employed to manipulate the behaviour of others rather than to tell them things about the world.

The contrast I would propose with the ape and monkey communication systems is that the extent of gestures and music-like vocalizations would have increased. To make this difference explicit, it is appropriate to refer to the early hominid communication system as 'Hmmmm' – it was Holistic, multi-modal, manipulative, and musical. While each of these features is found in the communication systems of modern apes and monkeys, I believe that they became integrated together among the early hominids. The result was a communication system more complex than that found now among non-human primates, but one quite different from human language.

10 Getting into rhythm

The evolution of bipedalism and dance

Dave Brubeck's 'Unsquare Dance': *Homo ergaster* playing with sticks and beginning to dance

By 1.8 million years ago, a new form of human ancestor had evolved: *Homo ergaster.* Some anthropologists designate this species as the first member of our genus, believing that *Homo habilis* and *Homo rudolfensis* should be categorized with the australopithecines.[1] Whether or not that is so, *Homo ergaster* may represent a turning point in human evolution.

A spectacularly well-preserved specimen of a male adolescent with a brain size of 880 cubic centimetres was discovered in 1984 at Nariokotome in Kenya. Dating to 1.6 million years ago, it showed that the ape-like, australopithecine body plan, with its suggestions of knuckle-walking and tree-climbing, had by this time been replaced by one that is distinctively modern. *Homo ergaster* walked upright on two legs in the same fashion as we do today. It is this evolutionary development – bipedalism – to which we must now turn, because it has significant, perhaps profound, implications for the evolution of human linguistic and musical abilities.

This may at first sound surprising, since so far we have been primarily searching for their origin in the evolution of the brain, vocal tract and social life. This is because I have largely been following a narrow definition of music, one that is concerned with sound alone. But, as I explained in chapter 2, neither music nor language can be usefully divorced from movement. Hence we must consider the whole of human anatomy in order to understand their evolution. Indeed, the impact of bipedalism on how we move and use our bodies, together with its impact on the human brain and vocal tract, may have initiated the greatest musical revolution in human history. As the ethnomusicologist John Blacking wrote in 1973, 'Many, if not all, of music's essential processes can be found in the constitution of the human body and in patterns of interaction of human bodies in society.'[2]

The anatomy of bipedalism

Although chimpanzees and gorillas can stand and take a few wobbly steps on two legs, this is as strenuously difficult and mentally challenging for them

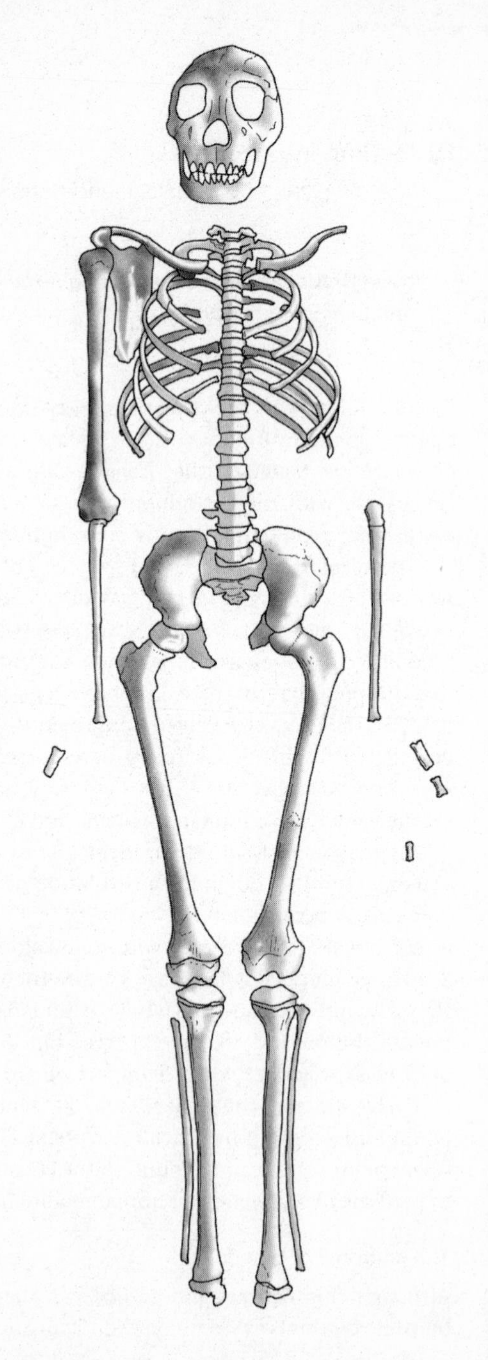

Figure 8 Skeleton of *Homo ergaster* from Nariokotome, Kenya. Specimen KNM-WT 15000, dating to around 1.6 million years ago. This individual is estimated to have weighed 67 kilograms and stood 1.6 metres tall.

as it is for us to run around on four limbs. The difference in locomotory preferences arises from contrasts between ape and human anatomy, notably in the length of the limbs, the nature of the hip, knee and ankle joints, and the form of the toes.[3] Because such anatomical adaptations pervade so many skeletal areas, palaeoanthropologists have been able to trace the gradual transition from four-legged to two-legged walking in our hominid ancestors, which culminated in the evolution of *Homo ergaster*.

Ape anatomy is in fact a compromise between adaptations suited for two different types of locomotion: a form of four-legged locomotion (quadrupedalism) known as knuckle-walking, and tree-climbing. While the human foot provides a platform and has the big toe aligned with the others, the ape foot has evolved for grasping. Its big toe is at a divergent angle from the other toes, in the same way as we have an opposable thumb. The toes themselves are curved to facilitate gripping, climbing and hanging from branches.

The key difference between human and ape knee joints is that the latter cannot be 'locked out' to produce a straight leg. Apes have to keep their legs slightly bent and hence need to employ muscle power to keep the body upright. Their thigh bones are vertical, rather than sloping inwards towards the knee like those of humans, causing their feet to be placed well apart when walking on two legs. As a consequence, apes have to shift their body weight from side to side so that their centre of gravity is brought over each weight-bearing leg in turn. We avoid this waddling gait by placing our feet relatively close together on the ground and using strong hip muscles to counter any side-to-side movement.

Anatomical adaptations for bipedalism continue up through the body: a broader pelvis, a curved lower spine, and a head held vertically. With such anatomy, humans have a striding, fluid gait, in which each leg alternately has a swing phase followed by a stance phase. The anthropologist Roger Lewin describes it thus: 'the leg in the swing phase pushes off using the power of the great toe, swings under the body in a slightly flexed position, and finally becomes extended as the foot once more makes contact with the ground, first with the heel (the heel-strike). Once the heel-strike has occurred, the leg remains extended and provides support for the body – the stance phase – while the other leg goes through the swing phase, with the body continuing to move forward. And so on'.[4]

The australopithecines and earliest *Homo*, which we considered in the previous chapter, habitually walked on two legs, but would have done so with markedly less efficiency than *Homo ergaster* because they possessed only a partial set of anatomical adaptations for bipedalism. These were combined with adaptations for arboreal life – tree-climbing and hanging – which had become lost by the time that *Homo ergaster* appeared. Pre-eminent among the known part-terrestrial/part-arboreal hominids is *Australopithecus afarensis*, which existed between 3.5 and 3 million years ago, and is best represented by Lucy, the unique collection of forty-seven bones from a single individual.[5]

Alongside the adaptations for tree-climbing already mentioned, these bones show that Lucy had several key anatomical features indicative of bipedalism: the upper leg bone (femur) has the same angled shape as that of modern humans, which would have resulted in the feet being placed close together when walking; the pelvis is short with its blades rotated inwards, far more similar to that of a modern human than of an ape; and the articulation of the big toe with the main portion of the foot indicates that it was parallel rather than divergent.

Dramatic proof of Lucy's bipedalism came in 1978 when a 27 metre-long trail of 3.6 million-year-old hominid footprints preserved in volcanic ash was discovered at Laetoli in Tanzania. *Australopithecus afarensis* is the only known hominid species from this region at this date, and is therefore assumed to have been responsible. The footprints were meticulously excavated by Mary Leakey, who revealed that three individuals had once walked across what was then a thick bed of wet volcanic ash. The footprints show a strong heel-strike, followed by the transfer of weight to the outside of the foot, then across to the ball of the foot, and finally concentrated on the big toe.

Although the anatomy of no other australopithecine species is as well known as that of Lucy, it appears that all such species were bipedal to a greater or lesser extent. This was, nonetheless, a quite different type of bipedalism from that of modern humans. Confirmation of this came in 1994 from a surprising source – the bony labyrinth of the inner ear.[6] This consists of three semicircular canals positioned at right angles to each other, which are crucial for enabling us to keep our balance when walking, running and jumping on two legs, and even hopping on one. Apes and monkeys also have a bony labyrinth, but the size and proportions of the three canals are very different from those in humans, reflecting their different locomotory methods.

Fred Spoor of University College London and his colleagues discovered that fossilized skulls still maintain the bony labyrinth and that this can be detected using the technology developed for making brain scans of living people, specifically CT scans. They scanned the inner ears of twelve fossil skulls coming from four different species: two types of australopithecine, *Homo habilis* and *Homo ergaster*.

The *Homo ergaster* specimens alone had an inner ear morphology like that of modern humans. That of the australopithecines was ape-like, while the

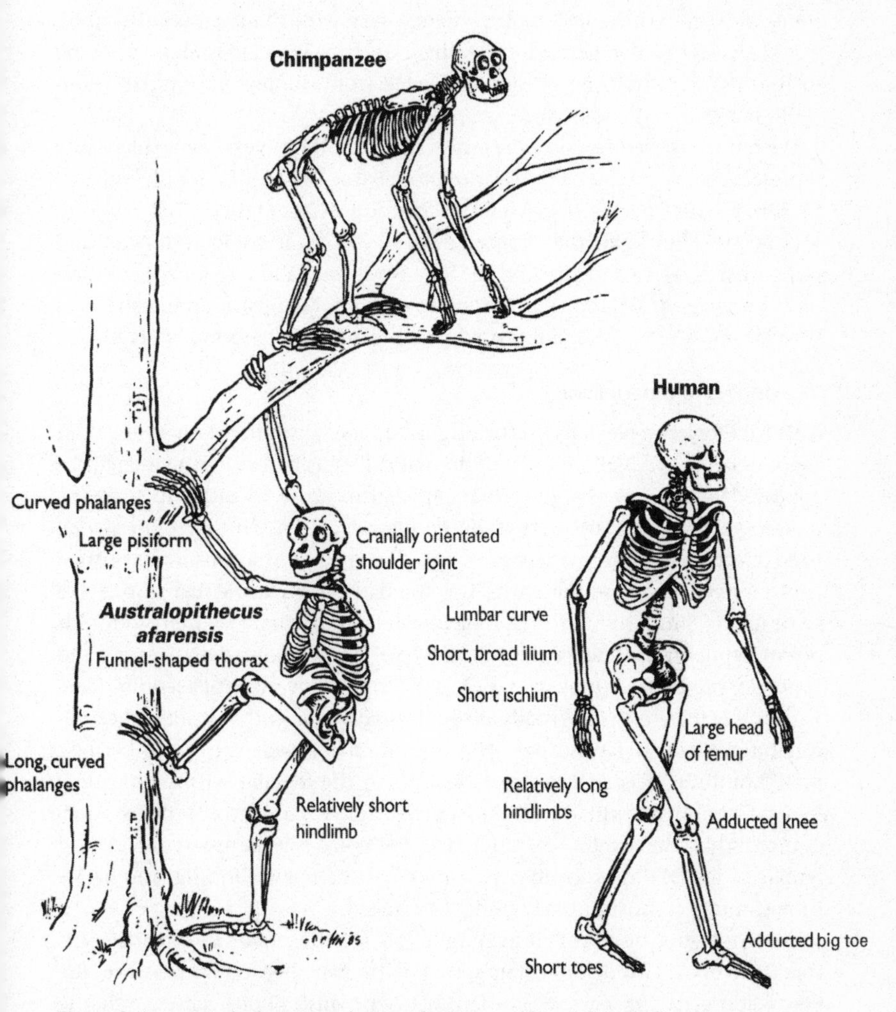

Figure 9 Anatomical adaptations for tree climbing and bipedalism in the chimpanzee, *Australopithecus afarensis* (Lucy) and *Homo sapiens*.

single *Homo habilis* specimen looked more like that of modern monkeys than of either apes or humans. Spoor and his colleagues concluded that although the australopithecines were able to stand and walk on two legs, more complex movements, such as running and jumping, would have been beyond them. *Homo habilis*, existing at a date contemporary with the australopithecines, was even less bipedal, perhaps spending more time knuckle-walking or living in the trees – something already suspected from the few postcranial (non-skull) bones that are known.

We return, therefore, to *Homo ergaster* at 1.8 million years ago, splendidly represented by the so-called Nariokotome boy, as the first fully bipedal hominid. This species possesses the anatomical adaptations necessary to produce the fluid gait that Roger Lewin describes. It could walk, run and jump, just as we can today. And to these we should add two more activities: body language and dance. But before we explore the musical implications of bipedalism, we must first address why this form of locomotion evolved.

The origins of bipedalism

Anthropologists have debated the origins of bipedalism for many years, their theories being gradually revised or dramatically falsified as new evidence has appeared. It was once argued that bipedalism arose in order to 'free the hands' so that stone artefacts could be manufactured. This hypothesis had to be rejected when the evidence for australopithecine bipedalism was pushed back far beyond 2.5 million years ago, the date at which the first stone tools were made. Similarly, the idea that bipedalism originated when hominids began living in savannah environments, perhaps to allow them to stand and watch for predators or prey over the top of long grass, could not accommodate with the evidence that partially bipedal *Ardipithecus* and australopithecines inhabited wooded landscapes. The notion that bipedalism arose because male hominids needed to carry foodstuffs to the females with whom they had pair-bonded similarly had to be rejected when anthropologists recognized that the extent of differences in body size between males and females – sexual dimorphism – indicated that such pair-bonding, and hence provisioning, is most unlikely to have occurred.

Currently, the most persuasive argument for the origin of bipedalism is that it involved two distinct stages, each with its own selective pressure, the first leading to the partial bipedalism of the australopithecines, and the second to the fully modern bipedalism of *Homo ergaster*.

As regards the first stage, Kevin Hunt of Indiana University has suggested that fruit may have played a crucial role.[7] He made a detailed study of the situations in which chimpanzees stand, and occasionally walk on two legs, and found that they do so principally in order to harvest fruit from small trees; using their hands for balance, they pick fruit and slowly shuffle from tree to tree. Standing on two legs permits both hands to be used, while the bipedal shuffling avoids the need to keep raising and then lowering their body weight, which delivers a major saving in energy expenditure. Hunt suspects that the australopithecines were behaving in a similar fashion, and argues that Lucy's anatomy was more suited to standing than walking. The broad pelvis, he suggests, made a stable platform, while the curved fingers and powerful arms should be seen as adaptations not for climbing trees but for hanging by one hand while standing on two legs and feeding with the other.

The second stage – the shift to a fully modern bipedalism – is most probably related to the spread of savannah-like environments in East Africa soon after 2 million years ago, owing to major climatic change. But rather than being driven by the need to see across tall grass, it probably relates to the need to reduce heat stress caused by increased exposure to the sun in landscapes that had markedly fewer trees.[8] Peter Wheeler of Liverpool John Moores University has proposed this in a theory that is neatly summarized as 'stand tall and stay cool'.[9]

Heat stress is a problem for all animals living in savannah environments, largely because brains begin to malfunction when raised 2°C above their normal temperature. When standing erect, a hominid would only absorb sunlight on the top of its head and shoulders, whereas when knuckle-walking its whole back would be exposed. Moreover, the air is cooler away from the soil itself, and wind velocities are significantly higher, both of which would improve the cooling effectiveness of bodily evaporation.

By using a range of computer and engineering models, Wheeler was able to demonstrate that by walking bipedally, savannah-living hominids would have been able significantly to reduce both heat stress and water consumption. This may have played a major role in opening up to them the new dietary niche of meat-eating, as such reductions would have allowed hominids to travel greater distances across savannah landscapes in search of carcasses, possibly at times of the day when competing carnivores would be resting in the sun.

Cognitive and vocal implications of bipedalism

While anthropologists have always been interested in the evolution of human intelligence and language, this reached a particular intensity during the mid-1990s. A succession of academic books, articles and conferences examined every conceivable aspect of brain size and shape, argued about the reconstruction of vocal tracts, and debated the implications for hominids of ape 'language' and intelligence. In 1996 came one of the most valuable contributions, both for its conclusions and for the way in which they were derived, because it looked at the hominid body rather than the brain.

Leslie Aiello, professor of anthropology at University College London, explained in a paper delivered to a joint meeting of the British Academy and the Royal Society that the origin of bipedalism had profound implications for the evolution of intelligence and language.[10] The fact that *Homo ergaster* had a larger brain than the australopithecines – not just in absolute terms but also relative to body size – could, she argued, be explained by the new demands on sensorimotor control that bipedalism both required and enabled. Standing or walking on two legs requires that the centre of gravity is constantly monitored and small groups of muscles frequently recruited and changed to correct its position; the movement of the legs has to be integrated with that of the arms, hands and trunk in order to maintain a dynamic balance. And once those arms and hands are free from a locomotory role, they can be used independently from the legs – for example, for carrying, signalling, throwing or making tools. Bipedalism requires, therefore, a larger brain and more complex nervous system just to attain this more complex degree of sensorimotor control. Once evolved for those reasons, the larger brain might then be used for other tasks, including the planning of foraging behaviour, social interaction and, eventually, language; intelligence may have been no more than a spin-off from walking on two legs.

The evolution of complex vocalization would have been additionally facilitated by the impact of a bipedal anatomy on another part of the body. As described in chapter 8, the human larynx is found much lower in the throat than that of the chimpanzee, and this enables a much wider array of sounds to be produced. Anthropologists traditionally assumed that strong selective pressures for spoken language had 'pushed' the larynx into the throat, even though this created a risk of choking on food. But Aiello argued that the low larynx was merely a consequence of the anatomical adaptations necessary for bipedalism. Because the spinal cord now had to enter the braincase from below rather than from behind (as shown by the position of the foramen magnum), there was less space between the spinal cord and the mouth for the larynx. This space had been further reduced by the changes in the hominid face and dentition that arose with a greater degree of meat-eating. Consequently, the larynx had to become positioned lower in the throat, which had the incidental effect of lengthening the vocal tract and increasing the diversity of possible sounds it could produce.

The claim that this increased the risk of choking, and hence would not have occurred unless there had been strong selective pressure for a low larynx, was shown to be false by Margaret Clegg, one of Aiello's graduate students, who examined vast numbers of medical records in order to document fatalities from choking.[11] She found that, with extraordinarily rare exceptions, death by choking only affects two classes of people: drunken young adult males who choke to death on their own vomit, and infants who are forced to eat solid foods before they are physiologically able to do so.

Aiello suggested that not only the position but also the make-up of the larynx itself may have been changed by the new bipedal anatomy, resulting in the production of a 'less harsh, more melodious sound' than in the australopithecines. The larynx of modern apes and humans is essentially a valve, which also comprises the vocal cords, which can be closed off to allow the air to build up and then opened to create a sound. This 'valvular larynx' also has a quite different function – one to do with movement. Air pressure behind a closed larynx stabilizes the chest and provides a fixed basis for the arm muscles – this is why we hold our breath when about to use our arms vigorously, such as when throwing a ball. Chest stabilization is really important for primates who use their arms for knuckle-walking or climbing. As a consequence, they have a relatively thick, cartilaginous larynx and vocal cords, which produce rather harsh-sounding vocalizations. The vocal cords of modern humans, however, are more membranous.

Bipedalism may have caused the change by relaxing the selective pressure on the locomotor function of the valvular larynx. If so, then by becoming less rigid, *Homo ergaster*'s vocal cords may have further enhanced the diversity of sounds of which this bipedal hominid was capable, even though there were no selective pressures for speaking or singing. These developments begin to look like no more than an accident of evolution.

Proto-language: compositional or holistic?

The social complexity of the early hominids and Early Humans, described in the preceding and following chapters, created selective pressures for enhanced communication. These were realized by the physiological changes I have described, especially those associated with bipedalism, which enabled *Homo ergaster* to produce a significantly larger array of vocalizations than either their immediate hominid ancestors or modern-day apes. It is quite legitimate, although not particularly helpful, to refer to the *Homo ergaster* communication system as proto-language.

In my opening chapter I referred to two differing conceptions of proto-language – compositional and holisitic – and firmly aligned myself with the latter. That alignment has already been partly justified by my interpretations of monkey and ape vocalizations in chapter 8, which I believe to be analogous to those of the common ancestor to apes and humans of 6 million years ago. These are holistic calls and provide a suitable evolutionary precursor for the type of holistic proto-language proposed by Alison Wray. But they provide no foundation for a 'words without grammar' type of proto-language as proposed by Derek Bickerton.

He acknowledges that his type of proto-language would have been a rough and ready form of expression and subject to large measures of ambiguity (as I noted in chapter 1). While it may have been adequate for communicating some basic observations about the world, it would have been unsuitable for what Alison Wray describes as 'the other kinds of messages' – those relating to physical, emotional and perceptual manipulation.[12] It would not, for instance, have been suitable for the type of subtle and sensitive communication that is required for the development and maintenance of social relationships. And as we have seen, the demands of social life appear to have been the principal selective pressure for the evolution of vocal communication in early hominids.[13]

It is important to appreciate that *Homo ergaster* would have lived in socially intimate communities within which there would have been a great deal of shared experience and knowledge about individual life histories, social relationships and the surrounding environment. There would have been relatively slight demands for information exchange compared with our experience today. We are familiar with using language to tell family members with whom we do not live, friends, colleagues, acquaintances and sometimes even total strangers, about ourselves, our children, work, holidays and so forth. We have different types of expertise and knowledge to share, and the majority of us even spend most of the day apart from our partners and children, as part of quite different social networks. We always have new things to tell each other. *Homo ergaster* would have experienced none of these demands for information exchange. There would have been limited, if any, selective pressure within their society for a 'creative language', one that could generate new utterances in the manner of the compositional language upon which we depend.[14]

Bickerton's version of proto-language is also problematic for *Homo ergaster* because it remains unclear why, if such proto-langauge had existed at 1.8 million years ago, it did not rapidly develop into fully modern language – and no linguist is arguing that the latter existed prior to *Homo sapiens* soon after 200,000 years ago. This is a particularly serious problem if simple rules did exist, such as Ray Jackendoff's 'agent-first' rule (see chapter 1), which would have provided clear precursors to grammar.[15] Could there have been almost two million years of a 'words without grammar' proto-language?[16] Probably not, if we accept the results of simulation studies recently conducted at the University of Edinburgh by the linguist Simon Kirby and his colleagues (reviewed in chapter 16). They have shown that some degree of grammatical complexity spontaneously emerges from the process of cultural transmission within a small number of generations. So if 'words' had existed within the utterances of *Homo ergaster*, then we should certainly expect the appearance of grammatically complex language long before the appearance of *Homo sapiens*.

It appears most likely, therefore, that the proto-language of *Homo ergaster* consisted of holistic utterances, each with its own meaning but lacking any meaningful sub-units (that is to say, words), much as Alison Wray has proposed. To illustrate such utterances, Wray provides some hypothetical examples. For instance, a string of syllables such as *tebima* might have meant 'give that to her', and another such as *mutapi* might have meant 'give that to me'. In neither case, nor in any other holistic utterance that one might imagine, would the individual syllables map onto any of the specific entities or actions in the meaning of the phrase. The closest types of utterance we have today are those like 'abracadabra', which might be translated as 'I hearby invoke magic'. Wray envisages hominids using such holistic utterances principally for the manipulation of other individuals, as commands, threats, greetings and requests.

Her notion of proto-language has come under extensive criticism from Derek Bickerton and Maggie Tallerman, a linguist from the University of Durham.[17] One of their arguments depends more on a point of principle than on evaluation of the evidence: they argue that because a holistic proto-language of this type is evidently on an evolutionary continuity with ape-like vocalizations, then it simply cannot be the precursor to modern language. This is based on their belief in a discontinuity between primate calls and human language. It is, however, a view that derives from misconceptions about the nature of primate calls.[18] As Richard Seyfarth, the anthropologist who studied the vervet alarm calls, has recently explained, 'many studies ... document continuities in behaviour, perception, cognition and neurophysiology between human speech and primate vocal communication'.[19]

Another criticism has been that holistic utterances would have been too short, too ambiguous and too few in number to have constituted a feasible proto-language. This criticism is weakened, if not entirely removed, partly by appreciating the nature of *Homo ergaster* society, with its relatively slight demands for information exchange, and partly by elaborating Wray's notion of proto-language by suggesting that the holistic 'Hmmmm' utterances of *Homo ergaster* would have been as much music-like as language-like. We should envisage each holistic utterance as being made from one, or more likely a string, of the vocal gestures that I described in the previous chapter. These would have been expressed in conjunction with hand or arm gestures and perhaps body language as a whole, as I will describe below. In addition, particular levels of pitch, tempo, melody, loudness, repetition and rhythm would have been used to create particular emotional effects for each of these 'Hmmmm' utterances. Recursion, the embedding of one phrase within another, is likely to have become particularly important in order to express and induce emotions with maximum effect. The key argument of this chapter is that both the multi-modal and the musical aspects of such utterances would have been greatly enhanced by the evolution of bipedalism.

The musical implications of bipedalism

When Leslie Aiello wrote about the impact of bipedalism on the evolution of language I doubt if she was making any differentiation between speaking and singing at this stage in human evolution – she often stresses the musical qualities of hominid vocalizations. The increased range and diversity of vocalizations made possible by the new position and form of the larynx, and changes in dentition and facial anatomy in general, would certainly have enhanced the capacity for emotional expression and the inducing of emotions in others. But the musical implications of bipedalism go much further than simply increasing the range of sounds that could have been made.

Rhythm, sometimes described as the most central feature of music, is essential to efficient walking, running and, indeed, any complex coordination of our peculiar bipedal bodies. Without rhythm we couldn't use these effectively: just as important as the evolution of knee joints and narrow hips, bipedalism required the evolution of mental mechanisms to maintain the rhythmic coordination of muscle groups.

The significance of such mental mechanisms becomes apparent from people who have either lost them because of cognitive pathologies or who have suffered mental disability from birth and always struggled to achieve fluid physical movements. When their lack of an internal rhythm mechanism is compensated for by provision of an external rhythm, marked improvements in walking and other physical movements occur.

The music therapist Dorita S. Berger gives a striking example in her account of Alonzo, an eight-year-old autistic non-verbal boy:

> [Alonzo] ran around the music therapy classroom aimlessly while the piano played and I sang a song describing and reflecting his behaviour back to him. Alonzo remained at the far end of the large classroom and did not approach the piano. Occasionally, he would stop and cast glances across the room toward the piano, but his physical behaviour remained chaotic, distant, out of control.
>
> Abandoning the piano and song, I opted instead for the simplicity of pure rhythm played on the conga drum. No singing; no melody; nothing but rhythm. In an instant, Alonzo's running halted, his body and mind frozen in a state of attention. He looked across the room at me and the drum. As suddenly

> as he had stopped he began moving in an organized, rhythmic, stepping demeanour, toward the sound. His steps accented the pulse of the conga's beat as he neared the drum.[20]

While such anecdotal accounts are provocative, their scientific value is questionable because they rely on a subjective interpretation of events. Fortunately, we have more than a decade of meticulously conducted scientific studies exploring the relationship between rhythm and movement conducted by the neuroscientist Michael Thaut, director of the Center for Biomedical Research at Colorado State University and himself an accomplished musician.[21]

One of Thaut's key studies has been with people who suffer from Parkinson's disease. This is an impairment of the basal ganglia of the brain, which, among other effects, disturbs the temporal aspect of motor control. Thaut explored the impact of rhythmic auditory stimulation (RAS) on Parkinson's sufferers, simply by playing them a regular beat as they were asked to walk on both flat and sloping surfaces. He had sufficient numbers of people to divide them into three groups. One received a daily thirty-minute 'gait-training' session with RAS for three weeks; another received the same training session but without RAS; the third group received no training at all. He found that significant improvements in walking emerged only in the RAS-trained group. Among these patients, their stride velocity increased by 25 per cent, their stride length by 12 per cent, and their step cadence by 10 per cent. Thaut extended his study to find that without any further training the improvements in gait gradually fell away and were no longer perceptible after five weeks.

When expressed as percentages, these improvements in walking achieved by the provision of a regular beat may not appear very impressive. But when one watches a video of Thaut at work the improvements in gait are really striking – a shift from cumbersome, shuffling movements to a quite unexpected degree of muscle control and fluid movement.

In fact, 'hearing' itself may not be necessary. In a further set of experiments, Thaut required his subjects to tap their fingers in rhythm with a beat. He found that as he changed the tempo, so they, too, changed the tempo of their finger-tapping. This is not altogether surprising – we are all able to match a rhythm to a greater or lesser extent. But Thaut discovered that if he altered the beat by an amount that was below the threshold of conscious perception, the subjects still altered their finger-beating in accordance with it – even though they were quite unaware that any change had occurred.

Music therapists can use the impact of rhythm on movement without worrying about the precise cause: if it helps those with mental and physical constraints, then use it. Those concerned with evolutionary issues, however, need to understand why such effects arise. Unfortunately, scientists are still some way away from understanding the link between auditory processing and motor control. Thaut has explained that there is some basic physiological data that indicates the presence of an auditory–motor pathway in the nervous system by means of which sound exerts a direct effect on spinal motor neuron activity. 'How could it be otherwise?' one might ask, when we all know the effect that listening to music can have on our own movements, as when without any thought we begin to tap our feet or sway our bodies. Thaut explains that he and other scientists remain unsure whether specific neural structures exist that are responsible for rhythmic motor synchronization, or whether neural excitation patterns in the auditory system are projected into the motor system circuits.

One hopes that with the further development and application of brain scanning technology we will gain a better understanding of how our auditory and motor systems are linked together. But following Thaut's studies, and accounts of the impact of rhythm on children such as Alonzo, we can conclude that efficient bipedalism requires a brain that is able to supply temporal control to the complex coordination of muscle groups. The loss of this cognitive ability can be partly compensated for by the use of an external beat.

The key point is that as our ancestors evolved into bipedal humans so, too, would their inherent musical abilities evolve – they got rhythm. One can easily imagine an evolutionary snowball occurring as the selection of cognitive mechanisms for time-keeping improved bipedalism, which led to the ability to engage in further physical activities that in turn required time-keeping for their efficient execution. Key among these would have been the use of the hands, now freed from their locomotory function. They would have been used for hammering: bones to crack them open for the marrow inside, nuts to extract the kernels, and stone nodules to detach flakes.

Music and motion

It is useful to reflect here on the frequent association that is drawn between music and motion.[22] In the Western tradition, music is described by words relating to movement, such as lento, andante, corrente (slow, walking, running), while some argue that both performer and listener must imagine inner movement in order fully to appreciate a piece of music. As I mentioned in chapter 2, John Sloboda, the distinguished music psychologist, argues that music is the embodiment of the physical world in motion.

Johan Sundberg, a musicologist based at the Royal Institute of Technology, Speech, Music and Hearing in Stockholm, has undertaken experimental studies to substantiate the music–movement association.[23] In one of these, he found strong similarities between the pattern of deceleration when people are stopping running and the changes in tempo as pieces of music are coming to an end, using the ritardandi in recordings of baroque music for his sample.

Sundberg does not speculate on why this association between music and motion should exist. It appears intuitively plausible, however, that it directly reflects how the changes in human mobility that accompanied the evolution of bipedalism had a fundamental impact on the evolution of human musical abilities. It may, indeed, be in this connection that the phenomenon of entrainment – the automatic movement of body to music – arose. Experimental work with chimpanzees seems essential since, according to this hypothesis, their lack of full bipedalism should mean that they also lack the phenomenon of entrainment to music.

Another issue is evidently tied up with the relationship between music and movement, that of emotion. Once again I will defer to the ethnomusicologist John Blacking. He recognized the very intimate link between music and body movement, so much so that he suggested that all music begins 'as a stirring of the body' and that 'to feel with the body is probably as close as anyone can ever get to resonating with another person'. So by getting into the 'body movement of the music', one can feel it very nearly as the composer felt it.[24] This leads on quite naturally to the phenomenon of dance. To address that, we must briefly return to the evolution of bipedalism itself.

Walking, running or dancing?

Our understanding of how bipedalism and human anatomy evolved was given a jolt in November of 2004 by a publication in the prestigious science journal *Nature*.[25] The anthropologists Dennis Bramble and Daniel Lieberman, from Utah and Harvard Universities, argued that running, and specifically 'endurance running', may have been far more important for the evolution of human anatomy than walking. The human ability to run long distances is unique among primates; and, when allowance is made for body size, humans can sustain running speeds that are comparable to the trotting of quadrupeds such as dogs and ponies.

The long, spring-like tendons that we have in our legs, most notably the Achilles tendon, have little impact on our walking ability but are vital to our capacity for running, making this 50 per cent more efficient than would otherwise be the case. The longitudinal arches of our feet have the same effect, as do our relatively long legs, compact feet and small toes. Also, the joints of our lower limbs, especially the knee, have larger surface areas than would be expected if we had evolved simply to walk on two legs; such joints are particularly well suited for dissipating the high impacts that arise from running.

Another anatomical adaptation that may be for running is one of the most noticeable but evolutionarily neglected aspects of our anatomy, big bottoms; humans have much larger behinds than any other primate. The bottom is a muscle, the gluteus maximus, which plays a limited role in walking but is essential for running. One of its functions is to provide balance, as our torsos have a tendency to pitch forward when we run. Our long waists and broad shoulders also play a role in stabilizing the upper part of the body to an extent that walking alone does not require.

When this suite of anatomical features is combined with our adaptations for reducing body heat, notably our sweat glands, minimal body hair, and the tendency to breathe through the mouth during strenuous activity, the human body appears to be adapted for running as much as for walking. Most of these adaptations were present in *Homo ergaster*, leading Bramble and Lieberman to suggest that running might have been essential for competing with scavenging wild dogs and hyenas on the African savannah.

Walking and running are not, of course, the only physical activities in which humans engage. We also climb, jump, skip and ... dance. Could it be that the springs within our legs, our long waists, big bottoms and small feet were all selected for neither walking nor running but for twirling, leaping, pirouetting, and perhaps even for performing an arabesque or an échappé beneath the African sun?[26] Probably not. But these would certainly have all been made possible by the bipedal anatomy of *Homo ergaster* – which goes far beyond what is required simply for walking on two legs.

Gesture, body language and dance

The new degrees of motor control, independence of torso and arms from legs, and internal and unconscious time-keeping abilities, would all have dramatically enhanced the potential for gesture and body language in *Homo ergaster*, hugely expanding the existing potential for holistic communication. This would have added to vocalization an invaluable means of expressing and inducing emotions, and manipulating behaviour. Given the complex social life of hominids, those who were most effective at gesture and body language would have been at a social and reproductive advantage – just as were those who were most effective at vocal communication.

The spontaneous gestures that accompany the spoken language of modern humans are quite unlike words because they are not composed with a grammar to create an additional level of meaning. In this regard they are holistic, just as a musical phrase is holistic, and are quite different from the intentional symbolic gestures that modern human uses, such as the V for victory or thumbs-up sign. These intentional gestures are symbolic, and there are no grounds for attributing them to species prior to *Homo sapiens*.

As mentioned above, the majority of spontaneous gestures used by modern humans are iconic, in the sense that they directly represent whatever is being verbally expressed, and I strongly suspect that these were adopted by *Homo ergaster* and later hominids as a central element of 'Hmmmm'. So, if I were describing something that is big, my gesture would most likely involve placing my hands together and then gradually moving them apart; whereas when describing something as small I might gradually move the finger and thumb of one hand together until they are almost touching.[27]

Geoffrey Beattie, a psychologist at Manchester University, has undertaken several studies of the spontaneous human gestures that are significant for understanding the evolutionary relationship between language and music, and are informative about the nature of 'Hmmmm'. One striking finding is that everyone appears to use a similar suite of spontaneous gestures, irrespective of what language they speak.[28] Beattie has confirmed that gestures play a complementary role to spoken utterances, rather than being merely derivative or supplementary. So gestures are not used simply to help the speaker retrieve words from his or her mental dictionary; they provide information that cannot be derived from the spoken utterance alone.[29] Beattie's experimental work indicates that gestures are particularly important for conveying information about the speed and direction of movement, about the relative position of people and objects, and about the relative size of people and objects.

Beattie's work has confirmed the critical role of gesture in human communication. This role is perhaps best expressed in the words of David McNeill, whose 1992 book, *Hand and Mind*, pioneered the notion that gesture can reveal thought.[30] McNeill explained that 'Utterances possess two sides, only one of which is speech; the other is imagery, actional and visuo-spatial. To exclude the gestural side, as has been traditional, is tantamount to ignoring half of the message out of the brain.'[31] Thus body movement appears to be as crucial to language as it is to music.

Although gestures often accompany speech, they can be used independently of speech, in either a spontaneous or a deliberate fashion. Indeed, it has been said that 65 per cent of human communication occurs through body rather than spoken language. We communicate in this manner quite unconsciously most of the time, and the popular books and TV programmes about reading body language are often both quite entertaining and informative on the subject of how much we unknowingly tell others about our innermost thoughts and feelings.[32] Just as we are often unaware of the signals we are giving out, we rarely appreciate how we base our judgements about other people on their gestures and postures rather than on what they say. Hence, when we get a hunch about somebody's untrustworthiness, or that they wish to develop a closer relationship, this is most likely to have come from an unconscious reading of their body language.

Several elements of body language today have both a universal and a cultural element. We all have an unconsciously intimate body space around us, which we only allow our family and closest friends to enter. Should a stranger enter that space, we experience the physiological responses of increased heart rate, adrenalin pouring into the bloodstream, and the tightening of muscles, all of which are preparations for a potential 'fight or flight' situation.[33]

Just as each of us has an unconsciously defined intimate zone, we also have personal, social and public zones around us, each at a greater distance, within which we feel comfortable about different types of people being present – friends, work acquaintances, strangers. Who is 'allowed' into each space can depend upon who else is watching: a work colleague may be kept at distance in the office but welcomed into the personal or even intimate zone in private.

Just like the facial expressions that we considered in chapter 7, body language can be faked. Indeed, learning about body language and using it to deceive is central to the actor's art.[34] Politicians also frequently attempt such deception, although the physiologically based 'micro-gestures', such as dilation of the pupils, may give them away. As a broad generalization, few would disagree that we are very poor at consciously attending to and using body language today. I suspect that this was quite different for our non-linguistic ancestors.

The examples of body language I have given above are all well known, and are complemented by subtle variations of posture and gesture that cannot be easily categorized. But their role in modern human communication should not be neglected. Rudolf Laban, perhaps the most influential writer and thinker about human movement in the twentieth century, understood this when writing his seminal work, *The Mastery of Movement* (1950). There he explained: 'The astonishing structure of the body and the amazing actions it can perform are some of the greatest miracles of existence. Each phase of movement, every small transference of weight, every single gesture of any part of the body reveals some feature of our inner life.'[35]

The true significance of body language can perhaps be appreciated by recognizing that whereas speaking is an intermittent activity – it has been estimated that the average person talks for no more than twelve minutes a day[36] – our body language is a continuous form of communication. Colwyn Trevarthen, emeritus professor of psychology at Edinburgh University, argues that appreciating the rhythmic and harmonious nature of human body movement is critical to understanding the origins of human music. He once described music as being no more than 'audible gesture' and has explained that:

> If we watch persons going about their ordinary business, working alone, mingling and chatting in a crowd, or negotiating and collaborating in a collective task, we see that, while the human body is built to walk on two legs to an inner drum, at the same time it is intricately juggling hips, shoulders and head as a tower of separately moving parts above the stepping feet. While walking, we freely turn and twist, glance with eyes jumping to left and right, extend waving limbs, make intricate gestures of the hands, talk, all in coordinated phrases of flowing rhythm. This movement has a multiplicity of semi-independent impulses, a potentiality for multi-rhythmic coordination that is surely richer than any other species possesses, inviting witty iteration and syncopation.[37]

Expressive body language for one might be dance for another. Rudolf Laban described how an actress playing the role of Eve might pluck the apple in more than one way, with movements of varying expression.[38] Her arms or her whole body might shoot out suddenly, avidly, in one direction towards the coveted object to express greed. Alternatively, she might be languid or even sensuous, with a nonchalant, slow lifting of an arm, while the rest of the body is lazily curved away from the apple. Our non-linguistic human ancestors might have used a similar variety of movements when miming or enacting different attitudes towards the butchery of a carcass or the sharing of food.

By so doing, they would have placed different nuances of intent or meaning onto the same basic holistic utterance/gesture. Laban gives the simple example of the expressive range of gestures that can accompany the word 'no'. He explains that one can 'say' this with movements that are pressing, flicking, wringing, dabbing, thrusting, floating, slashing or gliding, each of which 'says' 'no' in a quite different manner.[39] Once such gestures are integrated into a sequence of body movements and vocalizations, once some are exaggerated, repeated, embedded within each other, one has both a sophisticated means of self-expression and communication, and a pattern of movements that together can be observed as pure dance alone.

Today we are relatively desensitized to the subtleness of gestures and body movement, partly because we are so dependent upon language and hence have difficulty imagining the power of a gestural and movement-based communication system. Laban laments this loss: 'we have lost this language of the body, and there is little probability that we can rediscover it.'[40]

The musicality of *Homo ergaster*: the beginning of an evolutionary process

I began this chapter by suggesting that bipedalism may have initiated a musical revolution in human society. It did this by enhancing the diversity of vocalizations, body language and the use of rhythm. Whereas we should imagine the vocal communications of the australopithecines, *Homo habilis* and *Homo rudolfensis* as more melodious versions of those made by non-human primates today, those made by members of the *Homo ergaster* species, such as the Nariokotome boy, must have been very different, with no adequate analogy in the modern world. While bipedalism inadvertently extended acoustic diversity, its key consequence was to take the role of rhythm and movement in 'Hmmmm' to a new order of magnitude.

I must, however, be careful not to exaggerate the musicality and communication skills of *Homo ergaster*, as this species marks only the beginning of an evolutionary process. The holistic phrases used by *Homo ergaster* – generic forms of greetings, statements and requests – are likely to have been small in number, and the potential expressiveness of the human body may not have been realized until later, bigger-brained, species of *Homo* had evolved. Moreover, *Homo ergaster* certainly lacked the anatomical adaptations for fine breathing control that are necessary for the intricate vocalizations of modern human speech and song. In fact its ability to regulate breathing appears no different to that of the African apes today.[41]

While *Homo ergaster* may have been fully bipedal, it remained australopithecine-like in other characteristics. The Nariokotome specimen did have a relatively large brain compared to the 450 cubic centimetres of living African apes and australopithecines, but this is primarily a reflection of that specimen's large body size. It is not until after 600,000 years ago that the brain size of *Homo* increases significantly, exceeding that which can be accounted for by body size alone and reaching that of modern humans by 100,000 years ago at the latest, and possibly by 250,000 years ago.[42] There was, therefore, about one million years of effective stasis in brain expansion after *Homo ergaster* had first appeared.

The immense and rapid growth of the human brain that began after 600,000 years ago can best be explained by selective pressures for enhanced communication, resulting in a far more advanced form of 'Hmmmm' than that used by *Homo ergaster*. The next chapters in this book will explore these pressures and the manner in which 'Hmmmm' may have evolved, leading us towards the 'Singing Neanderthals' that were living in Europe by 250,000 years ago.

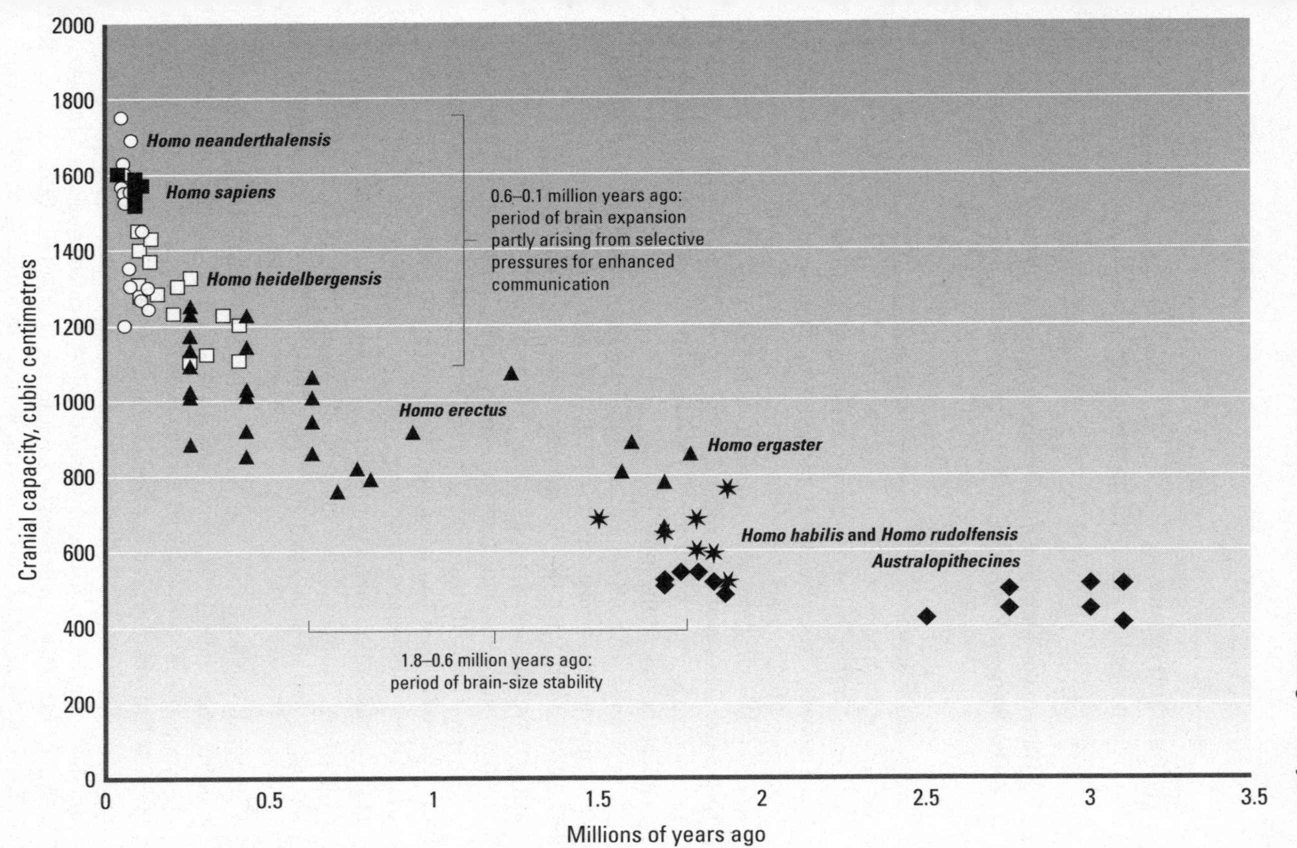

Figure 10 The evolution of human brain size.

11 Imitating nature

Communication about the natural world

'Watermelon Man' by Herbie Hancock: a group of *Homo heidelbergensis* returning to their campsite after a successful hunt

A remarkable archaeological discovery was made in 1995: three 400,000-year-old wooden spears. These were excavated by Hartmut Thieme at an archaeological site exposed by open-cast coal mining at Schöningen in southern Germany. Several notched pieces of wood were also found which, together with the spears, represent the oldest wooden hunting weapons ever discovered. Each spear had been made from the trunk of a thirty-year-old spruce tree. They had been carefully shaped from the tree trunks and look like modern javelins, with their tips carved from the particularly hard wood found at the base of the trunk.[1]

The Schöningen finds demonstrated that pre-modern humans were manufacturing wooden artefacts by at least 400,000 years ago. Few archaeologists had ever doubted that this was the case because wear traces found on stone artefacts had indicated woodworking. Many archaeologists, however, had doubted that pre-modern humans were big-game hunters; but the Schöningen spears proved them wrong.[2] They were found at a likely habitation site, one where fire had been used and where there were more than ten thousand well-preserved bones from horse and other large mammals. Many of the bones had cut marks from stone tools and fracture patterns indicative of butchery. To argue that the javelin-like spears had not been thrown to kill horses and deer, and perhaps even elephant and rhinoceros, would be perverse.

Although the Schöningen discoveries ended the debate about big-game hunting, they inflamed another – about the cognitive and communicative abilities of pre-modern humans. Robin Dennell, professor of archaeology at Sheffield University, argued that the Schöningen spears indicate 'considerable depth of planning, sophistication of design, and patience in carving the wood, all of which have been attributed only to modern humans'.[3] He also described himself as 'speechless' when learning of their discovery.

The Schöningen spears provide an appropriate introduction to this chapter because it will examine how the 'Hmmmm' communication system of early hominids would have been moulded by their interactions with the natural world between 1.8 million and 250,000 years ago. Both the previous chapters, and the two that follow, focus on the social world of pre-modern humans and how 'Hmmmm' expressed and induced emotional states and structured social interaction. My concern here is communication about animals, plants, rivers, weather and other features of the natural world. Our starting point is with the colonization of Europe and Asia soon after 2 million years ago.

Dispersal into Eurasia

Until very recently, the vast majority of archaeologists believed that human ancestors first dispersed out of Africa at around 1 million years ago. But a succession of new discoveries, and the redating of long-known fossils by new techniques, have doubled that estimate.[4] So it now appears that the date of initial dispersal virtually coincides with the appearance of *Homo ergaster* in the fossil record. That bigger-brained and fully bipedal Early Human species appears to have been the first to possess both the capacity and the motivation to disperse into new landscapes.

The first traces of pre-modern humans outside of Africa were discovered in Java at the end of the nineteenth century. Several more skull fragments, then believed to be around one million years old, were found between 1928 and 1937. These were assigned to *Homo erectus*, now thought to be an immediate descendant of *Homo ergaster*. In 1994, two of the fragments were redated to 1.8 and 1.6 million years ago, suggesting that *Homo ergaster* had not only dispersed from Africa as soon as it had evolved but had also undergone rapid evolutionary change. Indeed, if the earlier date is correct, then the initial dispersal may have been by *Homo habilis*.[5]

In 1995 a claim was published in *Science* for Early Human occupation at Longuppo Cave in China at 1.9 million years ago.[6] This had its problems: the supposed specimen, a tooth, may have come from an ape, whose presence in China at that date would not be unexpected. Moreover, chipped stone artefacts found at the site do not appear to have been directly associated with the sediments that were dated. More reliable evidence has since come from lake sediments in the Nihewan Basin in the north of China, where indisputable stone artefacts have been dated to 1.66 million years ago.[7]

Further evidence for dispersal from Africa prior to 1 million years ago has come from western Asia. The earliest occupation at 'Ubeidiya, an archaeological site in the Jordan Valley, has been dated to 1.5 million years ago.[8] This has large quantities of stone artefacts and some unquestionable Early Human fragments. It is also on a key dispersal route from Africa, as the Jordan Valley is merely an extension of the Rift Valley, where *Homo ergaster* evolved. In 2000, two *Homo ergaster* skulls, almost identical to that from

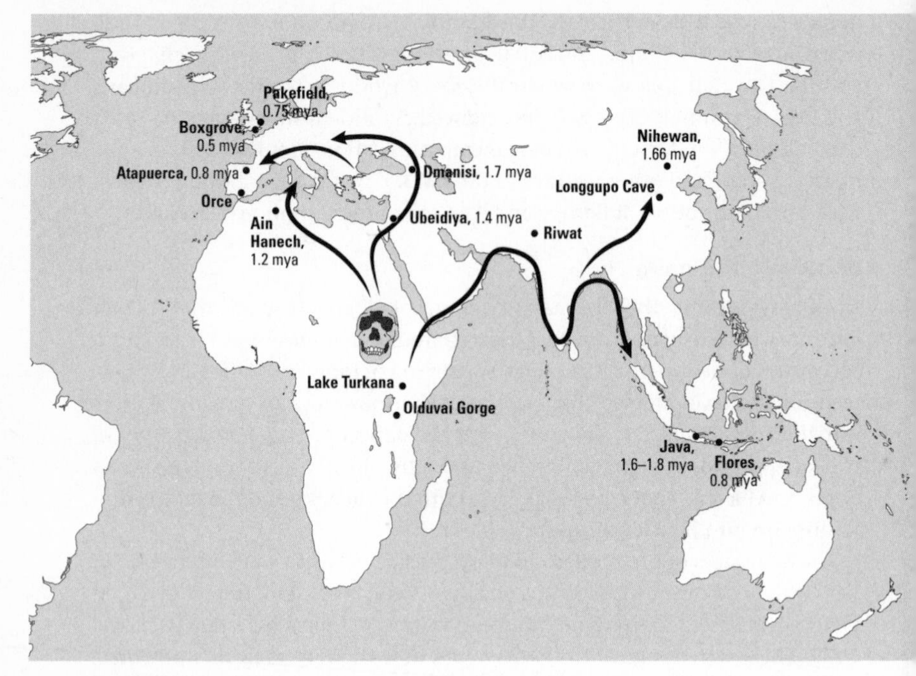

Figure 11 Early dispersals of hominids, most likely *Homo ergaster*, and location of key sites mentioned in the text.

Nariokotome in Africa, were found at Dmanisi, in Georgia, a further two thousand kilometres north of 'Ubeidiya, and dated to 1.7 million years ago.[9] Two further skulls and jaw bones have since been discovered, and the fossils are intriguing because they show a smaller brain (and body) size than the Nariokotome specimen, at a mere 650 cubic centimetres.

It appears, therefore, either that *Homo ergaster* dispersed out of Africa soon after it had evolved and rapidly spread throughout Asia, or that there had been an even earlier dispersal of *Homo habilis*.

Although the Dmanisi site has been described as being 'at the gates of Europe',[10] conclusive evidence for a European presence prior to 1 million years ago has remained elusive. Several sites have been claimed as having artefacts of a far older date, but these are almost definitely stones that have been fractured by frosts and collisions in rivers rather than by Early Human hands. The oldest authenticated site is TD6 at Atapuerca in northern Spain.[11] This is dated to 800,000 years ago and has a significant number of Early Human fossils, many of which have cut-marks from stone tools suggesting cannibalism. Orce, in southern Spain, with its supposed fossil fragment and stone artefacts dated to 1.9 million years ago, is another candidate.[12] But the fossil, a rather nondescript piece of bone, is more probably from a horse than from an Early Human. The date is also questionable. It is based on sedimentary evidence for one of the reversals of the earth's magnetic field that have occurred throughout the past.[13] But the reversal at Orce seems more likely to have been at 0.9 than at 1.9 million years ago.

By 0.5 million years ago, a substantial Early Human presence had been established in northern Europe, as demonstrated by the site at Boxgrove in southern England.[14] Although the evidence remains scanty and is still under analysis, recent discoveries of stone artefacts from Pakefield in Norfolk may imply occupation by 750,000 years ago.[15] The fossil fragments from Boxgrove and other European sites of a similar age are attributed to *Homo heidelbergensis*, another immediate descendant of *Homo ergaster* and the most likely direct ancestor of the Neanderthals. The bones of *Homo heidelbergensis* were large and thick-walled, indicative of a substantial body mass, which was probably a physiological adaptation for living in the cold landscapes of northern Europe. This was inherited and then further evolved by *Homo neanderthalensis*, a direct descendant of *Homo heidelbergensis*.

A quite different type of physiology evolved in part of the *Homo* lineage of south-east Asia. Excavations during 2003 in Liang Bang rock shelter on Flores Island in Indonesia produced the remains of seven individuals who had been no more than one metre high, even though they were fully grown adults. These were designated as a new species, *Homo floresiensis*.[16] In view of the date of stone artefacts found on the island, it is thought that a small population of *Homo erectus* made a short water crossing to Flores Island at around 800,000 years ago.[17] That population then became isolated and evolved into a dwarf form of *Homo*. This is a process known to happen to large terrestrial mammals isolated on small islands, especially if predators are absent. Dwarf elephants, hippos, deer and mammoths have been found on islands around the world. This is the only discovery of dwarf humans and shows just how physiologically diverse our genus can be, especially when compared to the relatively massive *Homo heidelbergensis* from Europe. Perhaps just as remarkable as their size is that *Homo floresiensis* survived on Flores Island until a mere 12,000 years ago.

Developments in stone technology

It is useful to refer to *Homo ergaster*, *Homo erectus* and *Homo heidelbergensis* as Early Humans rather than early hominids because this acknowledges their evolutionary status. These species are strikingly similar to *Homo sapiens* by virtue of their fully modern stature and bipedalism; they were large and robust individuals, with brain sizes of around 1000 cubic centimetres – about 80 per cent of that of modern humans. And those living after 1.4 million years ago had a stone tool technology significantly more complex than that which had gone before.

That is the date at which a new type of artefact appears in the archaeological record: the hand-axe. Unlike the Oldowan chopper tools made by *Homo habilis* and the australopithecines, hand-axes have a carefully imposed form. Many are pear-shaped, often about the size of one's palm; others are ovate; and some, technically known as cleavers, have a straight edge. Hand-axes from Africa, such as those found in the deposits of Olduvai Gorge or in vast numbers at the 0.5 million-year-old site of Olorgesailie, were made by chipping nodules of basalt, quartzite, chert and limestone. Many thousands have been excavated from sites in Europe, and in western and southern Asia, where flint was often the preferred raw material owing to its fine flaking qualities.

Meticulous studies of the scars left where flakes were detached, and the experimental knapping of stone, have given archaeologists a very good understanding of how hand-axes were made.[18] Stone hammers – spherical nodules of quartzite, for example – were used to detach large, thick flakes from a nodule of stone. After the approximate shape of the hand-axe had been created, other types of flakes were removed, notably 'thinning' flakes. These were often detached with a 'soft' hammer of antler or bone, which had to strike the stone at a very specific angle in order to remove flakes that would thin the hand-axe. Small preparatory flakes often had to be removed in order to create striking platforms; and throughout the manufacturing process sharp edges needed to be ground slightly so as to remove irregularities that might otherwise have deflected the force of the hammer blow.

Experimental studies have shown that hand-axes can be very effective butchery implements. Their association with animal bones at many archaeological sites, along with cut-marks on the bones and wear traces on the hand-axes, confirm their use for butchery – although they may have been used for other tasks as well. The site at Boxgrove in West Sussex provides clear evidence that hand-axes were used for butchering big game, including horse, deer and rhinoceros.[19]

Two other features of hand-axes must be noted. First, their longevity in the archaeological record: for more than a million years hand-axes remained effectively unchanged as the most advanced form of stone tool available. They were still being made by some of the last Neanderthals in Europe at just 50,000 years ago, although by that time they had become a relatively minor component of the technological repertoire. All in all, they are the most widespread and long-lasting tool type from the whole of human evolution.

Secondly, many hand-axes have a highly symmetrical form which appears to be quite unnecessary for their use as butchery implements. I will provide an explanation for such symmetry in the following chapter.

The process of dispersal

Although the evidence for Early Human dispersal from Africa remains both scanty and difficult to interpret, most palaeoanthropologists now agree that this occurred soon after 2 million years ago – more than half a million years before the first appearance of hand-axes and while only an Oldowan-like technology was available.

The process of dispersal might have been no different from that of the other large mammals that are known to have spread from Africa into Europe and Asia between 2 and 1 million years ago.[20] This period saw many changes in the earth's climate, which constantly altered the character of the vegetation cover and the foraging opportunities for many species, including Early Humans. In response, their populations grew and spread into new territories or else shrank and retreated, but without any conscious intention to do so. Accordingly, we would need no more than the techniques of biogeography and palaeoecology to explain the presence of stone artefacts and Early Human fossils at sites such as 'Ubeidiya, Dmanisi, Atapuerca and Flores Island.[21]

An alternative scenario, however, is that by 2 million years ago our human ancestors were already behaving quite differently from other large mammals, and this requires us to invoke concepts of exploration and discovery in order to explain their dispersal from Africa. But whether we should also attribute to them the 'Everest syndrome' – a conscious desire to explore new lands for no reason other than 'because they were there' – remains unclear.[22]

Another point of debate concerns the routes of dispersal. The conventional view has been that *Homo ergaster* left Africa via the far north-east, following the geographical route formed by the Rift Valley, the Nile Valley, and the Jordan Valley. According to this view, 'Ubeidiya is just where we should expect to find the earliest sites out of Africa. There were, however, two other possible dispersal routes. The first was to the east of the Rift Valley, from modern-day Ethiopia into modern-day Yemen. Today this requires a water crossing, but with even a minor reduction of sea level, as repeatedly happened during the Pleistocene, Early Humans could have spread from East Africa into south-west Asia on foot.[23] The second alternative would have involved a water crossing from the North African coast to Europe. The shortest route would always have been across the Straits of Gibraltar. But the strength of currents through the straits is likely to have made such a trip prohibitive for Early Humans, who are unlikely ever to have used anything more than very simple rafts. Thus the crossing from the coast of modern-day Tunisia to Sicily, which is likely to have been no more than fifty kilometres at times of low sea level, seems more probable.

The fact that the earliest dates for European settlement are found in the south rather than the north might support the idea of direct colonization from Africa. But the northern landscapes would have been more difficult for an African-evolved Early Human to master. Hence we should expect a delayed colonization of these areas regardless of whether Early Humans crossed the much-reduced Mediterranean Sea or arrived in Europe via an Asian route.

Moreover, it is most unlikely that there was only one dispersal route, or that the routes were used on just one occasion. Rather than envisaging the dispersal out of Africa as some type of one-off migration event which led to swathes of Early Humans gradually encroaching across the Asian and European land masses, we should think in terms of multiple minor dispersals, many of which would have led to nothing more than the deaths of those Early Humans that entered into new environments. So, although there may be traces of Early Humans out of Africa soon after 2 million years ago, it probably took another million years until a permanent and continuous Early Human presence was established. Indeed, the eventual appearance of sites in northern Europe at 0.5 million years ago might reflect a new-found ability to survive in northern latitudes that simply had not previously been present. That ability might have lain in the development of technology to improve their effectiveness as big-game hunters, as demonstrated by the Schöningen spears. But such technology, in turn, may have derived from a further evolution of their communicative abilities.

Communication about the natural world

By definition, dispersal means entering unfamiliar landscapes, in which new types of animals and plants would have been encountered, and where new distributions of fresh water, firewood, stone and other resources would have had to be found and the information transmitted to others. Such communication would have been essential regardless of whether we think of Early Human dispersal as more similar to that of other large mammals or to that of modern human explorers. Whichever is the case, Early Humans were dependent for their survival and reproduction on working cooperatively as part of a group, as will be examined further in chapter 14.

This would have been the case especially when they were engaged in big-game hunting. To achieve such cooperation, information about the natural world had to be transmitted from one Early Human to another. Modern hunter-gatherers frequently discuss the tracks and trails of animals, weather conditions, alternative hunting plans and so forth. Although it is always risky using modern analogies for Early Humans, a significant exchange of information about the natural world appears essential for big-game hunting.[24]

How could that have been achieved by 'Hmmmm' communication? First, we should recognize that Early Humans might have used objects to support their spoken utterances and gestures. By returning to one's community carrying some new type of rock or plant, the carcass of an unfamiliar mammal or the feathers of a bird, one could have communicated a great deal. Similarly, the expression of one's emotional state by vocalization and body language would have been very informative: happiness, excitement and confidence suggesting new foraging opportunities; withdrawal and submissiveness indicating that a long-distance hunting and gathering trip into a new landscape had been to no avail. Simple physiological indicators might also have been valuable: to observant eyes, a well-fed and healthy Early Human would have told quite a different story from one who had evidently been short of food and water.

What further means of information transmission could have been used? One might have been an elaboration of the number of holistic utterances, increasing to encompass more aspects of the natural environment and activities than the vocalizations relating to predators and food suggested for early hominids in chapter 8. Another could have been the addition of mimesis to the repertoire of communication methods within 'Hmmmm' – mimesis being 'the ability to produce conscious, self-initiated, representational acts that are intentional but not linguistic'.[25]

Mimesis and non-linguistic communication

Merlin Donald, a neuroscientist from Queen's University, Ontario, provided the above definition of mimesis in his 1991 book, *Origins of the Modern Mind*. In that work he proposed that Early Humans such as *Homo ergaster, Homo erectus* and *Homo heidelbergensis* used mimesis, and that this provides the bridge between ape and modern human forms of communication. The use of mimesis had, he argued, allowed them to create new types of tools, colonize new landscapes, use fire and engage in big-game hunting, all of which were beyond the capacities of earlier, non-mimetic hominids. Donald thought mimesis was so important that he characterized more than one million years of Early Human existence as 'mimetic culture'.

Donald distinguishes between mimesis, mimicry, and imitation. Mimicry is literal, an attempt to render as exact a duplicate as possible. So when mimicking another individual, of one's own or another species, one tries to reproduce exactly everything about it – or perhaps just the sound it makes or the way it moves. Imitation is similar, but not as literal. Donald suggests that offspring copying their parents' behaviour imitate but do not mimic. While mimesis might incorporate both mimicry and imitation, it is fundamentally different because it involves the 'invention of intentional representations'. So although the gestures of holding one's heart or covering one's face may have originated from the imitation of someone's grief reactions, when used to represent grief they would be described as mimetic. Personally, I find such fine distinctions difficult to comprehend and of limited worth, so my use of the term 'mimesis' is inclusive of both mimicry and imitation.

Donald explains that mimesis can incorporate a very wide variety of actions and modalities to express many aspects of the perceived world: 'Tones of voice, facial expressions, eye movements, manual signs and gestures, postural attitudes, patterned whole-body movements of various sorts, and long sequences of these elements'.[26] Although his view is that a linguistically based form of culture replaced mimetic culture, he recognizes that mimesis continued as a key ingredient of human culture throughout history and up to the present day. His examples include Ancient Greek and Roman mime, early Chinese and Indian dance, and the dances of Australian aborigines, in which individuals identify with, and act out the role of, a totemic animal.[27]

Mimicking animal movements

Acting out the roles of animals is likely to have been a key aspect of Early Human interactions – either the mimicking or the more sophisticated mimesis of animal sounds and movements, in order to indicate either what had already been seen or what might be seen in the future. To inform this proposition regarding Early Human behaviour, it is worthwhile briefly to consider some modern hunter-gatherers. Mimicking and performing mimes of animals are pervasive among all such societies as part of their hunting practices and religious rituals.

In the course of her anthropological studies of the 1950s, Lorna Marshall provided some detailed descriptions of how men and boys in the !Kung community of Nyae Nyae in the Kalahari Desert mimicked animals.[28] She explained that as a regular pastime they mimicked animal walks and the ways in which animals carry and throw their heads, catching the rhythms exactly. This was often done for nothing more than amusement; she described how delightful it was to watch one young man mimic an ostrich sitting down on its nest and laying eggs. Many of the !Kung musical games also involved the imitation of animals. Each animal was represented in the music by its own rhythmic pattern, which often seemed to catch the particular way it moved, such as the leap of a kudu (a type of long-horned antelope). Boys twanging bowstrings often provided the rhythms; while they did so, they moved their free arms and their heads so as to mimic the animals. Marshall described how the boys could even play their bows to mimic the sound of hyenas copulating. On some occasions everyone seemed to join in, crawling, pouncing, leaping and making animal sounds.[29]

But skill in mimicry also had practical value for the !Kung. During the hunt, stalking had to be undertaken in silence and therefore hunters used mimicry to indicate what they had seen. They also used conventional signs made with their hands; two fingers would be held erect from a gently clasped fist, the fingers chosen and positioned so as to indicate whether the horns were widely spread like those of a wildebeest or close together and curved like those of a hartebeest. When making the sign, they also moved the hand and forearm to represent the way in which the animal moved its head.

People were mimicked as well as animals, with the peculiarities of posture and movement of particular individuals so accurately copied that their identity could easily be guessed. Fighting was acted out in what Marshall described as 'pantomimes'. Arms would be thrown as if hurling weapons, accompanied by yelps and yells, and followed by mimicked axe strikes against fallen opponents. Stylized gestures were used, such as the flicking of a hand past an ear to pretend that a missile had just whizzed by. After a man had pretended to kill another he would appear to throw something over his shoulder, which was understood by those watching to mean that the fallen man was dead.

With the possible exception of this last gesture, none of the examples I have given above involved the use of symbolism, whereby an arbitrary gesture, movement or sound would represent that of an animal or another person. They were all iconic gestures, body movements or sounds – direct and intentional representations, which Donald would classify as mimesis. Neither were the communications compositional – that is, formed according to specific rules that could either refine their meaning or add an additional level of meaning. Mimesis is a holistic form of communication, and there is no reason why the non-symbolic Early Humans should not have used a similar type of mimesis in their 'Hmmmm' communication. The word 'similar' is critical here, for I do not believe that the Early Humans were capable of imagining themselves as the animals that they were imitating in the way that modern humans can.[30]

One can readily imagine that mime could have been used by Early Humans not just for animals, but also to indicate expanses of water, woodland or raw materials that had been observed.

Onomatopoeia and sound synaesthesia

As well as miming how animals move, Early Humans could have imitated their calls, along with the other sounds of the natural world. We know that traditional peoples, those living close to nature, make extensive use of onomatopoeia in their names for living things. Among the Huambisa people of the Peruvian rainforest, for instance, a third of the names for the 206 kinds of birds they recognize are clearly onomatopoeic in origin.[31] Although it is highly unlikely that Early Humans gave names to the animals and birds around them, mimicking their calls would have been a key feature of their 'Hmmmm' utterances.

The study of animal names provides another clue to the nature of Early Human 'Hmmmm' utterances, by virtue of the phenomenon of what has been called 'sound symbolism'.[32] This describes the way in which names often reflect some aspect of the animal concerned other than its call – most frequently its size. 'Sound symbolism' is not, in fact, a very good term, because it implies an arbitrariness between the name and the animal, whereas we are dealing here with precisely the reverse. 'Sound synaesthesia' is a better term, for it describes the mapping from one type of variable – size – onto another – sound.

Sound synaesthesia was recognized by Otto Jespersen in the 1920s – one of the linguists I referred to in my opening chapter because he also believed that language began from music-like expressions. Jespersen noted that the 'the sound [i] comes to be easily associated with small, and [u o a] with bigger things'.[33] He suggested that this is because the sound [i] is made by pushing the tongue forward and upward to make the smallest cavity between the tongue and the lips, while the [u o a] sounds result from a lowered tongue, which creates a large mouth cavity. In other words, such sounds are the product of physical gestures made by the tongue and the lips, which mimic the size of the object being named.

Jespersen's claim was little more than a hunch. But Edward Sapir, another distinguished linguist of the 1920s, undertook an intriguing and quite simple test. He made up two nonsense words, *mil* and *mal*, and told his subjects that these were the names of tables. He then asked them which name indicated the larger table and found, as Jespersen had predicted, that almost all of them chose *mal*.

Could this association between sound and size play a role in the naming of animals by traditional peoples in the modern human world? Brent Berlin, an ethnobiologist from Georgia University, recently examined the names that the Huambisa of Peru and the Malay of Malaysia use for fish. He chose fish so as to avoid any onomatopoeic influence over how the names might have originated. In both languages he found a very significant association between the size of the fish and the types of vowels used in their names.[34] Those that are small in size are more likely to have names using the vowel [i], while those that are relatively large are more likely to use the vowels [e a o u]. As the Huambisa and Malay live on opposite sides of the planet, it is very unlikely that such similarities in vowel use could have originated from a shared ancestral language.

The same association between size and vowel use was found among the bird names used by the Huambisa, as well as in three other, entirely unrelated languages, including that of the Tzeltal of Mexico. Berlin also investigated the words used for tapir, a large, slow-moving animal, and for squirrel, which is quick and small, in nineteen South American languages. In fourteen of these, the names for tapir involved [e a o u] vowels, while [i] was used in the squirrel names. The names of insects, frogs and toads appear to show the same patterns, although in these cases onomatopoeia plays a dominant role.

Onomatopoeia and sound synaesthesia may not be the only universal principles at work in the naming of animals. The bird names of the Huambisa tend to have a relatively large number of segments of acoustically high frequency, which appear to denote quick and rapid motion, or what Berlin calls 'birdness'. In contrast, fish names have lower frequency segments, which have connotations of smooth, slow, continuous flow – 'fishness'.

To develop his work further, Berlin tested whether English-speaking students would be able distinguish between the Huambisa bird and fish names. In one experiment, he took sixteen pairs of words – one the name of a bird and the other that of a fish – and asked 600 students to guess which was which. One such pair was *chunchuíkit* and *máuts*; another was *iyáchi* and *ápup*. Their guesses were correct at a significantly higher rate than would have been expected by chance alone. In some cases almost all of the students correctly guessed which was which; 98 per cent rightly thought that *chunchuíkit* denoted a bird and *máuts* a fish. The experiment was repeated on a much larger sample of bird–fish word pairs, with a similar result, and with the names for tapir and squirrel. In general, it appears that we can intuitively recognize the names belonging to certain types of animals in languages that are quite unfamiliar to us, by making an unconscious link between the sound of the word and the physical characteristics of the animal.

This finding challenges one of the most fundamental claims of linguistics: that of the arbitrary link between an entity and its name. The word 'dog', for instance, is often cited to demonstrate this arbitrariness because it neither looks nor sounds like a dog; and neither does the French word *chien*, the German word *hund*, or the Malay word *anjing*. Berlin's work, however, has shown that the names used for animals are frequently not entirely arbitrary but reflect inherent properties of the animals concerned, including the sounds they make (onomatopoeias), their size, and the way they move. The bird names of the Huambisa appear to derive from a rich mixture of all such properties.

The implications for Early Human 'Hmmmm' utterances are profound. Even though Early Humans are unlikely to have had names for the animals around them, they would have referred to such animals not only by mimicking their calls but also by using vocal gestures that captured their physical characteristics. When they were communicating about the slow-moving mammoths we can imagine them using acoustic utterances with low frequencies and [a], [o] and [u] vowels, in contrast to high-frequency utterances and the vowel [i] used to refer to animals such as small scurrying rodents or flitting birds.

If Early Humans were mimicking the movements and sounds of animals in the manner I have proposed, we should perhaps recharacterize their communication system from 'Hmmmm' to 'Hmmmmm': Holistic, manipulative, multi-modal, musical and mimetic. Its essence would have been a large number of holistic utterances, each functioning as a complete message in itself rather than as words that could be combined to generate new meanings.

Early Human holistic utterances

The type of communication system I am suggesting for Early Humans is one that still lacks words and grammar, and which continues to follow Alison Wray's arguments for the nature of proto-language. Each holistic message string that was spoken or gestured by an Early Human would have had a manipulative function, such as greeting, warning, commanding, threatening, requesting, appeasing and so forth.[35] It is, however, a rather more elaborate communication system than she proposed – one that uses not only gesture but also dance, and one with extensive use of rhythm, melody, timbre and pitch, especially for the expression of emotional states.

For hunting and gathering, 'Hmmmmm' might have included utterances that meant 'hunt deer with me' or 'hunt horse with me', either as two completely separate utterances, or else as one phrase of 'hunt animal with me' accompanied by mimesis of the particular animal concerned. Other phrases might have included, for example, 'meet me at the lake', 'bring spears', 'make hand-axes', or 'share food with ...' followed by a pointing gesture towards an individual or mimesis of that individual.

As Wray has explained, *Homo heidelbergensis* and other Early Humans could have had a great many holistic messages, ranging from the general 'come here' to the more specific 'go and hunt the hare I saw five minutes ago behind the stone at the top of the hill'. The key feature of such phrases is that they would not have been constructed out of individual elements that could be recombined in a different order and with different elements so as to make new messages. Each phrase would have been an indivisible unit that had to be learned, uttered and understood as a single acoustic sequence.

As Wray points out, the inherent weakness of a communication system of this type is that the number of messages will always be limited. Since each message must represent something different and be phonetically distinct, the number of messages within a holistic communication system is constrained by the ability of the speakers to pronounce, and of the hearers to perceive, sufficient phonetic distinctions. The memory capacities of Early Humans, which I assume were similar to ours, would have imposed additional constraints; their long-term memory would limit how many strings could easily be recalled, while their short-term memory would constrain the length of each string. If holistic phrases were used with insufficient frequency they would simply drop out of memory and be lost. Similarly, the introduction of new phrases would be slow and difficult because it would rely on a sufficient number of individuals learning the association between the new utterances and the objects and activities to which they related. The 'Hmmmmm' communication system would, therefore, have been dominated by utterances descriptive of frequent and quite general events, rather than the specifics of day-to-day activity.

In view of such constraints, Wray has argued that a holistic communication system would instigate and preserve conservatism in thought and behaviour in a manner that a language constituted by words and grammatical rules would not. It is this recognition that makes her proposals so applicable to Early Humans because those species did indeed show a marked lack of innovation throughout their existence between 1.8 and 0.25 million years ago. Their stone tools remain unchanged and extremely limited in range; the absence of bow and arrow technology and stone-tipped spears among Early Humans who relied on hunting in northern landscapes is striking. There is a complete absence of behaviours such as constructing huts and carving bone into tools, let alone engraving objects or painting cave walls. Although fire was used, evidence for the construction of even the most basic fireplace remains highly contentious.[36]

Performance at Bilzingsleben

To conclude this chapter we will look at one of the most interesting of Middle Pleistocene sites in Europe, that of Bilzingsleben in southern Germany.[37] This is approximately fifty kilometres from Schöninigen and was occupied by *Homo heidelbergensis*, living the same type of hunting and gathering lifestyle, at about the same date in human prehistory, give or take fifty thousand years, as those Early Humans who made the wooden spears. Excavated since the late 1970s by Dietrich Mania, this site has produced a vast quantity of animal bones, plant remains and stone artefacts, along with a few fragments of Early Humans. When occupied, the site was within oak woodland and close to the shore of a partly silted-up lake. The occupants appear to have hunted red deer and scavenged from the carcasses of rhinos and elephants. The locally available stone only came in small nodules. As a consequence, heavy-duty tools were made from the massive elephant and rhino bones that appear to have been readily available. These were smashed across wooden and stone anvils, and then chipped in order to produce robust edges that could be used for scraping skins and other tasks. Thus, rather than being carved, bone was treated as if it were stone – a further reflection of the conservatism of Early Human thought.

Bilzingsleben appears to have been a very attractive locale for *Homo heidelbergensis*, for the large quantity of animal bones they left behind have no gnaw-marks from carnivores, suggesting that such animals were absent from the vicinity. It was also attractive in 1999, when I was fortunate enough to visit Dietrich Mania and his excavations in the gentle rural landscape of southern Germany. He showed me some intriguing pieces of animal bones that carried near-parallel striations incised by stone blades, which he claimed were part of an intentionally made symbolic code. Mania also reconstructed one of the three circular arrangements of bone that he had excavated twenty years before. These intriguing features, approximately five metres across and made from bones such as rhino jaws and elephant thigh bones, are entirely unique for this period. Mania interprets them as the bases of windbreaks or, perhaps, huts. One contained a scatter of charcoal fragments, which may have been the remnants of a hearth. I wasn't convinced, and others have also questioned Mania's interpretations.

One critique came from Clive Gamble, professor of archaeology at Royal Holloway College, London, who questioned whether the bones that form the circular patterns really are contemporary with each other. Others have suggested that the rings were made by swirling waters from the lake and river at times of flood. Gamble also rejected the idea that those who occupied Bilzingsleben had intentionally incised bone, constructed huts and windbreaks, and built hearths; the scatter of charcoal was, he suggested, from a burnt tree.

Gamble thinks that Mania and many other archaeologists are too keen to find behaviour like that of modern humans in the archaeological record of Early Humans. As a consequence, they are overly ready to interpret archaeological remains on sites as huts, fireplaces and other features such as one finds on the campsites of modern hunter-gatherers such as the !Kung.

Rather than simply rejecting Mania's interpretation, however, Gamble used Bilzingsleben to illustrate the type of social life that he envisages for *Homo heidelbergensis* – one constituted by many minor encounters between individuals. At Bilzingsleben, Gamble believes that the most important social act was setting up the anvil:

> Carrying the anvil a few metres or a few hundred signified the commencement of structured activity, the start of rhythmic gesture and, for the archaeologist,

> the possibility of patterns that might survive. The gathering now began as a result of the material action that produced spatial and temporal consequences for social interaction. Sitting beneath the trees or beside their charred remains once they had fallen and been burnt, these were the places for transmitting skills, breaking bones on the anvils, knapping stones, stripping flesh, sharing, eating, rhythmically striking bones to form patterns and occasionally producing ordered cut marks on their surfaces.[38]

I find this a more persuasive interpretation of the Bilzingsleben evidence than huts, hearths and symbolic codes incised on pieces of bone. But Gamble's interpretation also epitomizes another weakness of current archaeological interpretations – it is much too quiet. What I would like to add to this scenario are the music-like, emotion-laden 'Hmmmmm' vocalizations used by Early Humans as part of their social interactions. I would also like to add body gestures, mimesis, and dance-like movements – in all, a highly evolved form of 'Hmmmmm' communication.

As both Mania and Gamble recognize, security from carnivores would have allowed for relatively large gatherings and long stays at a Middle Pleistocene locale. Bilzingsleben would have been a site from which individuals and small groups would set off for hunting trips, most probably using Schöningen-like spears. Hence a great deal of the 'Hmmmmm' communication is likely to have involved mime, vocal imitations and sound synaesthesia relating to animals, birds and the sounds of the natural world in general. It would also have been a site to which hunters returned to elicit help in locating a carcass or to share meat following a kill.

Once thought of in this light, the circular features become open to another interpretation. Rather than being the accidental products of swirling water or of the excavation methods of archaeologists, or the bases of windbreaks or huts, they simply become demarcated spaces for performance. Is it possible that Early Humans stepped into such spaces to sing and dance, to tell stories through mime, to entertain and enthral – and perhaps did so not only to elicit assistance in hunting but also to attract a mate? This last possibility, the use of music to attract members of the opposite sex, is the subject of my next chapter.

12 Singing for sex

Is music a product of sexual selection?

Vivaldi's Concerto in B flat major for trumpet and orchestra: a member of *Homo heidelbergensis* showing off a hand-axe he has made

This chapter is about singing, sex, and the social organization of Early Human communities. We have already examined how melodious vocalizations by australopithecines and the earliest *Homo* may have been important in creating and manipulating social relationships through their impact upon emotional states. We have also seen how bipedalism may have enhanced this function by providing greater temporal control, leading to more rhythmic vocalizations and a more elaborate use of the body as a medium for communication, while mimesis could have been crucial for the transmission of information about the natural world.

We now need to examine how the musical aspects of 'Hmmmmm' may have become further developed within Early Human society because they directly benefited those who were willing and/or able to engage in such communication. By 'benefit' in this context I mean reproductive success; the question we must address is whether those who could elaborate the musical aspects of 'Hmmmmm' gained advantage in the competition for mates in Early Human society.

The principles of sexual selection

Music has long been associated with sex, whether we are dealing with the works of Liszt or Madonna. Is this modern role of music telling us something about how the capacity for music evolved? Charles Darwin thought so: 'it appears probable that the progenitors of man, either the males or females or both sexes, before acquiring the power of expressing mutual love in articulate language, endeavoured to charm each other with musical notes and rhythm'.[1]

This proposition comes from Darwin's *The Descent of Man, and Selection in Relation to Sex* (1871), in which his theory of sexual selection was described – a theory that many believe to be as important as that of natural selection as described in *The Origin of Species* (1859). The essence of sexual selection is simply that mate choice is a key element in reproductive success, as your offspring will inherit some or all of your chosen mate's looks and behaviour. This is significant because if your genes are to spread into future generations, you need to have offspring who will themselves be able to reproduce; they must be fertile, healthy and be able themselves to succeed in the competition for mates. As the cost of reproduction is substantially higher for females than for males, and as they have relatively few chances to produce offspring during their lifespan, females should be far more choosy about whom they mate with than males. The power of these simple biological facts is that a great deal about the appearance and the courtship behaviour of animal species becomes explicable within the framework of evolutionary theory: these characteristics are there simply to make the animal, especially the male animal, attractive to members of the opposite sex.

In the 1930s, the geneticist R. A. Fisher developed Darwin's theory of sexual selection with the idea of 'runaway' sexual selection. Fisher suggested that if a heritable mate preference – for example, the preference for a larger than average tail – becomes genetically correlated with the heritable trait itself – in this case the larger tail – then a positive feedback loop will arise so that tails will eventually become far longer than would otherwise have been expected. This runaway selection may account for such features as the remarkably elaborate plumage of birds of paradise or the extravagant courtship displays of the lyre bird.

For many academics, the significance of sexual selection as an evolutionary force only became apparent in the late 1970s, when Amotz Zahavi introduced the 'handicap principle'. This was further developed in his 1997 book of the same name, which had the provocative sub-title *A Missing Piece of Darwin's Puzzle*.[2] Zahavi had recognized that for a physical or behavioural trait successfully to attract a member of the opposite sex, it had to impose an actual cost on the bearer. In other words, it must constitute a potential handicap to the bearer's own survival. Otherwise, the possession of such a trait could be faked, making it wholly unimpressive.

The peacock's tail that adorns the cover of Zahavi's book is the classic example. Its impressive size imposes an energy cost on the bearer and increases the risk of predation: the larger the tail, the more noticeable the bird, and the slower it will be at escaping from dangerous situations. Moreover, to possess an elaborate tail fan, the peacock has to maintain itself in a healthy condition; it has to be good at finding nutritious food and fighting parasites. So, in the parlance of sexual selection theory, a large and colourful peacock's tail is a 'reliable indicator' of particular good genes, since without such genes the bearer of this tail would either have been preyed upon or else would not have been able to sustain its elaborate nature.

In modern sexual selection theory, a distinction is drawn between 'indicator' and 'aesthetic' traits. The latter are physical or behavioural characteristics that exploit the perceptual biases of those looking for mates. For example, a particular species of bird may have a diet dominated by red berries and hence evolve eyes with a high sensitivity to red as well as a brain that is attracted to that colour. This perceptual bias may then predispose the bird to choose mates that have red, rather than green, yellow or blue, plumage. So red plumage would evolve within the bird population as a side effect of the predominance of red berries in the diet.[3]

When Darwin was writing about the significance of sexual selection in 1871 the notions of runaway selection, handicaps, and indicator and aesthetic traits had not been developed. Nevertheless he made a powerful argument that sexual selection explains much of the diversity within the natural world. He devoted a significant number of pages to birdsong, being well aware that this is principally undertaken by males during the breeding season. By the late nineteenth century, naturalists had realized that such singing must have a purpose, and they assumed that it represented a form of territorial competition between males. Darwin, however, argued that male birdsong had evolved via a mechanism of female choice: 'The true song ... of most birds and various strange cries are chiefly uttered during the breeding-season, and serve as a charm, or merely as a call-note, to the other sex.'[4] He went on to explain that 'unless females were able to appreciate such sounds and were excited and charmed by them, the persevering efforts of the males, and the complex structures often possessed by them alone, would be useless; and this is impossible to believe'.[5]

From birdsong to human music

Darwin followed his study of birdsong by applying the same logic of sexual selection to explain the origin of human music. Geoffrey Miller, an evolutionary psychologist at the University of New Mexico, has analysed Darwin's arguments step by step and claims that they are as perceptive and rigorous as those used by any modern evolutionary biologist. Darwin noted the ubiquity of music in all known cultures, the spontaneous development of musical abilities in children, and the manner in which it arouses strong emotions, before concluding: 'All these facts with respect to music and impassioned speech become intelligible to a certain extent, if we may assume that musical tones and rhythm were used by our half human ancestors, during the season of courtship.'[6]

Geoffrey Miller was not only impressed with Darwin's arguments but developed them further by drawing on the ideas of runaway selection and the handicap principle, and of indicator and aesthetic traits. In 2000 he published a provocative article with the explicit goal of reviving Darwin's suggestion that human music was shaped by sexual selection to function as a courtship display, an idea that he felt had been 'strangely neglected'.

The crux of Miller's argument is that if music-making is a type of biological adaptation, then it must relate to mate-attraction, since it provides no direct survival benefit.[7] We have already seen that Miller's last assertion is quite wrong: the musicality of our ancestors and relatives did have considerable survival value as a means of communicating emotions, intentions and information. In the following chapters I will explore further benefits, notably that of facilitating cooperation. Miller, failing to appreciate these, argues that music's biological value was just as Darwin originally proposed: a means to attract mates. This may, indeed, have been a further factor in its evolutionary history.

For Miller, 'music is what happens when a smart, group living, anthropoid ape stumbles into the evolutionary wonderland of runaway sexual selection of complex acoustic display'.[8] He believes that singing and dancing constituted a package of indicator traits for those choosing mates, predominantly females: dancing and singing revealing fitness, coordination, strength and health; voice control revealing self-confidence. Less certainly, he suggested rhythm might demonstrate 'brain capacity for sequencing complex movements reliably', and melodic creativity the 'learning ability to master existing musical styles and social intelligence to go beyond them in producing optimally exciting novelty'.[9]

Such explanations are needed, but they sound far less convincing than the general notion that music evolved as a form of courtship display. This is, indeed, the difficulty that both Miller and Darwin before him faced: trying to tie down precisely why music-making might have been subject to sexual selection is far more difficult than making the assertion that this was the case. When Miller turns from indicator to aesthetic traits he is even less persuasive, with vague claims about how rhythmic signals excite certain kinds of neural networks in mammalian brains – 'tonal systems, pitch transitions, and chords probably play on the physical responsiveness of auditory systems to certain frequency relationships' (a statement that lacks clarity of meaning, to say the least) – and how musical novelty attracts attention.[10]

Miller attempted to find evidence from the twentieth century that music is a product of sexual selection. He cites the example of Jimi Hendrix, who had sexual liaisons with hundreds of groupies, as well as two long-term relationships, and died at the young age of twenty-seven. Although only three children are known to have resulted from his sexual exploits, without the (assumed) use of contraception his genes would have flourished in later generations as a direct consequence, Miller argued, of his music-making. The link to his music is, of course, questionable; Hendrix's sexual attraction derived from a combination of good looks, style and being an anti-establishment figure as much as from any chords he played on his guitar.

In an attempt to find more substantive evidence, Miller examined the age and sex of the music-makers behind more than six thousand recent albums from the jazz, rock and classical genres. In each case, males produced at least ten times more music than females, and were most productive at around the age of thirty, which, Miller claims, is near the time of peak mating effort and activity. If this is the case, the evidence is compatible with what we would expect from a sexually selected behaviour. The sex and age bias, however, might be explained in many other ways relating to the particular structure and attitudes of twentieth-century Western society – women and older people have hardly had the same opportunities for musical expression and commercial success.

If, like me, you find Miller's idea about the origin of music attractive, his supporting arguments weak and his evidence fragile, then you, too, may be astonished at his complete dismissal of the archaeological and fossil records as a further source of evidence. He writes, 'It is just not very important whether music evolved 200,000 years or 2 million years ago.'[11] About this he is utterly wrong. As we will shortly see, society was structured very differently at those dates in human history, affecting the likelihood that music might have been used as a form of courtship display. Moreover, Miller himself recognizes one pervasive characteristic of music-making that sits uncomfortably with his theory: the fact that in all known societies music-making is frequently, if not always, a group activity.[12] This is an issue for chapter 14, which addresses the role of music in facilitating cooperative activities. But for now we must turn to the nature of hominid and Early Human social relations in order to examine if and when the sexual selection of musical abilities may have occurred.

Sexual selection in the African apes

When all that remains from our early prehistoric past is so scanty – fragmented fossil bones and scatters of stone artefacts – some degree of speculation is inevitable when we attempt to reconstruct hominid mating patterns. But there are, in fact, some features of the fossil record that suggest how early prehistoric social life was organized, and which allow us to evaluate Miller's sexual selection hypothesis for music. These features only become apparent because consistent relationships exist between primate body size and social organization. And so, before examining the fossils themselves, we must briefly consider the social organization of the African apes, our closest living relatives. Although a great deal of variation is present, arising from differences in body size, ecological conditions and group history, sufficient generalizations can be derived to allow palaeoanthropologists to make cogent suggestions about the social behaviour of extinct hominids.[13]

Initially, we must note that sexual selection can arise from two different aspects of primate mating behaviour, either or both of which may be operative in any particular group. First, males can compete with each other for opportunities to mate with females. This results in the selection of traits such as large male body size and large canines, and perhaps aggressive personalities. Secondly, females can choose their mating partners, leading to the selection of the indicator and/or aesthetic traits that make males attractive to females. These may overlap with those traits selected by male–male competition, such as large body size.

Chimpanzees live in groups composed of several males and females; the males tend to be resident, while the females leave their mothers to join other groups when they are sexually mature. Gorilla groups tend to have a single dominant male, the silverback, with younger males departing to form bachelor groups. Their groups do not shrink and enlarge in the manner of those of chimpanzees, which do so partly in response to changing distributions of resources and partly as a result of social conflict and the presence or absence of mating opportunities. Chimpanzee groups are often referred to as 'parties'; those groups that regularly exchange members constitute the 'community'. The parties themselves can be quite spread out, especially during foraging activity, when there can be as much as three hundred metres between one member and another.

Among all three species of African apes, the males and females build coalitions of three or more individuals who will forage and socialize together, and who often support each other during conflicts. Coalitions may form alliances. Social tensions and the potential for conflict are always present; these are alleviated by grooming, food sharing and, among bonobos, by sex. This last takes many forms, but is especially important between the females, who engage in the mutual rubbing of their genitals to reinforce friendships.

One of the key structuring principles of African ape communities is competition between males for mating opportunities. This usually results in the emergence of one dominant male who secures the majority of mating opportunities by virtue of his size and strength – the alpha male. African apes are, therefore, described as having a polygynous mating system – single males with multiple female mating partners – in contrast to the monogamous mating systems that are most frequently adopted by gibbons and by *Homo sapiens* in modern Western society.[14] Among the African apes, low-ranking males need to compete for whatever mating opportunities become available, usually only finding them when out of sight of the alpha male. But as alpha males are often dependent upon their allies to maintain their rank, they may reward their supporters with mating access to females.[15]

The role of female choice among the African apes is contentious, and is likely to vary with species and ecological conditions. Male gorillas certainly display to females, but the extent to which females are able to choose their partners is questionable. Nevertheless, the silverbacks often appear to be 'chosen' by the females within the group and to require their support to maintain rank. Among the bonobos at Wamba, the adult females are almost as dominant as the males and appear to choose their male breeding partners without meeting any objection from the rejected males. The female chimpanzees of Mahale also have some degree of choice. They often find themselves in highly dispersed parties and have to search actively for mates.

From a palaeoanthropologist's point of view, the most important feature of polygynous mating systems is a relatively high degree of sexual dimorphism, especially in terms of body size. Male gorillas, for instance, are twice the size of females, as are male orang-utans. Chimpanzees have an average male to female body size ratio of about 1.4:1. This pattern of relatively large male body size is most probably a consequence of sexual selection, operating through male–male competition, female choice, or a combination of both.[16] The equivalent value for *Homo sapiens* is 1.2:1, and for the monogamous gibbons 1.0:1.[17]

Further characteristics of the African apes are also sexually dimorphic. Male apes have large canines that are used in threat displays and are also most probably a consequence of sexual selection. In addition to size and teeth, males use vocalizations to intimidate other males. There are not, I believe, any recorded instances of males using vocalizations as a courtship display to females, and this is unfortunate for Darwin and Miller's theory. But it is not fatal. It may be the case that song and dance only became sexually selected traits in the *Homo* lineage. And so we must now turn to the fossil record.

Early hominid mating patterns

As sexual dimorphism is a correlate of polygynous mating systems and sexual selection, palaeoanthropologists may be able to infer the presence of these for early hominids by examining the skeletal evidence and estimating the difference between male and female body size. This is not quite as easy as one would wish. Body size must be estimated from individual and often fragmentary fossil bones, such as those from the forearms, legs, jaws and crania. Fortunately, palaeoanthropologists can make use of the correlations that exist between individual bone sizes and overall body size in primates and modern humans, although this will always leave some degree of uncertainty.

A more challenging problem concerns the comparison of two or more specimens. If a palaeoanthropologist finds two arm bones of different sizes, it must be decided whether they come from (1) the adult male and female of a single, sexually dimorphic species; (2) from an adult and a juvenile of the same species; (3) from two separate and differently sized species; or (4) from the same species living in different environments which have affected its development – perhaps by providing different levels of nutrition or different requirements for tree-climbing, which might have then influenced bone and muscle growth.

The fossils themselves may provide some help in reaching the appropriate interpretation: epiphyseal fusion will indicate that adulthood has been reached, while associated fossilized species may indicate whether the hominid was living in an open or wooded environment. But often there is no clear guidance, and this is why some palaeoanthropologists conclude that there were many hominid species in the past whereas others believe there were only a few morphologically variable species.

Henry McHenry, an anthropologist from the University of California, has struggled with such problems while trying to assess the extent of sexual dimorphism in our hominid ancestors and relatives.[18] He has measured a large number of limb bones, jaws and crania, and then used these measurements in order to estimate body weight. He has also examined variation in the size of canines and forelimbs. McHenry concluded that the sexual dimorphism of the 3.5 million-year-old *Australopithecus afarensis*, represented by Lucy, was greater than that present in modern humans and slightly more than that of chimpanzees, but well below that found in gorillas and orang-utans.

For *Australopithecus afarensis*, McHenry has the footprint trails from Laetoli to work with, in addition to the collection of fossil specimens. Those footprints came from three individuals, the largest estimated as weighting 40.1 kilograms and the smallest 27.8 kilograms – a size ratio of 1.44:1. Unfortunately, it remains unclear whether this contrast in size reflects the difference between a male and a female, or between an adult and a juvenile.

Further key evidence regarding *Australopithecus afarensis* comes from the fossil locality known as 333 at Hadar in Ethiopia. This contained the remains of at least thirteen individuals who appear to have died at the same time, perhaps in a flash flood – the 'first family' mentioned in chapter 9. Of the five clear adults within this group, three were relatively large and two small. They ranged from an estimated 50 kilograms to 33.5 kilograms, confirming a likely male to female size ratio of 1.4:1. A similar degree of difference was apparent from the dental remains, but McHenry found a striking contrast when he compared the size of the forelimbs. These suggested a body size range between 30 and 60 kilograms, equivalent to that found among gorillas today, which have the largest degree of sexual dimorphism of any living primate.

McHenry concluded that *Australopithecus afarensis* lived in groups with a polygynous mating system in which the degree of male–male competition was less intense than that seen among modern-day chimpanzees and gorillas. He suggested that the large forelimbs might be explained by bipedalism – now that they were no longer being used for locomotion they may have taken over the role of threat and aggression previously played by the canines.

McHenry studied the skeletal remains of other australopithecine species and the earliest *Homo*. He found that each had male to female size ratios around the 1.4:1 mark, suggesting a polygynous mating system. In all of their societies, therefore, male–male competition and/or female choice is likely to have existed, resulting in sexual selection. This raises the possibility of sexually selected vocal displays – we must remember that the semi-bipedal australopithecines and earliest *Homo* are likely to have had an enhanced range of vocalizations compared with that found in the African apes today (see chapter 9). Moreover, these were set to become even more elaborate with the evolution of the fully bipedal *Homo ergaster*. But that species brought a dramatic change in the extent of sexual dimorphism, bringing into question whether sexual selection remained a potent evolutionary force.

Early Human mating patterns

If we accept that *Australopithecus afarensis* was sexually dimorphic, the shift to the modern human body size ratio of around 1.2:1 came rather suddenly at 1.8 million years ago, when *Homo ergaster* appears. Nariokotome and other relatively well-preserved specimens allow for more confidence in the estimation of body size than the fragmented remains of australopithecines.

In general – and exceptions are present – body size increases in both sexes: *Homo ergaster* males were up to 50 per cent larger than their australopithecine counterparts; females show an even greater increase, being up to 70 per cent larger than female australopithecines. What might this tell us about the mating patterns of *Homo ergaster* and its immediate descendants, and hence about the likelihood of sexual selection for musical ability?

On the basis of the comparative study of living primates, as described above, *Homo ergaster*'s relatively low degree of sexual dimorphism appears indicative of a monogamous rather than a polygynous mating system. If so, this may have taken the form either of lifetime pair-bonding between males and females, or of short-term relationships in the form of serial pair-bonding.[19]

A second reason to think that monogamous mating systems may have appeared with *Homo ergaster* relates to the body and brain size of this species, and the consequent energy demands placed on pregnant and nursing females. We have already noted that *Homo ergaster* male body size increased by up to 50 per cent and female body size by 70 per cent, compared with their australopithecine forebears. Brain size effectively doubled in the large-bodied specimens: ranging between 400 and 500 cubic centimetres for australopithecines, and between 800 and 1000 cubic centimetres for *Homo ergaster.*

These increases in size would have placed substantial energy demands on pregnant and nursing females, while also inhibiting their ability to acquire food for themselves.[20] Such demands would have become particularly severe owing to the phenomenon of 'secondary altriciality', which must have first arisen at this time. This simply means giving birth to infants that are still undergoing fetal rates of development and consequently need constant care. This arose because bipedalism requires a relatively narrow pelvis and hence puts a severe constraint on the width of the birth canal. Human babies are literally forced along this, and often have quite misshapen skulls for a period afterwards. To be born at all through the narrow bipedal pelvis, infants effectively had to be born premature, leaving them almost entirely helpless for their first eighteen months of life – some consequences of which will be explored in the next chapter.[21]

With the energy costs of feeding the developing fetus and then infant, along with the constraints on their own activity that this imposed, female *Homo ergaster* seem likely to have required substantial support. Where did this come from? One possibility is that males began provisioning for and protecting their mating partners and offspring. It may have been in their interest to do so, as their own reproductive success would have been dependent upon their offspring surviving to reproduce. Pair-bonding would therefore have developed, as the males needed to ensure that the infants they were supporting were those that they themselves had fathered.

This scenario appears to fit neatly with the reduction in sexual dimorphism found among *Homo ergaster,* because male size would no longer have been the key factor in mating success. There is, however, a major problem with this interpretation, which has led several palaeoanthropologists to question any change in mating patterns at this date in human evolution. This is that the key reason why sexual dimorphism was reduced in *Homo ergaster* was simply the relatively greater increase in female than in male body size.[22] This increase in body size can be explained by the bipedal adaptation to dry savannah environments, which required larger distances to be travelled on a daily basis in the search for food and water, and a dietary change involving increased meat-eating. Since no larger types of humans have ever evolved, it seems likely that a size threshold had been reached in human evolution such that males simply could not get any bigger. Henry McHenry thinks this was the case, arguing that males who exceeded 60–70 kilograms were selected against because of physical failures such as back injuries – something that continues to plague humans of large stature today.

Another reason to question a change in mating patterns with the appearance of *Homo ergaster* is that the females may have been supported not by males but by the other females within the group, especially those to whom they were related. Female cooperation is widespread among mammals in general, sometimes involving the suckling of each other's young and 'babysitting'.[23] Even though female apes move between groups and will not be with their biological kin, they often cooperate together, sometimes to protect themselves against aggressive males.

Food sharing among female apes and other non-human primates is rare, however, and the provisioning of young is hardly known at all. Professor Leslie Aiello and Cathy Key, her colleague at University College London, explain this last point simply by the lack of any need for it: plant foods are sufficiently abundant and easy to procure for all adults and young to feed themselves. But if the diet changed to include greater amounts of meat from either hunting or scavenging, as is likely to have happened with the appearance of *Homo ergaster*, then food sharing and provisioning may have become key factors of female cooperation.[24]

The grandmothering hypothesis and female choice

The notion of mutually supportive female kin networks has gained considerable popularity in recent palaeoanthropological studies, partly through the studies of Aiello and Key, and partly due to the influence of the grandmothering hypothesis. This arose from fieldwork by the anthropologists James O'Connell, Kristen Hawkes and Nicholas Blurton-Jones among the modern Hadza hunter-gatherers of East Africa.[25] They found that although males spend a great deal of time hunting they provide only limited amounts of food for the women and children.[26] Of far greater importance was plant gathering by post-menopausal women, and the manner in which such women would care for their grandchildren, thus enabling the physically stronger young mothers to forage for food. By providing such support, the grandmothers are improving their own inclusive fitness, although it should be noted that post-menopausal women in these communities also provide support to unrelated individuals. O'Connell and his colleagues believe that this kind of female–female support is not only applicable to *Homo ergaster* society, but is directly related to the evolution of post-menopausal lifespans within the *Homo* lineage – something quite absent among the African apes.

The grandmothering hypothesis suggests that the necessary support for pregnant and nursing *Homo ergaster* females may have come from their female kin rather than their male mating partners. Would such female support have been sufficient? Cathy Key has attempted to estimate the increase in reproductive costs for females that arose from increases in body and brain size during human evolution, and to assess when it would have become to the advantage of males for them to begin provisioning for and protecting their mates.[27] *Homo ergaster* seems likely to have fallen just below a body and brain size threshold that triggers a necessary transition from female–female to male–female cooperation. That point was most probably reached at some time between 600,000 and 100,000 years ago, when a further increase in brain size occurred, this most likely being selected to enable enhanced communication.

It appears, therefore, that male–male competition continued in polygynous *Homo ergaster, Homo erectus* and *Homo heidelbergensis* society, much as it had done in australopithecine and *Homo habilis* society. A key difference, however, was that size could no longer have been a significant variable in mating success, because of the biomechanical threshold that had been reached. In consequence, there would have been a greater emphasis on female choice, and males would have needed traits other than physical stature to indicate 'good genes'.

So female choice is likely to have begun to play a far more significant role than it did among the australopithecines and than it does among present-day African apes. Whereas female gorillas often suffer aggression from males that might be twice their size, and almost always respond submissively to male advances, *Homo ergaster* females would have been in a stronger position. Not only were they closer to the males in size and strength, but they would most probably have had far stronger female alliances to help protect them against unwelcome advances. As a consequence, males would have needed to invest more time and energy in their displays so as to attract female attention and interest; they could no longer rely on brute force against one another and against the females in order to achieve their reproductive ends.

What types of display might have been adopted? Singing and dancing appear to be ideal candidates, especially since the capacity for both had been enhanced by the anatomical changes associated with full bipedalism. So here we return to Darwin's hypothesis as elaborated by Miller: that singing and dancing may have provided both indicator and aesthetic traits for females when choosing mates.

My review of the fossil and comparative primate evidence has shown that this hypothesis is more likely to be correct than it may have initially seemed. And while direct evidence for male singing and dancing may be lacking, there is, in fact, substantial evidence for what may have been male display. It comes in a form that archaeologists possess in abundance: stone tools.

Hand-axes and the problems they pose

As I described in the previous chapters, hand-axes appear in the archaeological record at 1.4 million years ago and remain a key element of Early Human technology until a mere 50,000 years ago. It is not only hand-axes' longevity and their prolific numbers at many localities that are so striking; many have a very high degree of symmetry that has been deliberately imposed by skilful and often delicate chipping of the stone. Such symmetry is often found simultaneously in three dimensions – in plan, in profile and end-on. Hand-axes are often attractive to the modern eye and they have been called the first aesthetic artefacts.

This is intriguing because, as far as archaeologists understand, hand-axes were used simply for butchering animals, cutting plants and working wood. Plain stone flakes, or those minimally shaped by chipping, could have performed such tasks quite effectively.[28] So why did Early Humans invest so much time in making highly symmetrical artefacts? And when made, why were they so often discarded in a near-pristine condition?[29]

Before proceeding with some answers, a few caveats must be noted. First, some degree of symmetry will arise quite unintentionally when a stone tool is made by alternately flaking the two sides of a nodule to produce a sharp edge; this may have been how symmetry first arose in the earliest hand-axes, before it became a feature that was deliberately imposed. Secondly, some nodules of stone, such as flint pebbles that have been heavily rolled in water, are symmetrical to begin with, and these may have produced a high degree of symmetry with little effort or even intention on the part of the stone-knapper. In other cases, however, the raw material nodules might have been such that even the most expert stone-knapper could never turn them into symmetrical hand-axes.[30] These, though, are the extreme cases. In the vast majority of instances the degree of symmetry of a hand-axe is a direct consequence of the intention and skill of the stone-knapper. The archaeological evidence shows the entire range, from hand-axes in which symmetry is hardly present at all to those that are practically perfect in all three dimensions.

The sexy hand-axe hypothesis

Could it be that many hand-axes were made primarily to impress members of the opposite sex? If so, they would provide hard evidence (quite literally) that sexual selection was a potent force in Early Human society, and would support the notion that music may have been used to the same end, as Darwin originally proposed. Investing time and energy in shaping a finely symmetrical artefact would certainly have been a handicap on the stone-knapper, and its production would certainly have indicated the kinds of mental and physical capacities that any mother would wish to be inherited by her offspring. This is the essence of what has become known as the sexy hand-axe hypothesis, proposed in 1999 by myself and the evolutionary biologist Marek Kohn.[31]

We argued that the ability to make a finely symmetrical hand-axe acted as a reliable indicator of cognitive, behavioural and physiological traits providing the potential for high reproductive success. Hence females would preferentially mate with those males who could make such tools. The converse would also be true: males would be attracted to females who were adept at producing symmetrical hand-axes. Because high-quality raw material is necessary, making such a tool would indicate the possession of environmental knowledge: the ability to locate stone with good flaking characteristics would imply a similar ability to locate sources of high-quality plants, carcasses, shelters and water. More generally, it would demonstrate the kinds of perceptual and cognitive skills for environmental exploitation that a parent would 'wish' their offspring to inherit.

The making of a symmetrical hand-axe requires considerable technical skill as well as the ability to conceive and successfully execute a plan. Moreover, that plan needs to be modified continually as contingencies arise – for example, when unexpected flaws are encountered in the material and when miss-hits occur. Planning, flexibility and persistence would be important qualities in many aspects of hunter-gatherer life; a potential mother would wish for them in her offspring, and hence would be attracted to those males who displayed such qualities in their tool-making. Similarly, making hand-axes would have been a reliable indicator of health, good eyesight and physical coordination.

The production of any other relatively difficult-to-make artefact could function as an indicator trait, and the argument thus far does not address the one special feature of hand-axes, their symmetry. Kohn and I argued that this is likely to have functioned as an aesthetic trait, exploiting a perceptual bias in the opposite sex towards symmetry. Many studies of living things have shown that symmetry in physiological features is a good indicator of genetic and physical health, since genetic mutations, pathogens and stress during development are all liable to create asymmetries. Studies have shown, for instance, that swallows with symmetrical tail feathers and peacocks with symmetrical eyespots gain enhanced reproductive success.[32] Studies of modern humans have shown that both men and women are preferentially attracted to members of the opposite sex whose faces and bodies have high degrees of symmetry.[33]

Since attraction to symmetry exists in such a wide range of animals,

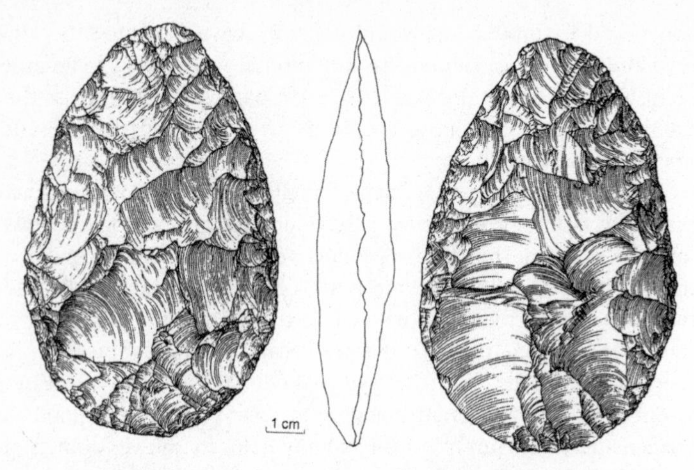

Figure 12 Hand-axe from Boxgrove, made by *Homo heidelbergensis* around 500,000 years ago. This is typical of the many fine symmetrical specimens discovered at this site.

including modern humans, it is reasonable to assume that it was also present in *Homo ergaster, Homo erectus* and *Homo heidelbergensis*. In fact, hand-axes must have appealed to members of the opposite sex then for the same reasons that we find them attractive today.

Some argue that the sexy hand-axe hypothesis leaves an unresolved problem: which sex made the hand-axes? There has been a long history in archaeology of assuming that males produced hand-axes because they are associated with hunting and scavenging activities. But there is no evidence as to which sex made them. My view is that if we are dealing with a polygynous mating system involving sexual selection, then we must conclude that males would have been making the most symmetrical hand-axes. This does not, however, preclude females from also making hand-axes of a utilitarian nature. If the mating system involved some degree of male choice, then females may also have been engaged in manufacturing hand-axes with a high degree of symmetry for display purposes. This would have involved the transition from a strictly polygynous system to a form of serial pair-bonding in which both males and females had some degree of mate choice.

Whether or nor hand-axe display was a male preoccupation alone, the sexy hand-axe hypothesis provides answers to many of the long-standing problems that hand-axes have presented to archaeologists. As well as issues of manufacture and symmetry, it explains their abundance in the archaeological record. The simple possession of a hand-axe would be quite insufficient as an indicator of good genes because it could have been stolen from another individual; a female (or male) observer would need to see the hand-axe actually being manufactured in order to ensure that he/she was not being cheated. So it is not surprising that we should find so many hand-axes in the archaeological record, often several hundreds discarded together in pristine condition. Once made, they were of limited further use.

Another peculiar feature of the archaeological record also becomes explicable: the hand-axe 'giants'. Although not great in number, several specimens exist that would have been far too large for use. The classic example comes from Furze Platt in southern England – a flint hand-axe no less than 39.5 centimetres in size and weighing 3.4 kilograms. Similarly, quartzite hand-axes almost 30 centimetres in length have come from Olduvai Gorge – described by Professor Derek Roe of Oxford University, a world authority on hand-axes, as 'dramatic objects'.[34] This is precisely what the sexy hand-axe hypothesis suggests they were intended to be – objects to catch the attention of potential mating partners and impress them with the knowledge, skill, physical strength and mental characteristics required for their manufacture.

The sexy hand-axe hypothesis is important for our consideration of music. If correct, it provides evidence that polygynous mating systems continued to be a dominant feature of hominid society after the emergence of *Homo ergaster* and the loss of significant sexual dimorphism. Owing to this, and to enhanced degrees of female cooperation, the opportunities for female choice of mating partners may have increased, requiring males to invest more time and energy in display. Biomechanical constraints on body size required that means other than stature alone had to be found to advertise 'good genes'. If making sexy hand-axes was one, another might have been singing and dancing – just as Charles Darwin argued in 1879 and Miller in 2000. Although this would not constitute the origin of music-making, it would certainly have spurred its evolution by providing an additional selective pressure on the existing musical elements of 'Hmmmmm'.

The presence of sexy hand-axes and sexy singing among our Early Human ancestors provides a quite different perspective on the past from the scenarios traditionally offered by archaeologists. To see how different, I will begin the next chapter by considering what hominid life might have been like half a million years ago at one of the most famous of archaeological sites, Boxgrove in southern England.

13 The demands of parenthood

Human life history and emotional development

'Kind of Blue' by Miles Davis: *Homo heidelbergensis* at dusk in the security of trees after feasting on horse meat

Imagine a group of hominids – members of the *Homo heidelbergensis* species – clustered around the carcass of a horse on the banks of a wet and muddy lagoon. Two of them are expertly cutting tendons with hand-axes; others watch them at work, and another is returning from a nearby cliff-face with a flint nodule from which he will make another hand-axe. As they work, the Early Humans communicate using 'Hmmmmm' – gestures, body language and mime combined with holistic utterances, many of which have a highly musical nature. The adult males are the loudest and have the most extrovert movements; in the background a gentler form of communication is under way, between mothers and their infants. That is the subject of this chapter: how the need for mother–infant communication was a further selective pressure for the evolution of 'Hmmmmm', and ultimately for the capacities for music and language in modern humans.

Boxgrove: discovery, excavation and interpretation

Boxgrove is a truly remarkable site because so many of the animal bones and stone artifacts remain exactly where they were discarded half a million years ago, having been rapidly but gently buried in the mud of the lagoon.[1] This has allowed archaeologists to reconstruct Early Human activities in far more detail than they can elsewhere; the vast majority of sites of this period were long ago washed away by rivers, leaving the stone artifacts rolled and battered in gravel deposits just as if they were unworked pebbles. At Boxgrove, the scatters of flakes show precisely where the flint-knappers sat and shaped their tools; numerous flakes have been refitted to show exactly how they were removed from amorphous nodules to shape almost perfectly symmetrical ovate hand-axes.[2]

Boxgrove is one of the earliest archaeological sites in Britain – indeed in the whole of northern Europe. It was occupied during an interglacial period when the sea level was low and Britain was no more than a peninsular of the continent. With the retreat of the ice sheets, *Homo heidelbergensis* had spread northwards, living on animals acquired by either hunting or scavenging, and exploiting plant foods from the woodland that now flourished in the warmer and wetter climate. Indeed, much of Britain at 0.5 million years ago would have been thickly covered in trees, which were useful for collecting berries and nuts, and sleeping secure from carnivores, but not for hunting and scavenging. For those activities the coastal plain and marshes would have been preferred, as is apparent from the excavations at Boxgrove. The site provides not only an insight into the behaviour of *Homo heidelbergensis* but also a vivid glimpse of the interglacial environment, as a great many other animals lived and died in its vicinity. The bones of elephant, rhino, horse, deer, lion, wolf and hyena have been excavated, along with those of smaller animals such as beavers, hares, voles, and frogs. It had been a vibrant, heavily populated landscape, a northern-latitude Serengeti.

Many of the horse, deer, and rhino bones have cut-marks from stone tools. Some of them can be pieced together to reconstruct the butchery process, as the cut-marks cluster around the major muscle attachments, showing how the carcasses were taken apart.[3] The same bones are also covered in carnivore gnaw-marks, showing how hyenas had come scavenging after the Early Humans had left. Their presence indicates that Boxgrove was a dangerous locality, not a place for lingering and one where constant vigilance would have been essential. In this regard, Boxgrove was quite different from the near-contemporary site of Bilzingsleben where no traces of carnivores were found, an indication of the safety provided by the surrounding trees. The emotions and 'Hmmmmm'-ings of the Early Humans at the two sites must have been quite different: those at Boxgrove coming from people who were anxious, adrenalin-fuelled, and working quickly, and some of whom may have been wishing to keep their infants calm.

Mark Roberts from the University of London discovered and excavated the site during the 1990s. It was a remarkable achievement: starting with no more than a hunch that a few previously found tools marked the location of an important site, then working with hardly any funds to prove himself right, and within a few years running a multi-million-pound excavation – an archaeological rags-to-riches story. To his own disappointment, and perhaps even more so that of English Heritage who funded the project, very few remains of the Early Humans themselves were discovered, no more than part of a chewed leg bone and a few teeth. Nevertheless, these, and the date of the site, were sufficient to identify those who butchered the animals at Boxgrove as *Homo heidelbergensis*.[4]

Even though the evidence is so well preserved, archaeologists disagree about what precisely happened beside the Boxgrove lagoon. Clive Gamble, whose comments on Bilzingsleben we reviewed earlier, suspects that the carcasses had not been hunted but scavenged. As several of them appear to have been complete when the butchery begun, the carnivores must have been chased away soon after making their kill. Gamble also believes that the Early Humans were not 'saying' much to each other as they worked, because the spatial patterning of the debris from tool-making, in isolated and discrete clusters, is quite different from that which we find in the campsites of *Homo sapiens* hunter-gatherers, whether recorded in the recent archaeological record or observed by anthropologists. These typically have large palimpsests of debris in the vicinity of fireplaces, reflecting how people positioned themselves to chat while making tools and sharing food.[5]

Mark Roberts has a quite different view from Gamble, believing that the Boxgrove hominids were far more like modern humans. He argues that a hole found in a horse's pelvic bone was caused by a spear that pierced the hindquarters of the animal.[6] Roberts, and his colleague Simon Parfitt, made a detailed study of the cut-marks on the horse bones and decided that the animals had been skinned and defleshed in a manner that appears inconsistent with scavenging. Any carnivore gnaw-marks arose, they argued, only after the hominids had left the site and abandoned the carcasses. Roberts is far more ready than Gamble to attribute spoken language to the Boxgrove hominids. Consequently he provides a picture of *Homo heidelbergensis* as cooperative, talkative and efficient big-game hunters – masters of their landscape.

To my mind the truth lies somewhere between Roberts's and Gamble's interpretations. The finds from Schöningen give strong support to the argument that the Boxgrove hominids were big-game hunters, while the anatomical evidence suggests that *Homo heidelbergensis* was capable of more complex spoken utterances than Gamble is prepared to accept. But there can be no question that the spatial patterning of debris at Boxgrove adds to the enigma of the hand-axes themselves by suggesting behaviour and thought quite different from that of modern humans. What is missing from the site is as important as what is present: there are no hearths and no traces of any structures.

Roberts seems too quick to write modern behaviour into the past, while Gamble correctly tries to understand how our ancestors differed from modern humans. By attributing those who occupied Boxgrove with 'Hmmmmm' communication, we can allow them a sufficient level of communication to be successful big-game hunters, while constraining their thought to produce the behavioural conservatism that is evident from Boxgrove and all other sites of this period.

Singing and dancing at Boxgrove

While Roberts and Gamble may differ in the specifics of their interpretations, the underlying picture of Boxgrove that they provide is nonetheless quite consistent. It is one focused on butchery by Early Humans who are anxious to take away joints and fillets of meat to eat in safe locations, hidden from prowling carnivores. The hominids use not only their hand-axes and flakes but also their brute strength to separate the carcasses. They may be communicating to each other or working independently, but either way it is the skinning and the cutting of tendons that command their attention. When tools become too blunt or greasy, they are discarded and new hand-axes are made from nodules of stone brought from the cliff-base nearby. It is a busy, bloody and rather brutal scene.

The arguments forwarded in my previous chapters, those about rhythm, mimesis and sexual selection, suggest that we can embellish this vision of Boxgrove. We should envisage the Early Human males choosing to spend more time than was strictly necessary when making their butchery tools, investing time and effort to create a finely proportioned and pleasing artefact. Doing so at a site like Boxgrove would have been a true handicap because it would have deflected attention from the carcasses, the other members of the group and the surroundings. Perhaps the males sat with their backs to the carcass, oriented towards a cluster of young females while they worked; their minds as much on who was watching them flint-knap as on how the hand-axe was to be used for skinning and cutting tendons.

If the manufacture of hand-axes was undertaken partly for social display, then we might also follow Geoffrey Miller and envisage the Early Humans making sexually enticing melodic and rhythmic utterances while they chipped at the stone. Moreover, although the movements of the Early Humans around the carcass and around each other are likely to have been energetic, they would have also been graceful and expressive – a display reflecting physique, fitness and coordination, rather than the clumsy movements that are generally attributed to our prehistoric ancestors and relatives.

As the availability of a complete horse carcass to be butchered was most probably a rare event for *Homo heidelbergensis* – whether or not they were hunters – we can readily imagine that those at Boxgrove were relieved, happy and excited by their work. Hence, we should envisage rhythmic Early Human movements around the carcass and melodic vocalizations merging into something close to what we would recognize as dance and song. And throughout the proceedings, there would have been holistic, manipulative utterances: 'cut those tendons', 'sharpen the axe', 'take meat to [gesture indicating person]', and so forth. With all of this display, expression and communication, no wonder *Homo heidelbergensis* brains exceeded 1000 cubic centimetres in size.

I suspect that there would have been two further aspects to the Boxgrove 'Hmmmmm', ones that would have been found wherever hominids were grouped together. The second is the subject of the next chapter. Here, I am concerned with females singing to their infants. In addition to the butchery, the tool-making, the ostentatious noise and movement, we must imagine a young mother cradling and gently singing to her newborn baby. Another woman may be making a hand-axe or cutting meat while her infant sits near by. As the mother works, she maintains eye contact with her child; she smiles, raises her eyebrows, gestures and makes gentle, melodic utterances.

Singing to infants

In chapter 6 I summarized the evidence that infants are born musical and used this to underline music's biological basis. The most crucial elements of that evidence are the universal nature of infant-directed speech (IDS) and the fact that this is used long before infants have begun the process of language acquisition. IDS has been of interest to a wide range of academics, from Anne Fernald to Colin Trevarthan; another is Ellen Dissanayake, a child psychologist from Seattle. She believes that the need for enhanced mother–infant interactions during human evolution caused such communication to become intense and to take on a musical nature. Dissanayake suggests that this may have been the ultimate source of music-making among adults and throughout society as a whole.[7]

We began to cover one of the key aspects of her theory in the previous chapter – the physical helplessness of human infants at birth. As I explained, the evolution of bipedalism resulted in a relatively narrow pelvis and hence birth canal, which limits the size of infants at birth, especially their brain size. To compensate, infants were effectively born premature and continued rapid foetal growth rates for the first year of life outside of the womb. Human infants double in weight during their first year, growing an average of 25 centimetres.

As we saw, the consequence for species from *Homo ergaster* onwards was that childcare became a costly activity of relatively long duration, compared with the situation for earlier hominids and our existing primate relatives. From this point on, considerable pressure was placed on nursing mothers in terms of increased energy costs and constraints on their own movement and activities – the rationale behind the grandmothering hypothesis introduced in the previous chapter.

Dissanayake believes that the musical aspects of IDS evolved as a direct response to the increasing helplessness of human infants as early hominids evolved into Early Humans, dismissing the Darwin/Miller hypothesis that male competition and adult courtship provided the original selective pressures for music. And she follows Colin Trevarthan in believing that music should be characterized as a 'multimedia' package, in which facial expression, gesture and body language are equal in importance with vocal utterances.

According to Dissanayake, the musical characteristics of mother–infant interactions arose because they provided substantial benefits to both parties. Her arguments hinge on the role of musical sound and movement in expressing and inducing emotional states, as explored in chapter 7, ultimately achieving concordance between the emotions experienced by parent and infant. Such concordance was essential, she argues, for developing their relationship and, ultimately, the enculturation of the infant.

In Dissanayake's words, the solution was accomplished 'by coevolution in infants and mothers of rhythmic, temporally patterned, jointly maintained communicative interactions that produced and sustained positive affect – psychobiological brain states of interest and joy – by displaying and imitating emotions of affiliation, and thereby sharing, communicating, and reinforcing them'.[8] The mechanism behind this coevolution was natural selection: those mothers whose physiological and cognitive make-up made them more able to provide infant care, and those infants who were receptive to such care, ultimately gained a reproductive advantage.

As I described in chapter 6, compared with normal speech IDS involves exaggerated vowels, repetition, higher pitches, a greater range of pitch, and a relatively slow tempo. These are used in concert with facial expressions and gestures – raised eyebrows, smiles, head nods and uplifts, hand waves – and physical contact such as tickles and cuddles. Through such stimuli we elicit emotions in prelinguistic infants and reveal our own feelings; we influence the infant's behaviour, maintaining alertness, soothing, showing displeasure and so forth.

In her theory, Dissanayake places great emphasis on the microdynamics of modern mother–infant interactions – how each partner tracks the emotional expressions written in the body, face and voice of the other. Each is able to enter, she claims, 'the temporal world and feeling state of the other'. By this means the infant has its level of attention and arousal controlled, gains emotional regulation and support, has its neural structures for emotional functioning reinforced, has the development of cognitive abilities enhanced, and ultimately is supported in the acquisition of culturally specific rules of social behaviour and 'language'. Today, of course, as our infants mature, IDS changes so as to support the acquisition of language, which has overwhelmed the significance of music-like utterances and gestures in adult modern humans.

Dissanayake's ideas, based as they are on the uttered and gestural communication of emotion and mood, provide a valuable contribution to the 'Hmmmmm' theory of Early Human communication. Even in the absence of language, *Homo ergaster* infants had to learn how different utterances related to different moods, and 'infant-directed Hmmmmm', with its exaggerated prosody, would have facilitated such learning.

The big baby problem

Dean Falk is a distinguished professor of anthropology at Florida State University who specializes in the evolution of the human brain. One of her latest theories looks at the very first stages of language, what she calls 'pre-language', from both a developmental and an evolutionary perspective.[9] Like Dissanayake, and Leslie Aiello before her, Falk believes that the appearance of bipedalism was a key factor in the evolution of human language and cognition. She, too, has stressed how bipedalism led to what were effectively helpless infants. But while Dissanayake addressed the consequent emotional and social needs of the babies, Falk's concern is largely with the mothers who had to carry them around.

Female monkeys and apes rarely put their babies down. Chimpanzee infants are unable to cling properly for the first two months of their life and are carried underneath their mothers' bodies. When placed on the ground, infant chimpanzees are prone to emit a 'hoo' sound as a means of re-establishing physical contact with their mother. This may be repeated and become part of a whimpering sequence. The mothers frequently emit a similar 'hoo' sound when they wish to pick up their infant or remove it from a dangerous situation. Bonobos are very similar. They are sensitive to the 'screams' of their infants and emit 'barks' or 'hiccups' in alarm situations, which elicit an immediate response from the infants.

While such ape mother–infant turn-taking vocalizations are similar to those found in humans, they differ dramatically in their frequency and diversity. When compared with the constant sing-song chattering of a human mother, chimpanzees and bonobos are practically silent.

When a couple of months old, chimpanzees and bonobos are sufficiently developed to climb onto their mothers' backs and to cling to their body hair for a free ride. They can do this for up to four years, after which the juveniles become entirely independent.

Falk has recognized that the infants of *Homo ergaster*, and perhaps even the australopithecines, would have had a more difficult time than those of the great apes today. As their period of physical helplessness lasts considerably longer, modern human infants also take longer than chimpanzees to gain control over their own posture and locomotion – approximately three months to lift the head and nine months to sit unaided. A similar, although somewhat shorter, period of physical incapacity is likely to have been experienced by the infants of our Early Human ancestors. As a consequence, *Homo ergaster, Homo erectus* and *Homo heidelbergensis* mothers were required to support and carry their infants around for prolonged periods of time, unless they could find another strategy to care for them.

The vertical posture of bipedalism would have further increased the burden on mothers, as would their large body size. We know that some members of *Homo ergaster* had already reached modern human stature, while the Boxgrove leg bone and other skeletal remains indicate that by 0.5 million years ago body size was often substantial – six feet tall, with plenty of muscle and fat. The infants would accordingly have been at least as large as those of modern humans, so carrying around a one-year-old would have been just as exhausting 1.8 million years ago as it was at Boxgrove and as it is today.

It may have been even more so. It seems likely that hand in hand with the evolution of bipedalism came the loss of body hair, leaving just the few patches that we have today. Such loss would have been another physiological adaptation for keeping cool when foraging on the open savannah.[10] The Early Human infants are likely to have had a grasping reflex, but this would have become increasingly limited in value as their parents became less hairy. Although human infants are still born with a grasping reflex today, they never develop the ability to cling unaided to their mother's bodies.

Body hair, fire and clothing

We must be cautious, however, as there is no direct evidence as to when we evolved to become effectively hairless. Indeed, the idea itself is an illusion because we have the density of hair follicles on our bodies that would be expected of an ape of our size. The difference between ourselves and the chimpanzee is that our body hair is so fine and short that it is practically invisible. Such a light covering would have been beneficial for *Homo ergaster* when foraging on the hot savannah during the day, but it would have been a considerable disadvantage at night when needing to keep warm.

Keeping warm may have been achieved by using fire or by wearing some form of clothing. Indeed, the evolutionary biologists Mark Pagel and Walter Bodmer have argued that the loss of body hair occurred only after our ancestors had invented artificial means to keep their bodies warm.[11] Once that had been achieved, it would have been advantageous for them to lose their body hair – not only for heat regulation during the day but because thick hair harbours a great number of parasites, some of which would have spread disease. All primates today spend a considerable amount of time grooming each other to remove such parasites, time that could be usefully spent doing other things, especially if social bonding was being achieved by 'Hmmmmm'.

While there are no traces of clothing in the archaeological record before relatively recent times, which may be no more than a factor of preservation, there is evidence for the use of fire at the 1.6 million-year-old site of FxJj50 at Koobi Fora in Kenya.[12] This is a patch of heavily oxidized soil amid a scatter of stone flakes and fragmented animal bones, interpreted as a campfire that had been kept burning for several days. Since none of the bones show signs of burning, the fire was most probably used to provide warmth and deter predators rather than to cook food. It is possible, therefore, that by 1.6 million years ago our ancestors had lost their body hair and gained the advantages not only of reduced heat stress during the day but also of reduced parasite load. They would then have faced the problem of how to carry their infants around with nothing for them to grasp, and, if they could not be carried, of how else to care for them.

It seems unlikely that *Homo ergaster* had begun wearing 'clothes', such as hide wraps, capes or cloaks, which might have provided something other than hair for their babies to grasp. One ingenious piece of research has attempted to date the origin of clothing by identifying when human body lice evolved – the lice that feed on skin but live within clothes. Mark Stoneking and colleagues from the Max Planck Institute in Leipzig used DNA to reconstruct the evolutionary history of lice and deduced that human body lice first appeared around 75,000 years ago, give or take at least forty thousand years.[13] This would, they claim, associate the origin of lice, and hence the wearing of clothes, with modern humans, leaving all of our ancestors and relatives naked but hairy. This seems most unlikely; by 250,000 years ago, for instance, Neanderthals were living in ice-age Europe and could hardly have survived without clothing.

My guess is that simple clothes, nothing more complex than a hide wrap, would have been used by Early Humans in northern landscapes, such as the *Homo heidelbergensis* who butchered animals at Boxgrove. The very first garments, however, might not have been for warmth but for carrying babies. Dean Falk has noted that in the vast majority of modern and recent societies, infants are carried around in slings, and she suggests that these were the first type of 'tool' made from hide or vegetable matter. I can readily imagine these also being used at Boxgrove. But Falk is surely correct to suggest that there is likely to have been a period after the appearance of bipedalism and loss of body hair but before such supports were in use, which would most probably equate with *Homo ergaster*. How, then, would they have coped with their big, helpless babies?

'Putting down baby'

One possibility was proposed in the previous chapter: the use of non-nursing females, especially grandmothers. But such support would have been principally for weaned infants, as babies most frequently remain with their mothers and, of course, require regular feeding. Demand-feeding – feeding whenever the baby cries for food – is pervasive in all traditional societies and requires close contact between mother and infant all day and night; its approved absence in modern Western society is quite peculiar.

A second possibility is that the babies were 'parked' – left in a safe place such as the crown of a tree – while the mother undertook her foraging, toolmaking, courtship and other activities at some distance from the 'parking' spot, perhaps for lengthy periods of time. This is a tactic used by many small primates, such as langurs. But it is very rarely seen in the higher primates. This is not surprising, because it risks letting one's infant injure itself or be killed by predators, or even by another member of one's own group (such as a male who is not the father). Also, because infants require frequent feeding, lactating mothers would have been constrained as to the distance they could travel from the 'parking' spot. In general, the hunter-gatherer-scavenger lifestyle on the Africa savannah, with intense social competition and prowling carnivores, was the exact converse of what is amenable to baby 'parking' – at least if one wished to find one's baby again alive and in a single piece.

A third alternative solution to the big, helpless, demanding baby problem is the frequent 'putting down' of the infant for short periods within sight and sound of the mother, unlike 'parking' where the possibility of such contact is lost. This would allow the mother to use both hands to pick fruit, butcher a carcass, drink from a river or knap stone; without the baby she could reach, bend, run and stretch, and simply gain some relief from carrying its weight around. When she was ready to move on, perhaps to a new patch of berries or to use a newly chipped stone flake, she could simply scoop the baby up from the ground, as she could whenever the child became distressed.

Dean Falk suspects that such 'putting down' did indeed occur and was essential to the development of 'prelinguistic communication'. For, once the baby is 'put down', the mother would still have eye contact, gestures, expressions and utterances to reassure the infant, these being substitutes for the physical contact that the infant would desire. The emotionally manipulative prosodic utterances that we associated with IDS would, Falk suggests, have been a 'disembodied extension of the mother's cradling arms'. In this regard, a *Homo ergaster* or *Homo heidelbergensis* mother would have been little different from a *Homo sapiens* mother today.

The origin of lullabies

Such utterances would have been essential because human babies do not like being 'put down' any more than ape babies. Like laughter, crying by modern human infants has a different respiratory nature from that of chimpanzees, occurring as short breathy exhalations alternating with long intakes of breath. By three months of age, human infants use different types of crying to express different emotions, such as anger, hunger or pain. The extent of crying by modern human babies is directly related to the extent to which they are left alone; crying increases the grasping reflex, which supports the arguments of psychologists that the major reason why infants cry is to re-establish physical contact with their mother.

As Dean Falk speculates, Early Human babies are unlikely to have been any happier at being separated from their mothers than are those of modern-day apes and humans. Similarly, it is reasonable to suppose that infant crying created the same feelings of distress for Early Human mothers as we feel today when our own babies cry. When we are busy and cannot pick them up, we comfort our babies with the utterances and gestures of IDS – a rapid peal of these is almost guaranteed to stop the crying. This use of IDS is precisely what Falk believes was selected in that evolutionary gap between the origin of bipedalism and the invention of baby slings.

Some of the periods of 'putting down' could have been longer if the baby fell asleep – which, as every parent knows, is a time of immense relief and often furious activity. Falk speculates that the precursors of the first lullabies are likely to have been sung by the first bipedal Early Humans. They may have cradled, stroked and gently rocked their infants, perhaps after feeding, when, if they were anything like modern human babies, they would have been naturally sleepy.

In summary, Falk argues that those mothers who had a biologically based propensity to attend to their infants when they had been 'put down', by making use of vocalizations, expressions and gestures, were strongly selected; their genes, and hence such behaviours, would have spread in the population. She argues that the IDS that evolved was initially prelinguistic and was constituted by types of melodic and rhythmic utterances that lack symbolic meaning but that have the emotional impacts we discussed in chapter 7, and which I have placed as a central feature of 'Hmmmmm'. But, just as the nature of modern human IDS itself changes as the infant matures, Falk argues that over evolutionary time 'words would have emerged in hominids from the prelinguistic melodies and become conventionalized'.

First words: 'Mama' or 'Yuk!'?

Although the emergence of language from 'Hmmmmm' is an issue that I will address in a later chapter, I will finish this review of Falk's 'putting down' baby theory with her idea for the first word, and then offer what I think is a more likely alternative.

Falk suggests that 'it does not seem unreasonable to suggest that the equivalent of the English word "Mama" may well have been one of the first conventional words developed by early hominids. After all, wouldn't maturing paralinguistic infants, then as now, be inclined to put a name to the face that provided their initial experiences of warmth, love and reassuring melody?'[14] Well, maybe, and the word 'Mama' can readily be imagined as the sound emerging from the gestures of the baby's mouth when it wishes to breastfeed and is looking at its mother.

Other than the gentle melodies to provide comfort and care, what else might Early Human mothers have been 'Hmmmmm'-ing to their babies? 'Yuk!' is a good guess, this being used to dissuade their babies from touching or eating substances, such as feces or rotting meat, that would make them ill.

The expression 'Yuk!', and closely related sounds such as 'eeeurrr', are found in all cultures of modern humans, accompanied by the characteristic facial expression of wrinkling the nose and pulling down the corners of the mouth. This is the expression of disgust, which since the time of Charles Darwin has been recognized as a universal human emotion.[15] It arises when people are faced with bodily excretions, decaying food, and certain types of living creatures, notably maggots. 'Yuk!', 'eeeurr' and related sounds are vocal consequences of the facial expressions they accompany and are further examples of sound synaesthesia – they are slimy, sticky, smelly, oozy noises, sounding the way that vomit, feces, spilled blood and maggots look.

Valerie Curtis, of the London School of Hygiene and Tropical Medicine, has made cross-cultural studies of what people find disgusting and concluded that disgust can be best understood as an evolved mechanism for defence against infectious disease.[16] By having this inbuilt, we don't need to learn that any of the items named above should not be touched or eaten – we know that automatically from their look and smell. Actually, some teaching does appear necessary, as the disgust reaction only develops in modern human infants between the ages of two and five years old; younger infants are quite happy to eat whatever they find lying on the ground. And so parents in all cultures are frequently saying 'Yuk!' to their babies while making the appropriate facial expression and placing the offending material well out of reach.

Imagine the challenge of being a caring *Homo ergaster* parent on the Africa savannah. Localities such as FxJj50 appear to have been frequently reused, leaving rotting meat, dried blood, human and hyena feces for your baby to chew upon. If you were a *Homo heidelbergensis* mother at Boxgrove, how would you have prevented your infant from crawling into the guts of a recently butchered animal? Without the 'Hmmmmm'-ing of 'Yuk!', infectious disease may have knocked out not only these species but the whole of the *Homo* genus just as it had got going!

I must note here that I have at least one critic of the 'Yuk!' proposal, a critic whose views I must respect, not because of her knowledge about human evolution but on account of her experience in raising children: Sue, my wife. She believes that 'Yumyumyumyumyummm' – the sound a parent makes when trying to persuade a baby to eat some food – would have had an evolutionary priority over 'Yuk!'. But as I have explained to her, 'Yumyumyumyumyummm' is not a word but a holistic phrase; it initially means something like 'eat this lovely food, my darling', but can alter its meaning as the intonation changes when the child stubbornly refuses to open its mouth.

Back to Boxgrove

Both Ellen Dissanayake and Dean Falk have provided compelling ideas about how the needs of Early Human infants would have created selective pressures for the development of vocal and gestural mother–infant interactions, which would have been of a music-like nature. This would have provided a further dimension to 'Hmmmmm' and the evolutionary precursor of the first stages of the IDS used by *Homo sapiens* today.

In the above scenario for life at Boxgrove 0.5 million years ago, the mother singing to her infant was just one of the two additional 'musical' happenings with which I wished to embellish the scene of butchery. The other is of a group of hunters arriving at the lagoon carrying a freshly killed carcass while singing together, or perhaps a communal dance by a group of women. Or even a song that all the males and females joined in with after they had eaten their fill of horse meat. Whatever the specifics, it is difficult to resist the idea that some form of communal singing and dancing would have occurred at sites such as Boxgrove and Bilzingsleben. Making music together is, after all, one of the most compelling features of music today. Why is that?

14 Making music together

The significance of cooperation and social bonding

Beethoven's Choral Fantasia: Neanderthals when the winter ice begins to thaw

Why do people like making music together? Whether we are talking of choirs, orchestras, football crowds, children in the playground, church congregations, or !Kung Bushmen, the members of all such formal and fleeting groups sing and dance with each other. For some groups it is their *raison d'être*, for others it is supplementary; for some the music they make is carefully rehearsed, for others it is spontaneous and may be entirely improvised. The music might be made to worship a divine being or 'just for a laugh'. Overriding all such diversity is a common thread: music-making is first and foremost a shared activity, not just in the modern Western world, but throughout human cultures and history.

When I've asked members of my family and my friends why this is so, they readily provide the answers: 'because it is a fun thing to do', 'because it makes me feel good', 'because it builds friendships and makes us bond together', and so on. Such responses are not surprising; common sense and our own experience tell us that this is the case. Common sense, however, is an inadequate explanation and can often be wrong – remember, the earth is not flat. Why and how does music-making create social bonds? Why should our minds and bodies have evolved so that we find group music-making so pleasurable?

Other pleasurable things – eating and sex – have obvious pay-offs. In previous chapters I have explained why music-making by individuals may have evolved as a pleasurable activity: it can be used to manipulate those around us, to transmit information about the natural world, to advertise one's availability to potential mating partners, and to facilitate the cognitive and emotional development of one's child. But none of these factors explains why music-making in modern human society is predominantly a group activity.

To answer that question, this chapter follows a chain of argument that takes us from croaking frogs to marching soldiers, and finishes at the site of Atapuerca in northern Spain at 350,000 years ago. There we will imagine communal singing and dancing by a group of *Homo heidelbergensis* as they dispose of their dead.

Is it about sex (again)?

Group music-making involves synchronizing vocalizations and movements. Behavioural synchrony is extremely rare in nature; the few recorded examples include fireflies 'flashing', frogs croaking, and fiddler crabs waving their claws in time with each other.[1] In each case, the explanation appears to be sex: the synchronous flashing, croaking and waving are undertaken by groups of males seeking to attract females for mating. Two explanations have been put forward as to why the males cooperate in this manner, rather than simply seeing who can flash, croak or wave the 'loudest'. The first is that the synchronized 'signal' may attract less attention from potential predators than if all the males of a large group were to perform independently. The second is that the synchronized 'signal' is 'louder' than would be the case if all the males were to perform independently, and this may be of value when competing for females against other multi-male groups.

This second possibility has been favoured by Björn Merker, of the Institute for Biomusicology in Ostersund, Sweden, as a possible explanation for hypothesized synchronous music-making by our prehistoric ancestors, and hence for our musical abilities today. Merker assumes that early hominids – he is unspecific as to which – had a social organization similar to that of chimpanzees today. That is, the mature females left the group into which they had been born and joined another for the purposes of mating and rearing their young. In such circumstances, Merker suggests, the males in any one group would have had a shared interest in attracting 'mobile' females and hence may have synchronized their vocal displays as a means of out-competing the males in another group – although such behaviour is not found in chimpanzees or any other primate today.

The loudness of the call would be a direct reflection of two factors, both of interest to females choosing which group to join. First, it would indicate the number of males within the group, which would provide a measure of resource availability, such as the abundance of fruiting trees. Secondly, the ability of males to synchronize their calling would, Merker suggests, be a measure of the extent to which they were able to cooperate together. As a consequence, loud, synchronized calls would not only function to attract females but also serve as a deterrent to males from other groups who might be tempted to intrude upon their territory.

Modern-day chimpanzees appear to be quite unable to synchronize their vocalizations. Unlike humans, they cannot keep to a beat, even with training. Bonobos may be able to do a little better in view of the following observation made on a captive group by the anthropologist Franz de Waal: 'during choruses, staccato hooting of different individuals is almost perfectly synchronized so that one individual acts as the "echo" of another, or emits calls at the same moments as another. The calls are given in a steady rhythm of about two per second.'[2] As Merker notes, this is rather different from the simultaneous calling that he describes as true synchrony and which is so characteristic of human music.

Nevertheless, Merker argues that such multi-male synchronized calling evolved from noisy bouts of cooperative calling – similar to those made by chimpanzees when they find an abundant food source – among ancestors common to humans, bonobos and chimpanzees before 6 million years ago. In his words: 'Just as chimpanzee pant-hooting displays at a newly discovered large fruiting tree attract mixed groups of males and females to the site of the commotion, we should picture these hypothetical hominid display bouts as key social gatherings with potential participation by all given members of a given territorial group and attended by considerable excitement.'[3] Once females had been attracted to a group of males engaged in synchronized singing and dancing, they would have had an opportunity to choose between individual males for mating partners – just as Geoffrey Miller, and Charles Darwin before him, argued.

When I reviewed Miller's sexual selection arguments in chapter 12, I noted how he himself had found the propensity for people to sing and dance together a challenge to his theory for the origin of music. Merker's ideas would help to resolve Miller's dilemma if they were feasible for our earliest hominid ancestors, the australopithecines and *Homo habilis*. But I find that rather doubtful.

The problem with Merker's ideas is that synchronous calling by hominids in order to attract mates would also attract predators, as would long-distance calls by lone hominids. We know that hominids on the African savannah were competing for carcasses with carnivores, and that they often became the prey themselves. The idea that they would have synchronized their calls to attract wandering females and to deter groups of other hominid males seems most unlikely, especially when the relatively open landscape constrained their ability to escape from predators by climbing trees. A far more likely strategy for such hominids would have been to keep quiet and hope that the prowling carnivores would pass them by.

Whereas synchronous calling would not have been a feasible method for attracting females in environments inhabited by predators, we might nonetheless generalize Merker's arguments into that of building trust between males, and more particularly females, who need to engage in a variety of behaviours that require mutual support and reliance.

Singing and dancing together

This is the most likely interpretation for the duetting by gibbons that was described in chapter 8. Indeed, among modern humans, those groups who are dependent upon each other are particularly prone to make music together. William McNeill collated the relevant evidence in his 1995 book entitled *Keeping Together in Time: Dance and Drill in Human History.*[4] As the title implies, the book is primarily about rhythmic body movement. This falls, however, under the broad definition of music I have adopted, and McNeill's arguments are as applicable to singing together as they are to dancing together.

McNeill starts with an anecdote about his experience of military life in the 1940s, which is worth repeating. He describes how he was drafted into the US Army in 1941 and underwent basic training with thousands of other young men. This involved a great deal of marching about on a dusty Texas plain: 'a more useless exercise would be hard to imagine ... we drilled, hour after hour, moving in unison and giving the numbers in response to shouted commands, sweating in the hot sun, and every so often, counting out the cadence as we marched: Hut! Hup! Hip! Four!' Having explained how pointless the drilling initially appeared to be, he then describes how in time it 'somehow felt good ... a sense of pervasive well-being is what I recall; more specifically, a strange sense of personal enlargement; a sort of swelling out, becoming bigger than life, thanks to participation in collective ritual'. McNeill goes on to explain that a feeling of emotional exaltation was induced in all of his fellow recruits: 'Moving briskly and keeping in time was enough to make us feel good about ourselves, satisfied to be moving together, and vaguely pleased with the world.'[5] This is the essence of the argument that McNeill develops throughout his book: communal music-making is actively creating, rather than merely reflecting, that pleasing sense of unity.

In some cases music itself aids the performance of a collective task by rhythmically facilitating physical coordination. But in the majority of cases it appears to be cognitive coordination that is induced by the music, the arousal of a shared emotional state and trust in one's fellow music-makers. McNeill provides many examples from traditional societies. He cites anthropologists describing Greek villagers whose communal dancing makes them feel 'light, calm and joyful', or Kalahari Bushmen for whom 'being at a dance makes our hearts happy'.

One of the reasons for such feelings might be simply the release of endorphins in the brain, chemicals that have a primary role in pain control. They are triggered by any physical activity, hence playing sport not only improves our physical fitness but also puts us in a good mood. The evolutionary psychologist Robin Dunbar has proposed that communal music-making leads to endorphin surges within the brains of the participants, resulting in them feeling happy and well disposed towards each other.[6] This has yet to be formally tested, and it also begs the question of why we should bother with communal music-making if a self-induced endorphin fix can be achieved by singing or dancing alone. One reason might be that the endorphin surge is significantly greater when making music with others. But then we must ask why evolution should have 'designed' our brains in this manner: why is communal activity so important?

Here I am reminded of John Blacking's studies, to which I have referred in earlier chapters. He undertook one of the most informative studies of communal music-making when he studied the Venda people of South Africa during the 1950s. He described how they performed communal music not simply to kill time, nor for any 'magical reasons' such as to create better harvests, nor when they were facing periods of hunger or stress. In fact, they did the reverse: they made communal music when food was plentiful. Blacking believed they did so, at times when individuals were able to pursue their own self-interest, precisely in order to ensure that the necessity of working together for the benefit of society as a whole was maintained as their key value through the exceptional level of cooperation that was required in their music-making.[7]

McNeill is keen on the notion of 'boundary loss' – the manner in which group music-making leads to 'a blurring of self-awareness and the heightening of fellow feeling with all who share in a dance'. The origin of such feelings, McNeill argues, lies in our evolutionary past, 'when our ancestors had danced around their camp fires before and after faring forth to hunt wild and dangerous animals'. Those individuals who practised the hunt and enhanced their levels of coordination via dance would, he suggests, have been reproductively more successful, leading to the genetic transmission of their capability, which eventually came to be tapped in a great variety of ways in the modern world – not least by drill sergeants.

McNeill's characterization of prehistoric life as dancing around campfires is simplistic, to say the least. He does not really answer the question of why group music-making should lead to enhanced reproductive success, beyond his general assertions about enhanced group cohesion. But his notion of 'boundary loss' is of considerable value. To understand why, we need to look rather more closely at the significance of cooperative behaviour in human evolution.

Cooperate or defect?

The importance of cooperation in hominid society has frequently arisen in early chapters: for predator defence, for hunting and scavenging, for childcare, and, indeed, for almost all other behaviours in which our ancestors would have engaged.

While it is easy to appreciate that cooperative behaviour would have been pervasive in Early Human society, just as it is today, explaining how it evolved and has been sustained is rather more difficult. When couched in the language of 'selfish genes', Darwinian theory tells us that to be reproductively successful individuals should be looking after their own interests. Perhaps that is simply what we are doing when we cooperate: we provide help to others in the expectation that the favour will be returned. The classic means by which academics explore this is the 'prisoner's dilemma' model – a simple but extraordinarily powerful way of understanding complex social situations.[8]

This model describes a great many of the social situations in which we find ourselves as individuals, and it can also be applied more widely to interactions between groups and, indeed, between entire nation states. It enables us to focus on the key issues involved in cooperation, which in turn will lead us to the inherent value of making music together. The name derives from one particular scenario in which we imagine two prisoners who have been arrested and charged with a crime. The prisoners, call them A and B, are separated, and each is offered the same deal: 'If you confess and your partner doesn't, you will get a light sentence (say, one year) and your partner will be imprisoned for a long time (ten years); if neither of you confess, you'll both receive a medium sentence (two years); if both of you confess, you will serve a lengthy sentence (seven years).'

What should each prisoner do? Their dilemma is that if Prisoners A and B both confess, then they will each receive a longer sentence than if they had both remained silent – seven years rather than two. But if Prisoner A remains silent while Prisoner B confesses, then the former will face an even longer sentence – ten years.

The prisoner's dilemma model is an aspect of game theory, which is used to analyze behaviour in many disciplines from economics to ecology. The choices are usually described in abstract terms of 'cooperation' and 'defection'; in the above scenario, remaining silent is a form of cooperation with one's partner, while confessing is a form of defection. Both prisoners have most to gain if they both cooperate, but most to lose if one attempts to cooperate while the other defects.

If this situation arises just once, and the prisoners are never to see each other again, then clearly the most sensible strategy is to confess, because this minimizes the maximum length of the sentence that might arise. But if the prisoners have past experience of a similar situation, and expect that it might happen again, their decision is more difficult to make. If, for instance,

Prisoner A were to remain silent on this occasion, would that make it more likely that Prisoner B would also remain silent on the next occasion, thus reducing the possible prison sentence? But if Prisoner B had confessed, causing Prisoner A to serve a long prison sentence, then on the next occasion Prisoner A may be out to seek revenge, or at least will have little confidence in his partner remaining silent.

The delight of the prisoner's dilemma model is that it captures the essence of real-life situations. Occasions do arise when we have to decide whether to help out a stranger, someone whom we will never meet again. Far more frequently, however, we have to decide whether or not to help those we will definitely meet again – members of our family, friends or colleagues. For each of these we will have views about their trustworthiness, their loyalty and their own willingness to be helpful; hence we base our decisions on the way they have treated us (and others) in the past.[9]

How to behave?

Should one always be cooperative and risk being repeatedly taken for a sucker? Or should one always defect and risk losing out on the best returns which would come if everyone else were to choose to be cooperative?

Robert Axelrod asked this question in the early 1980s, and found the answer by one of the most novel and entertaining of scientific methods – a computer prisoner's dilemma tournament. He invited experts in game theory to submit computer programs that would play against each other, deciding whether to cooperate or defect on the basis of previous interactions in the game. Each game had 200 'moves' – opportunities to cooperate or defect – and each program played against all the other programs, including one that made entirely random decisions, as well as against itself.

Fourteen programs were submitted by groups including economists, psychologists, mathematicians, sociologists and political scientists. Some programs were generous, frequently cooperating with their opponent – and were often exploited. Others were unforgiving: if their partner ever defected against them, then they would defect for the rest of the game. One of the simplest was called 'TIT for TAT', and it did exactly that. In the first game, 'TIT for TAT' attempts to cooperate. Subsequently, it copies whatever its opponent did in the previous game; if the opponent defects, then 'TIT for TAT' will defect on the next game, and similarly for cooperation.

Several of the programs pitched against 'TIT for TAT' built up a history of cooperation and then tried occasionally to cheat their opponent by defecting; others tried to 'understand' their opponent by assessing the probability that it would cooperate or defect and then responding so as to maximize their own returns. Indeed, some of the programs were able to beat 'TIT for TAT' in any one game. But when all programs played all others, 'TIT for TAT' was the most successful. The game seems a reasonable model for real-life social situations in which we have to interact with many different kinds of people – some generous, some forgiving, some entirely predictable in their behaviour, and some completely unfathomable.

The real world is, of course, far more complex than the prisoner's dilemma in a great many ways. Any situation in which we have to choose whether to cooperate or defect with someone else will never be identical to one that has arisen before (as they were in Axelrod's computer tournament). We also face the problem of trying to assess whether someone else did or did not cooperate; we may have to take their word for it, or they may claim that they attempted to cooperate but made a mistake, or we may misinterpret what they were attempting to do.

Theory of mind, self-identity and group identity

Because the real world is so very 'messy', we are in a constant state of trying to assess what those around us will do in any situation: will they cooperate or defect? In today's world, and ever since material symbols were invented, we short-circuit this decision-making process by wearing particular types of clothing and ornaments that provide extra clues as to our status, our personality, and those with whom we wish to cooperate – epitomized by the old school tie and including such items as uniforms, heraldry, designer clothes, team colours and so forth. But such cultural supports were unavailable to our non-symbolic hominid ancestors.

Many anthropologists believe that the challenge of surviving in a complex social environment, in which decisions about cooperation were paramount, gave rise to the theory of mind, our most important mental attribute other than language. In fact, it is widely thought that the evolution of language was dependent upon the possession of this capability.

I introduced this notion in chapter 8 when discussing how we often engage in 'mind-reading' when deciding how to interact with another person. Often this is quite easy, as they may be behaving in an overt way, wearing particular types of clothes, or using body and/or spoken language to tell us that they are angry, sad or happy, whether they are to be trusted, avoided or exploited. In such situations we don't really need a theory of mind at all, we just need to be a good 'behaviourist'. But on other occasions we do need to think what they are thinking; and that involves thinking what they are thinking that we are thinking.

The fact that we reflect on our own minds, that we examine our own beliefs and desires and appreciate how they contrast with those of the people around us, is integral to our sense of self. Indeed, I would argue that such mental self-reflection is fundamental to recognizing ourselves as unique individuals; having a theory of mind is synonymous with consciousness. Consequently, when we are in a situation similar to a prisoner's dilemma, we consciously reflect on what is going through our own minds and what might be going through the minds of the other 'players' in the game. When we get this wrong, we can make social gaffes, such as causing offence by telling an inappropriate joke. More seriously, the proverbial second-hand car salesman, who appears so trustworthy when in fact he/she is quite the opposite, can exploit us.

The consequence is that choosing to cooperate with others becomes an extremely difficult undertaking. This is exacerbated because the majority of situations involve cooperating not with just one other person but with several. This would certainly have been the case for our hominid ancestors too. Imagine a scenario in the ice-age landscapes of northern Europe at 500,000 years ago in which several individuals or small groups set off from their base camp in the morning to search for plant foods and animal carcasses. With a reciprocal food-sharing arrangement, all or some of the acquired food will be returned to the base camp for distribution among the group. Those who found nothing to share will be tolerated because their contribution will be made on a future foraging trip, when it would be their turn to be successful and that of others to return empty-handed.

In this scenario – and many others that would have arisen, such as communal hunting and warning about predators – each individual would be in a prisoner's dilemma situation with all other individuals: should one cooperate and bring back food, or simply eat it all oneself and come back empty-handed? Those who choose the latter are termed 'free-riders'; they benefit themselves by defecting and going unnoticed while everyone else is cooperating.

How does making music together help?

The dilemma facing our human ancestors, and which continues to face us today, is how to ensure that cooperation occurs to everyone's mutual benefit, rather than pervasive defection, to everyone's loss.

When Robert Axelrod described his computerized prisoner's dilemma tournament in his 1984 book *The Evolution of Cooperation*, he specifically addressed the question of how cooperation can be promoted in real-world situations. Two of his proposals seem to describe precisely what is achieved when people make music together. The first is to 'enlarge the shadow of the future'. By this, he meant making any future interactions appear relatively important when compared with those of the present. Making music together seems to be an ideal way of doing this, because it is a form of cooperation with very little cost. If I were to start singing and you were to synchronize your voice with mine, that would be a form of cooperation, and I would take it as a willingness on your part to cooperate with me again in the future. I might change the melody or rhythm to test whether you would follow my lead; you might do the same. We might also both be aware of a third person who refuses to join our song, or who deliberately sings another, and interpret that as a sign of their unwillingness to cooperate with us both now and in the future.

Axelrod's second proposal is simply that 'frequent interactions help promote stable cooperation'. This can also be achieved by music-making, which is not only 'cheap' to do but can be embedded into other activities. So when I am writing in my study I may sing along (very badly) to the music playing in the background; my wife, perhaps working elsewhere in the house, might join in (singing very nicely) – it is a cheap interaction undertaken while we are primarily engaged in other tasks. One can readily imagine the same for hominids at a campsite on the savannah, or Early Humans on the tundra – some are cutting carcasses, others scraping skins or knapping stone; while doing so they can sing together, forging a bond between themselves while engaged in their own individual activities.

Thus music-making is a cheap and easy form of interaction that can demonstrate a willingness to cooperate and hence may promote future cooperation when there are substantial gains to be made, such as in situations of food sharing or communal hunting. It can be thought of as the first move of a 'TIT for TAT' strategy that is to always cooperate, one that can be undertaken at no risk because there is nothing to lose if the other members defect – that is, if they do not join in the song or dance. However, it is precisely because there are very few costs and benefits associated with joint music-making that free-riders could exploit the situation. They could easily join in with the music-making, because they have nothing to lose, and then simply defect whenever they perceive a short-term gain for themselves.

A sense of self

It is because of this that McNeill's notion of 'boundary loss' is so important. I explained earlier that one's self-identity is intimately related to the feeling that one has a suite of beliefs, desires and feelings that are quite unique to oneself.

When people join together for a group activity – a family meal, a meeting with work colleagues, a football team assembling for a match – they typically arrive from quite different immediate experiences. I might arrive late for a meeting at work feeling angry because another meeting overran; some of those waiting for me are frustrated by my absence, others are pleased; some are feeling happy because they have had a good day, others are feeling gloomy. In such a situation we have a very profound sense of self, and rapidly put our minds into action to assess the feelings of others before deciding what to say – at least, we do if the meeting is going to be a success.

All group activities start in a similar way. When five or six hominids or Early Humans set out together to hunt or to look for carcasses, one of them might have been feeling hungry, another fearful; some of them may have wanted to go one way, while others believed the opposite direction was best. When each individual begins a group activity in a different emotional state, the situation is ripe for conflict, defection and free-riding.

The different emotions experienced by different members of any group reflect different states of mind, and will directly affect how the world is perceived and what are thought to be the costs and benefits of different actions, including those of cooperating or defecting. As I explained in chapter 7, 'happy' people tend to be more cooperative. Trying to achieve consensus and widespread cooperation when everyone is in a different mood is particularly difficult, as many of us will be all too well aware.

Making music together will help in such situations, as it will lead to the diminution of strong feelings of self – or, to use McNeill's term, 'boundary loss'. This is the process in which football crowds, church choirs and children in the playground are all engaging. In some cases it is by mutual consent and understanding, as, perhaps, in the case of Japanese factory employees who sing the company song before starting work. In other cases people are manipulated, as by the chanting at Nazi rallies.

Those who make music together will mould their own minds and bodies into a shared emotional state, and with that will come a loss of self-identity and a concomitant increase in the ability to cooperate with others. In fact, 'cooperate' is not quite correct, because as identities are merged there is no 'other' with whom to cooperate, just one group making decisions about how to behave.

Indeed, when psychologists have examined the results of experiments in which people are placed in prisoner's dilemma-type situations, they have concluded that cooperation is fostered by the 'extent to which players come to see themselves as a collective or joint unit, to feel a sense of "we-ness", of being together in the same situation facing the same problems'.[10]

The 'we-ness' is more formally known among psychologists as an 'in-group biasing effect', and this formed the centre of a recent discussion about the nature of human cooperation in the eminent journal *Behavioral and Brain Sciences*. This publication allows academics from many disciplines to contribute comments following its lead articles. Linda Caporael, a professor of psychology, and her colleagues had proposed that their experimental evidence about the willingness of people to cooperate with each other was incompatible with the view that people always attempt to maximize their own self-interest.[11] They correctly characterized this as the prevailing belief of those academics who consider humans from an evolutionary perspective and have become heavily influenced by the notion of 'selfish genes'. Although not explicitly stated, there was an evident attempt by Caporael and her colleagues to undermine the evolutionists' view of human nature.

They argued that group behaviour is not merely the sum of individuals acting to maximize their own returns but is 'mediated by a redefinition of self in terms of shared category membership or social identity'. This received considerable support from many commentators, although most of the evolutionists argued that this is quite compatible with their own view of human behaviour as a product of natural selection. Several of the contributors noted the importance of identifying the mechanisms that support the development and maintenance of group membership, and observed that we should be 'on the look-out' for cues that prompt group feeling. Listening, rather than just looking, will also be useful, because making music together is the key method for the creation of group identity.

A proponent of this is the jazz musician, cognitive scientist and writer William Benzon, who devoted much of his 2001 book, *Beethoven's Anvil*, to the proposition: 'Music is a medium through which individual brains are coupled together in shared activity.'[12] This recall's McNeill's use of the term 'boundary loss' in explaining the feeling of group unity that he felt arose from long periods of rhythmic marching with other army recruits, and which he generalized into a consequence of group music-making. Benzon also draws on some of his own experiences of playing music, which he describes as musicians sharing in the creation of common sounds, rather than sending musical messages to each other. He explains this by stating simply, 'we were coupled'.

For Benzon, 'coupling' arises when people synchronize the sounds and movements they make, enter similar emotional states, and hence attune their nervous states to one another. Most of us have actual experience of this when we sing with others and are no longer able to distinguish our own voice from all the rest.

The neurobiologist Walter Freeman is an equally strong proponent of music-making's role as a means of social bonding, and he brings to this his knowledge of brain chemistry.[13] He has suggested that during group music-making the hormone oxytocin may be released into the basal forebrain and that, by loosening the synaptic connections in which prior knowledge is held, this 'clears the path for the acquisition of new understanding through behavioural actions that are shared with others'.[14] Freeman describes music as the 'biotechnology of group formation'.

As far as I know, this hypothesis has not been fully tested. Animal experiments have, however, demonstrated that the suppression of oxytocin in the brain inhibits the development of social bonds. When this was tried with female prairie voles, for instance, they continued to mate with their sexual partners but failed to develop the normal patterns of attachment.[15] So one can readily imagine that a surfeit of oxytocin would do the opposite, and that group music-making may indeed facilitate the release of the chemical into the brain.

The specific role of oxytocin, if any, need not detain us; the fact that music creates for us a social rather than a merely individual identity is one that many of us have personally experienced. And it is something that we might envy. After John Blacking's experiences studying the Venda people during the 1950s, he reflected on the experience of growing up in Venda society as compared with his own years in public school: 'When I watched young Venda developing their bodies, their friendships, and their sensitivity in communal dancing, I could not help regretting the hundreds of afternoons I had wasted on the rugby field and in boxing rings. But then I was brought up not to cooperate, but to compete. Even music was offered more as a competitive than a shared experience.'[16]

At Atapuerca 350,000 years ago

That joint music-making forges social bonds and group identity is the 'common-sense' understanding with which I began this chapter. We can now understand why this is the case, and why it would have been so important for our earliest ancestors, the australopithecines and first *Homo*, living in large, complex social groups within which there was a great deal of competition for food and mates. To thrive, cooperation with others was as essential as successful competition. Our hominid ancestors were faced with prisoner's dilemma-like situations; social life was difficult and taxing, and there was a constant risk of being taken for a sucker. Moreover, the hominids lacked both the cues of language and material symbols to help resolve their social dilemmas over whom to trust and whom to exploit.

Hominids would have frequently and meticulously examined the likely intentions, beliefs, desires and feelings of other members of a group before deciding whether to cooperate with them. But on other occasions simply trusting them would have been more effective, especially if quick decisions were necessary. As a consequence, those individuals who suppressed their own self-identity and instead forged a group identity by shared 'Hmmmmm' vocalizations and movements, with high emotional and hence musical content, would have prospered.

Joint music-making served to facilitate cooperative behaviour by advertising one's willingness to cooperate, and by creating shared emotional states leading to 'boundary loss'/'we-ness'/'coupling'/'in-group bias'. With the evolution of *Homo ergaster* and full bipedalism, 'Hmmmmm' gained additional musical qualities, while further selective pressures for its evolution arose from the need to transmit information about the natural world, to compete for mates and to care for infants. As Early Humans colonized northern latitudes, developed big-game hunting, and coped with the dramatic environmental changes of the Pleistocene, the need for cooperation became ever greater. Hence communal 'Hmmmmm' music-making would have become pervasive in Early Human society.

To complete this chapter, we can consider one location where this would have occurred – Atapuerca in northern Spain at 350,000 years ago.[17] This is a limestone hill riddled with caves, tunnels and crevasses, one of which is known to archaeologists as the Sima de los Huesos – the cave of bones. It is a fitting title because it is packed with human bones dating to at least 300,000 years ago and attributed to *Homo heidelbergensis*. I really do mean packed: since the first samples were discovered in 1976, more than two thousand have been recovered by the excavation of a small area to only a shallow depth. These bones represent at least thirty-two individuals, which may be just a fraction of those present within the pit.

There is no indication that the pit was a carnivore den and that the human bones are merely a reflection of an animal's favoured food. Cave bears are certainly represented among the bones, but these appear to be from animals that had accidentally fallen to their deaths, since the pit is at the bottom of a vertical shaft almost forty-six feet deep. This explanation cannot be applied to the human bones owing to their great numbers. And in view of their excellent preservation, it is most unlikely that they had been washed into the pit within sediment flows from elsewhere. The fact that all the bones of the skeleton are represented, including some of the smallest and most fragile, suggests that complete bodies were deliberately dropped into the pit. Juan Luis Arsuaga, one of the three directors of the Atapuerca excavations, describes the Sima de los Huesos as the first evidence for human funerary behaviour.

A key question is how the collection of bodies accumulated. Was this in a piecemeal fashion, one or two each year as members of the *Homo heidelbergensis* group died, or does the large number of bodies reflect some catastrophic event? One means of answering this is from the age distribution of the bodies. Arsuaga and his colleagues have been able to ascertain the age

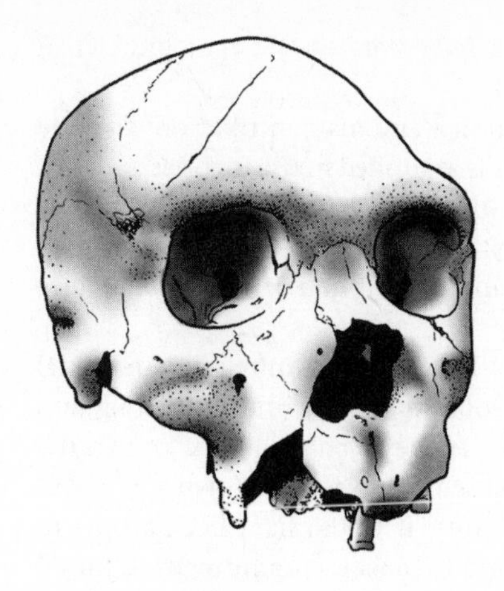

Figure 13 Near-complete skull of a *Homo heidelbergensis* individual from the Sima de los Huesos at Atapuerca, dating to around 350,000 years ago.

of death for many of the individuals they have identified, or at least to place them into an age group. Only one of the thirty-two individuals was a child, who died at the age of four or five. Whether the fact that children are otherwise absent is significant or not is difficult to assess – although children's bones are especially fragile, preservation in the Sima de los Huesos is excellent and more bones of children might have been expected.

The large majority of those represented are young adults – the fittest and strongest members of any community. These may have accumulated slowly over many generations; when someone died in the vicinity of Atapuerca, their body may have been deliberately deposited within the crevasse. But because of the similarity in their state of preservation, their density and the frequent bone articulations, Arsuaga and his colleagues think it much more likely that the Sima de los Huesos humans died either together or within a very short time of each other.[18]

One possibility is that the community was stricken down with disease, an epidemic like the plagues of medieval Europe. But such epidemics usually require large and dense populations living in unhygienic conditions, and this seems inherently unlikely for Early Human hunter-gatherers living in northern Spain at this date. Also, such epidemics tend to strike down predominantly the very young and the very old, precisely the age groups absent from Sima de los Huesos.

As an alternative, Arsuaga suggests that the population living in the region of Atapuerca might have been hit by an ecological catastrophe, such as persistent drought or harsh winters, leading to famine.[19] The young and old might have died, while the surviving members of the group(s), the young adults, took refuge in the vicinity of the Atapuerca hill. The thirty-two bodies may have accumulated over one or a few years as the ecological crisis persisted.

Such ideas are no more than speculation – and my contribution is to add further speculation, but speculation that is grounded in the theories and data that have been provided throughout this book. It is simply that as those bodies were thrown into the crevasse, either all together or one by one over a number of years, the sounds of communal singing and dancing would have echoed around the cave.

The deposition of such bodies would have been an intensely emotional event for the survivors. Indeed, they would hardly have needed song and dance to create a shared emotional state, as they would all have known the dead and would have understood the consequences for their own survival of losing the young adults of their community. It would have been a time to consolidate and confirm social bonds, and to make a commitment to future cooperation. Communal 'Hmmmmm' singing and dancing would have achieved this: sound and movement expressing grief and facilitating 'boundary loss' among the individuals who wept 350,000 years ago in the cave at Atapuerca.

15 Neanderthals in love

'Hmmmmm' communication by *Homo neanderthalensis*

The Spanish folk song 'Nana' by Manuel de Falla played by Pablo Casals: *Homo neanderthalensis* burying a baby in Amud Cave

Most anthropologists are tempted to equate the large brain of *Homo neanderthalensis* with a capacity for language. This is not unreasonable: what else could all that metabolically expensive grey matter have been used for, when the smaller-brained *Homo ergaster* and *Homo heidelbergensis* had already demonstrated such skills as tool-making, living in complex social groups and adapting to ever-changing environments? But the temptation must be resisted; the Neanderthals who inhabited Europe and south-west Asia had brains as large as those of modern humans but behaved in a quite different fashion, one that indicates the absence of language – an assertion I will justify below. So, what were the Neanderthals doing with such large brains?

This book has been providing the answer: the Neanderthals used their brains for a sophisticated communication system that was Holistic, manipulative, multi-modal, musical, and mimetic in character: 'Hmmmmm'. While this was also the case for their immediate ancestors and relatives, such as *Homo ergaster* and *Homo heidelbergensis*, the Neanderthals took this communication system to an extreme, it being a product of the selective pressures for enhanced communication discussed in the previous chapters. They utilized an advanced form of 'Hmmmmm' that proved remarkably successful: it allowed them to survive for a quarter of a million years through dramatic environmental change in ice-age Europe, and to attain an unprecedented level of cultural achievement. They were 'singing Neanderthals' – although their songs lacked any words – and were also intensely emotional beings: happy Neanderthals, sad Neanderthals, angry Neanderthals, disgusted Neanderthals, envious Neanderthals, guilty Neanderthals, grief-stricken Neanderthals, and Neanderthals in love. Such emotions were present because their lifestyle required intelligent decision-making and extensive social cooperation. So now is the time to look at the Neanderthals themselves and to gain a new perspective on their behaviour and society.

The evolution of the Neanderthals

Homo neanderthalensis had evolved in Europe by 250,000 years ago and survived throughout the latter part of the ice ages before becoming extinct soon after 30,000 years ago.[1] Genetic evidence indicates that our species, *Homo sapiens*, shared an ancestor with the Neanderthals somewhere between 300,000 and 500,000 years ago, and then evolved quite separately in Africa. *Homo sapiens* spread into Europe at 40,000 years ago and may have played a causal role in the Neanderthal extinction. The identity of the shared ancestor is uncertain because the fossil record between 1 million and 250,000 years ago is particularly difficult to interpret.[2]

Homo heidelbergensis, a descendant of *Homo ergaster*, is the most favoured candidate. But some anthropologists consider this 'species' to be no more than a catch-all term for various fragmentary fossil specimens whose identity and integrity as a group is questionable. A case in point is the collection of specimens from the site known as TD6. Like the Sima de los Huesos, this site is at Atapuerca in Spain, but it dates to 800,000 years ago and its *Homo* fossils are the earliest known from Europe.[3] The excavators have used these specimens to define an otherwise unknown species, *Homo antecessor*, which they claim was not only the first hominid species to colonize Europe but also the common ancestor to modern humans and the Neanderthals. But few anthropologists accept their interpretation, preferring to classify the specimens as *Homo heidelbergensis*.

The Cambridge anthropologists Robert Foley and Marta Lahr provide an alternative view for Neanderthal evolution. They suggest that the common ancestor to the Neanderthals and modern humans is a species that evolved from *Homo heidelbergensis* in Africa and which they call *Homo helmei*.[4] Because the stone-tool technology of the Neanderthals and of the earliest modern humans is strikingly similar – both using the Levallois technique, which will be described below – Foley and Lahr argue that this must have been invented by a common ancestor, as the chances of independent invention are minimal. As this technology does not appear until after 300,000 years ago, that is the earliest possible date for the separation of the two lineages. In this scenario, a population of *Homo helmei* dispersed into Europe where they evolved into *Homo neanderthalensis*, while another population remained in Africa where they evolved into modern humans.

But who knows? The fossil record is so sparse before 100,000 years ago that several evolutionary scenarios can be proposed. All that we can be confident about is that: (1) the Neanderthals and modern humans shared a common ancestor between 300,000 and 500,000 years ago; (2) the Neanderthals evolved in Europe and modern humans evolved in Africa; (3) modern humans dispersed out of Africa to colonize the whole world, while the Neanderthals became extinct soon after 30,000 years ago.

The Neanderthal brain, body and lifestyle

Modern humans have brain sizes of between 1200 and 1700 cubic centimetres, with an average of around 1350 cubic centimetres.[5] That of the Nariokotome boy (*Homo ergaster*), 1.6 million years ago, was around 880 cubic centimetres, and would have reached 909 cubic centimetres had the boy survived to become an adult. By 1 million years ago, average brain size had only enlarged to about 1000 cubic centimetres, reflecting a period of evolutionary stability that appears to have continued for another half-million years. The first European *Homo* specimens, from TD6 at Atapuerca, are too fragmentary to allow an estimation of brain size, but two well-preserved *Homo heidelbergensis* specimens from the Sima de los Huesos dated to around 300,000 years ago, have brain sizes of 1225 and 1390 cubic centimetres. Neanderthal brains were larger still, with a female average of 1300 and a male average of 1600 cubic centimetres.

Measurements of brain size only become meaningful when they are compared to body size; animals with bigger bodies have bigger brains irrespective of any changes in intelligence. Consequently, anthropologists compare brain sizes by using a mathematical relationship between brain weight and overall body weight for a closely related set of species, and then assess the extent to which the brain of any particular animal exceeds what would have been expected from its body size alone. If, for instance, we were to take mammals as a whole, then the *Homo heidelbergensis* specimens from the Sima de los Huesos have brains between 3.1 and 3.9 times larger than we would expect from their body size. This figure is called their encaphalization quotient, or EQ. That of *Homo sapiens* is 5.3, the largest of any known species, living or extinct.[6]

Although the biggest known brain capacity belongs to a Neanderthal from Amud Cave – 1750 cubic centimetres – the Neanderthals had a larger body mass than *Homo sapiens*, arising from thicker bones and more muscle. In consequence, they have a smaller EQ: 4.8. The difference from 5.3 may not translate into a difference in cognitive abilities, and we must appreciate that there is a great deal of brain size variation within both *Homo sapiens* and *Homo neanderthalensis*. (Psychologists have been unable to find any correlation between brain size and intelligence among modern humans, and so within our species, and one assumes within any species, a bigger brain does not make you any smarter.) Although below that of *Homo sapiens*, the *Homo neanderthalensis* EQ is impressive. Their large brains were metabolically expensive and could only have evolved under strong selective pressures. As

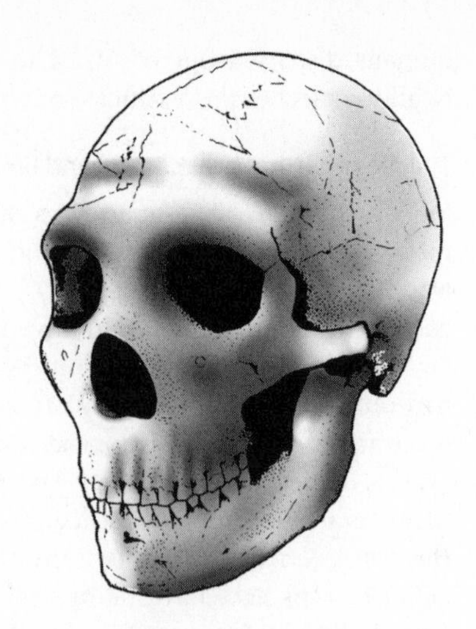

Figure 14 Skull of *Homo neanderthalensis*, the Amud I specimen, dating to around 45,000 years ago.

explained above, and as will be further argued in this chapter, the principal selective pressure was that for enhanced communication.

The Neanderthals have a suite of distinctive physical features which become particularly marked after 125,000 years ago. These include a stout body with relatively short arms and legs compared with modern humans. The Neanderthals had large, barrel-like chests and carried substantial muscle, as is evident from the muscle scars on their bones. They had large, projecting noses, swept-back cheeks, and lacked the bony protuberance on the base of the mandible that forms the human chin. As a whole, these anatomical features are similar to those we find among modern humans living in cold environments: they represent physiological adaptations for minimizing heat loss from the body.[7]

Just as we know a great deal about the Neanderthal body from their skeletal remains, archaeologists have gained many insights into Neanderthal lifestyles from the debris left behind in caves and open-air occupation sites.[8] Like their forebears, the Neanderthals were hunters, focusing on the large mammals of ice-age Europe. Scavenging may have been employed, and plant gathering is likely to have been important whenever the opportunity arose. We know that the Neanderthals controlled fire, from thick deposits of ash and occasional hearths within their caves, although no stone-built fireplaces are known.

The cave deposits frequently contain dense accumulations of discarded stone artefacts and fragmented animal bones, suggesting intense occupation. Caves would have provided valuable shelter, especially during the ice-age winters; we have no evidence that the Neanderthals ever constructed huts of any kind. They would also have been key loci for social life. It seems likely that the Neanderthals lived in relatively small, intimate communities, not significantly different from those of their ancestors. In contrast to the modern humans who followed them into Europe after 40,000 years ago, there are no signs of social differentiation, specialized economic roles, sites where large numbers of people aggregated, trading relations, or long distance gift exchange.[9] Just like *Homo ergaster*, it is probable that the Neanderthals had detailed knowledge about the life histories, social relationships and day-to-day activities of all the other members of their group, and rarely came into contact with 'strangers'.

Given such social intimacy, it is not surprising that we see a very marked pattern of cultural transmission in stone-tool technology; each generation appears to have copied almost exactly the techniques of manufacture used by the previous generation. Their techniques required considerable skill, which is evident not only from their manufacture of hand-axes but also from their use of what archaeologists call the Levallois method, a technique for producing stone flakes with a predetermined shape and size.

This involved the careful flaking of a stone nodule to create a convex surface. A striking platform was prepared by removing tiny chips from one end of the nodule, then a single skilfully aimed blow would remove a single large flake. That was the most basic form of the technique; there were variants, some of which enabled numerous flakes to be removed from a single core and some of which enabled the production of long, thin blades.[10] One of the most frequently used variants produced a pointed flake that could be attached to a spear shaft without further modification.

While spears were used for hunting, they appear to have required the Neanderthals to get close to their prey; we know from their anatomy that their lives were physically demanding and that they were frequently injured. We know more about Neanderthal anatomy than that of any other type of pre-modern human, owing both to the number of fossils and to the fact that several of them are excellently preserved. That is most probably a consequence of the fact that the Neanderthals buried their dead, or at least some of them. The motivation for such burials remains unclear, but a fortunate outcome is that the bodies were protected from prowling carnivores.

One skeleton is of particular significance, the Kebara I specimen, dating to around 63,000 years ago and excavated in 1983 from Kebara Cave in Israel.[11] This burial was of an adult male whose skeleton had survived intact, apart from the leg bones which had been destroyed by acidic soils in the cave. It provided us with the first complete set of Neanderthal ribs, vertebrae and pelvis. The cranium was missing, perhaps taken by a carnivore if the head had been left partially exposed, or perhaps removed beforehand.

Vocal and auditory capacities: did the Neanderthals have language?

The Kebara burial also provided anthropologists with the first Neanderthal hyoid bone – in fact, the first from any pre-modern human. The hyoid bone is attached to the cartilage in the larynx and anchors the muscles necessary for speech. The Kebara hyoid is effectively identical in form to that of a modern human, and has been used to argue that the Neanderthal vocal tract was also essentially the same as that which you or I possess.[12] It was certainly not identical, and contrasts in facial morphology probably led to nasal-like vocalizations. But the earlier claim that the Neanderthal larynx was positioned high in the throat, like that of a modern human baby or a chimpanzee, has now been thoroughly dismissed.

That claim was originally made by Phillip Lieberman and Ed Crelin in 1971, on the basis of the La Chapelle-aux-Saints Neanderthal specimen.[13] It was highly influential for many years, but it has been rejected not only because of the Kebara find, but also because the La Chapelle-aux-Saints specimen is now recognized as having been too distorted and incomplete for a reliable reconstruction. So, although the Neanderthal vocal tract may have been unable to produce exactly the same range of sounds as a modern human, they would certainly have been sufficiently diverse to enable speech if the neural circuitry for language was present in the Neanderthal brain.

Two further pieces of anatomical evidence support the claim for vocal capabilities equivalent to those of modern humans. The first is the dimensions of the hypoglossal canal, which carries the nerves from the brain to the tongue. The width of this is greater in modern humans than in chimpanzees and gorillas, reflecting the larger supply of nerves required for the complex motor control necessary for speech. The canal is partially preserved at the base of the cranium on fossil skulls. When Richard Kay and his colleagues from Duke University measured its width on a variety of specimens, they found that those from *Australopithecus africanus* and *Homo habilis* fell into the chimpanzee and gorilla size range, while those from Neanderthals had a canal size equivalent to that of modern humans.[14]

A second nerve-carrying canal measurable from fossilized remains is that of the thoracic vertebrae, through which the nerves for controlling the diaphragm, and hence breathing, pass. Anne MacLarnon and Gwen Hewitt, of Roehampton University, measured the dimensions of this canal for a wide variety of living primates, modern humans, and a sample of hominid and Early Human fossils. They found that it has had a similar evolutionary history to the hypoglossal canal: the vertebral canals of australopithecines and *Homo ergaster* are similar in size to those of African apes, while those from Neanderthals are larger and match the size of those from *Homo sapiens*.[15] MacLarnon and Hewitt considered and rejected explanations for its increase in size relating to bipedalism, breathing for running, or the avoidance of choking on food. Their conclusion was that the increased number of nerves passing through the thoracic vertebrae arose in order to provide increased control of breathing for speech. The Neanderthals had the level of fine respiratory control that is required for modern speech.

The evidence from the hypoglossal and vertebral canals indicates that Neanderthals possessed motor control over their tongues and breathing similar to that of modern humans. This also appears to be the case for the perception of sound. In chapter 9, I described how a study of the bones of the inner ear from the early hominid fossil Stw 151 indicated that it had been more sensitive to higher-frequency sounds than modern humans. A similar examination of five *Homo heidelbergensis* skulls from the Sima de los Huesos, by Ignacio Martínez and his colleagues, showed that by 300,000 years ago sound perception had become equivalent to our own today.[16] As *Homo heidelbergensis* was most probably a direct ancestor of the Neanderthals, we can conclude that they, too, had a similar auditory capacity to that of modern humans.

Unlike earlier anatomical developments relating to vocal communication, such as the descended larynx, the Neanderthals' enhanced motor control over the tongue and breathing, and their auditory capacity matching that of modern humans, are best explained as having been specifically selected by evolution for vocal communication.

Does this imply language, as those undertaking the above studies have assumed? That would, indeed, be a convenient explanation for the cultural achievements of the Neanderthals. We must never forget that they survived for more than two hundred thousand years, living through periods of major climatic change.[17] To do so, they would have required extensive cooperation in hunting and gathering, which would in turn have required the careful fostering and maintenance of social relationships. Such relationships would also have been important in sustaining their technological traditions. Few modern flint-knappers can replicate the techniques used by the Neanderthals to make their tools, skills that had been accurately and consistently transmitted from generation to generation. It might seem extraordinary that they could have been so accomplished in hunting, socializing and tool-making without the use of spoken language. Indeed, perhaps this is too extraordinary to believe.

While the evolution of language would conveniently explain the large

Neanderthal brain, the vocal tract, auditory capacity, and motor control over the tongue and breathing, there is nonetheless overwhelming evidence that language had not yet evolved within this lineage of the *Homo* genus. So we must look to an advanced form of 'Hmmmmm' to explain these anatomical developments and cultural achievements.

Three features of the Neanderthal archaeological record provide the compelling evidence for advanced 'Hmmmmm' rather than language. The first has already been described: the evidence that Neanderthals lived in small, socially intimate communities. Just as I argued for *Homo ergaster,* members of Neanderthal groups would have shared knowledge and experience about their social world and their local environment, and would have had a common purpose. With the possession of an enhanced form of 'Hmmmmm' there would have been limited, if any, need for the type of novel utterances that could only have been produced by compositional language. To put it plainly, they didn't have much to say that had not either been said many times before or could not be said by changing the intonation, rhythm, pitch, melody and accompanying gestures of a widely understood 'Hmmmmm' utterance.

The two other sources of evidence for advanced 'Hmmmmm' rather than compositional language need further discussion: the absence of symbolic artefacts, and the continued presence of immense cultural stability.

The absence of symbolic artefacts

Symbolic artefacts are those that have been either shaped or decorated so that they have a meaning that is quite arbitrary in relation to their form. Pieces of art are obvious examples, especially those that are non-representational. A problem archaeologists face is that modern hunter-gatherers – indeed, modern humans in general – frequently attribute symbolic meanings to entirely unmodified found objects and to natural features of the landscape. We can never be sure that the Neanderthals did not do the same. But the absence of any objects that have been intentionally modified and lack a feasible utilitarian or other non-symbolic interpretation suggests that we should err on the side of caution.

The connection between symbolic artefacts and spoken language is simple, although to some it is contentious: if Neanderthals were able to use words – discrete utterances with symbolic meanings – they would have also been able to attribute symbolic meanings to objects.

Objects that have symbolic meanings provide an invaluable aid to social interaction; we use them continually and are entirely surrounded by them. And so it is inconceivable (to me) that Neanderthals could have had the capacity to produce symbolic objects but did not discover or choose to use that capacity for more than two hundred thousand years, while living in the most challenging of environments and often on the very edge of survival itself. The absence of symbolic objects must imply the absence of symbolic thought, and hence of symbolic utterances. Without these, by definition, there was no language.

There are a few objects made by Neanderthals and their immediate ancestors that have been claimed to have symbolic significance. These are so rare, so varied in nature, and so unconvincing, that basing an argument for symbolic thought and language on their existence – as some do – is frankly bizarre.

There is, for instance, the so-called Berekhat Ram figurine.[18] This is a piece of volcanic stone, no more than three centimetres in size, found in a 250,000-year-old deposit at a site in Israel. Some claim that the stone was deliberately modified into a female form with head, bosom and arms. Others, including myself, think that any similarity between the stone and a female form is entirely coincidental: it is equivalent to the faces we sometimes see in the clouds or the moon – it is in the eye of the beholder. A microscopic study of the stone has provided strong evidence that a stone blade did indeed modify it. But was most probably for an entirely utilitarian purpose, perhaps to make a wedge to support an anvil. Alternatively, the incision might have been made when blunting a razor-sharp flint flake so that it could be used by children or for a task that required a blunt edge, such as scraping fat off a skin.

Another contentious artefact is the incised fragment of bone from Bilzingsleben, the marks on which Dietrich Mania believes were part of a symbolic code.[19] Although a member of *Homo heidelbergensis* rather than a Neanderthal would have been responsible, verification of Mania's interpretation would suggest that the Neanderthals, as probable descendants, were also capable of making symbolic artefacts. A few lines, however, do not make a symbolic code. An alternative and more likely interpretation of the Bilzingsleben lines is that they derived from using the bone as a support when cutting grass or meat, or perhaps even for beating out rhythms.

Those who cannot bear the idea that Neanderthals lacked symbolic thought claim that they lived such a long time ago that few, if any, symbolic artefacts would have survived.[20] Time is not, however, the only determinant of preservation. There are numerous, extremely well-preserved Neanderthal sites which have provided many thousands of artefacts and bones, in addition to well-preserved burials. And yet all that can be found at such sites are a few pieces of scratched bone and stone.

A more challenging argument for the presence of symbolism comes from the possibility of Neanderthal body painting. Stone nodules containing the mineral manganese dioxide, which have been scraped with stone tools, have been found at several Neanderthal sites.[21] Powdered manganese dioxide can be mixed with water, or other liquids such as blood and tree sap, to make black paint, as used by *Homo sapiens* when they painted cave walls after reaching Europe at 40,000 years ago. Numerous specimens of worked manganese dioxide nodules have come from the excavations at the Neanderthal-occupied cave of Pech de l'Aze and are currently under analysis by Francesco D'Errico in his laboratory in Bordeaux. He believes that the Neanderthals may have made substantial use of manganese dioxide pigment, with the evidence having been 'missed' at many sites simply because the excavators did not expect to find it.

As the Neanderthals have left no traces of pigment on cave walls or artefacts, the most likely explanation is body painting. This need not imply the creation of symbolic images. We can guess that the Neanderthals were white-skinned, having evolved in high latitudes, and we know that they were big-game hunters. It seems entirely plausible that the paint was used simply to camouflage their bodies. Alternatively, or perhaps in addition, it may have been used for cosmetic reasons – to emphasise an aspect of one's appearance as a means of sexual attraction.

Had the Neanderthal pigment use been for symbolic purposes, I would expect to see a wider range of pigments represented at their sites, especially nodules of ochre used to create red paint. This is the colour that dominates the earliest symbolic activities of modern humans in southern Africa (as I will describe in the next chapter), and it is a colour that has far more evocative connotations than black – as the evolutionary psychologist Nicholas Humphrey once wrote, red is the 'colour currency of nature'.[22]

Immense cultural stability

The second major argument against the idea that Neanderthals possessed language is the immense stability of their culture. The tools they made and the way of life they adopted at around 250,000 years ago were effectively no different from those current at the moment of their extinction, just after 30,000 years ago. As we know from our own personal experience and from a moment's reflection on human history, language is a force for change: with it we can exchange ideas so that our technology is improved and new ways of living are introduced. Indeed, many psychologists, philosophers and linguists claim that language is a medium not merely for exchanging ideas but for complex thought itself – a view I endorsed in my *Prehistory of the Mind* and will further support in my next chapter. So, if the Neanderthals had possessed language, how could their culture have remained so stable and so limited in scope? Well, it simply could not have done so.

To claim that Neanderthal culture was stable is neither to dismiss an impressive degree of cultural diversity nor to deny the complexity of their behaviour. In addition to their hand-axes and Levallois flakes, the Neanderthals made stone artefacts using a variety of other techniques and were evidently responsive to the types of raw materials available to them.[23] They had very high levels of knapping skill, and we must assume that they also made tools from bone, wood and other plant materials, even though these have not survived. Nonetheless, the absence of innovations throughout the time of Neanderthal existence is striking; they simply relied on selecting from a repertoire of tried and tested tool-making methods.

As with the lack of symbolism, this gains added significance because the Neanderthals were teetering on the edge of survival, as I will describe below. If ever there was a population of humans that needed to invent bows and arrows, the means for storing food, needles and thread, and so forth, it was the Neanderthals. But all of these came only with language-using modern humans, who then went on to invent farming, towns, civilizations, empires and industry. In contrast, the Neanderthals showed immense cultural stability, and so we are still looking at a species that had a holistic rather than a compositional form of communication.

The Chatelperronian problem

The only significant change in Neanderthal culture appears to have come towards the end of their existence in Europe, at 35,000 to 30,000 years ago. A small number of Neanderthal occupation sites in France have revealed bone objects and tooth pendants that suggest new forms of behaviour – behaviour of the type that, when undertaken by modern humans, involves symbolic thought. These objects are part of the so-called Chatelperronian industry, which seems to combine the traditional Neanderthal manufacturing techniques for stone artefacts with those characteristic of the first modern humans in Europe.[24] Most archaeologists believe that the Neanderthals saw these new arrivals in their landscapes and attempted to copy the stone tools they were using and the beads and pendants they were wearing, perhaps without understanding their symbolic dimension. Others believe that, after more than two hundred thousand years of living without any of these tools and body decorations, the Neanderthals finally began using them immediately before modern humans arrived in Europe with an almost identical set of artefacts.

These interpretations have been vigorously debated within the pages of academic journals over the last few years, without any resolution of the problem. Francesco D'Errico is the leading proponent of the independent invention theory and has carried out a meticulous examination of the relevant dating and stratigraphic evidence from the French caves in order to ascertain which came first: Neanderthal pendants or modern humans in the landscape. He has decided that it was the former, but the archaeological evidence may simply be insufficiently detailed to distinguish between the two scenarios. For many archaeologists, myself included, the coincidence of Neanderthals coming up with the use of beads just before modern humans appeared wearing them is just too great to be believed.

So my view is that the final Neanderthals in Europe were imitating the symbol-using modern humans without understanding the power of symbols. Imitation was, after all, at the centre of Neanderthal culture as the key means by which tool-making traditions were passed on from one generation to the next. We should not be at all surprised that they imitated the behaviour of the new, black-skinned modern human arrivals – and perhaps painted their own skins to look like them – even if they were unable to appreciate the manner in which arbitrary meanings could be attached to physical objects such as beads.

The Neanderthal mind

Having considered the complete lack of evidence for Neanderthal language, we must now turn to the Neanderthal mind, before considering their use of 'Hmmmmm'. The underlying reason why the Neanderthals have caused so much controversy since their discovery in 1856 is that in some respects they appear to be modern, like us, while in others they appear little different from the much earlier forms of *Homo*, which possessed considerably smaller brains. They appear so modern – so intelligent – because they were able to make complex stone artefacts, to hunt big game, and simply to survive for more than two hundred thousand years in harsh northern landscapes. Conversely, they appear to be so 'primitive' – so lacking in intelligence – because of the stasis in their technology, the absence of symbolic thought, and their lack of language.

My first attempt to resolve this paradox was published in my 1996 book, *The Prehistory of the Mind*.[25] Drawing on theories from evolutionary and developmental psychology I argued that the Neanderthals had a 'domain-specific' intelligence. By this I meant that they had very modern-like ways of thinking and stores of knowledge about the natural world, physical materials and social interaction, but were unable to make connections between these 'domains'. So, for instance, they had the technical skills for making artefacts as sophisticated as those of any modern human, and they undoubtedly also had complex social relationships which had to be continually monitored, manipulated and maintained. But they were unable to use their technical skills to make artefacts to mediate those social relationships, in the way that we do all the time by choosing what clothes or jewellery to wear, and as do all modern hunter-gatherers through their choice of beads and pendants.

Another example concerns hunting technology. Although the Neanderthals must have had an extensive and detailed knowledge of animal behaviour, they were unable to design specialized hunting weapons because they could not bring their technical and natural history intelligence together into a single 'thought'. Helmut Thieme and Robin Dennell have stressed the javelin-like qualities of the Schöningen spears; but while we must believe that the Neanderthals had similar weapons as well as flint-tipped thrusting spears, neither matched the sophistication of the multi-part spears used by modern human hunter-gatherers. There is no evidence to suggest that the Neanderthals created specific types of hunting weapons to target different types of animals in different situations.

This domain-specific intelligence partly explains the Neanderthals' large brain: they had neural circuits equivalent to ours in number and complexity, enabling tool-making, socializing and interaction with the natural world. What they lacked were the additional neural circuits that made connections between these domains, which may have required very little extra capacity. The collaborative team of archaeologist Thomas Wynn and psychologist Frederick Coolidge have suggested that these circuits related to working memory. Their view is that the Neanderthal capacity to hold a variety of information in active attention was less than that of modern humans.[26] Whether or not enhanced working memory is the key, the extra neural circuits that modern humans possess provide them with what I term 'cognitive fluidity'. This is, in essence, the capacity for metaphor, which underlies art, science and religion – those types of behaviour that go unrepresented in the Neanderthal archaeological record.

While the domain-specific, rather than cognitively fluid, mentality that I attributed to the Neanderthals in my 1996 book goes a long way towards explaining their behaviour and the archaeological record they left behind, it cannot entirely account for their large brains. For that we need to understand their communicative behaviour and emotional lives – topics which were neglected in my previous work. It is this omission, of course, that this book has been attempting to rectify; the Neanderthals may not have had language, but they did have 'Hmmmmm'.

'Hmmmmm'-ing Neanderthals

In my previous chapters I have examined several functions of music-like vocalizations and argued that *Homo habilis*, *Homo ergaster* and *Homo heidelbergensis* would have employed these to varying extents within their communication systems. *Homo habilis* most likely used vocalizations and body movements to express and induce emotions in a manner that far exceeded any modern non-human primate but that was quite restricted in comparison with modern humans, owing to the possession of a relatively ape-like vocal tract and limited muscular control. Both vocalizations and movement would have been substantially enhanced following the evolution of bipedalism, so that later species of *Homo* were able to engage in extensive mimesis in order to communicate about both the natural world and their emotional states. In both *Homo ergaster* and *Homo heidelbergensis*, singing may have been a means of attracting mates and a behaviour that was subject to some degree of sexual selection. Singing would have also been a means to reassure infants when 'put down', and to facilitate their emotional development and the acquisition of an adult communication system. Finally, as cooperation was essential in those Early Human societies, singing and dancing would most probably have been used to engender social bonds between individuals and groups.

By having a larger body size, by living in more challenging environments, by having particularly burdensome infants, and by having even greater reliance on cooperation, the Neanderthals would have evolved a music-like communication system that was more complex and more sophisticated than that found in any of the previous species of *Homo*.

We so often emphasize what the Neanderthals lacked in comparison with *Homo sapiens* – symbolic objects, cultural change, language, cognitive fluidity and so on – that we forget that they, too, had as much as five hundred thousand years of independent evolution. The selective pressures to communicate complex emotions through advanced 'Hmmmmm' – not just joy and sadness but also anxiety, shame and guilt – together with extensive and detailed information about the natural world via iconic gestures, dance, onomatopoeia, vocal imitation and sound synaesthesia, resulted in a further expansion of the brain and changes to its internal structure as additional neural circuits were formed. We must also recall from earlier chapters the evidence for and interpretation of perfect pitch. Without the development of language, it is most likely that Neanderthals maintained the capacity for perfect pitch with which we must assume they were born, and this would have enhanced their musical abilities in comparison with those found in both earlier *Homo* and modern humans.

The Neanderthals would have had a larger number of holistic phrases than previous species of *Homo*, phrases with greater semantic complexity for use in a wider range of more specific situations. I think it is also likely that some of these were used in conjunction with each other to create simple narratives. Moreover, the Neanderthal musicality would have enabled the meaning of each holistic utterance to be nuanced to greater degrees than ever before, so that particular emotional effects would have been created. But however sophisticated a communication system Neanderthal 'Hmmmmm' was, it remained one of relatively fixed utterances that promoted conservatism in thought and cultural stasis.

This interpretation of the Neanderthal brain and communication system allows us to look at some of the key fossil finds and archaeological sites from a new perspective. These provide no more than snapshots into what would have been Neanderthal communities distributed across ice-age Europe and the Near East, living in relative, but not complete, isolation from each other. Some contact and exchange of members would have been essential in order to maintain genetic and probably demographic viability. But we should envisage each group as developing their own version of 'Hmmmmm', some aspects of which may have been relatively unintelligible to members of other groups, although there would have been a great many common features. With this view of their communication system, we can gain a more comprehensive view of Neanderthal lifestyles than has previously been available.

The struggle for survival

By examining their skeletal remains, anthropologists have established how tough the world was for the Neanderthals; with a slight increase in mortality or a decrease in fertility, their communities could easily have dwindled and disappeared.[27] It appears that very few Neanderthals lived beyond the age of thirty-five, leaving their populations only marginally viable.[28] This conclusion is confirmed by the high incidence of disease and trauma evident from the skeletons: Neanderthal life was hard, often short, and frequently involved considerable pain. The so-called 'old man' of La Chapelle, who had not reached forty when he died, had suffered from degenerative joint disease in the skull, jaw, spinal column, hips and feet, a rib fracture, extensive tooth loss, and abscesses. Many skeletal remains display evidence of similar diseases and fractures.[29]

Some of the skeletons show sufficient healing to suggest the presence of social care. The classic example is the Shanidar 1 man, Shanidar being a cave in Iraq from which a particularly important collection of Neanderthal remains have been excavated.[30] This specimen shows evidence of head injuries and extensive crushing to the right side of the body. His injuries were probably caused by a rock fall; the man also suffered from blindness in at least one eye, infections and partial paralysis. Yet he had survived for several years, and he is most unlikely to have done so without substantial care – having food and water brought to him, being kept warm, and receiving emotional support and perhaps medicinal plants.

I think it most likely that 'music therapy' was used as a means to reduce stress, improve healing and facilitate the return of motor coordination. We saw in chapter 7 how effective this can be, and I can readily imagine Neanderthals singing gently to calm the distressed and those in pain, and tapping out rhythms to help the movements of those who were injured or disabled.

Whether such care, involving music therapy or not, arose simply from affection and respect, or whether even a disabled man had some practical contribution to make to the group, such as knowledge about how to survive during infrequently occurring but severe conditions, will never be known. What we can be confident about, however, is that even when the Neanderthal groups were struggling for survival in an ice-age climate, they found the time and energy to look after the ill and injured.

In ice-age conditions, making decisions was a matter of life or death; and Neanderthal life was full of decisions – from what to hunt, where to hunt, who to hunt with, what tools to make and so on to decisions about who to mate with and whether to care for an injured member of the group. The males had decisions to make regarding whether to provision not only females but also infants. Providing food and protection to an infant of whom they were definitely the father was perhaps an easy choice to make; more contentious would have been decisions involving infants whose paternity was unclear, or even those whose fathers had died or were injured.

As was considered in chapter 7, to arrive at rational decisions, human minds require the capacity not only to make judgements and to process considerable information, but also to feel emotion. Without feeling happy or sad, getting angry or elated, anxious or confident, the rational, 'optimal' choices cannot be made. Therefore, since making the right decision was of such consequence for the Neanderthals, they must have been highly emotional people, and that would have found expression in their utterances, their postures and their gestures.

This is why cooperation – the sharing of information and resources, working as a team during a hunt, caring for each other's well-being – would have been even more essential for their survival than it had been for the species that preceded them. There is unlikely ever to have been a population of humans – modern, Neanderthal or otherwise – for whom the creation of a social identity to override that of the individual was more important. For that, music is likely to have been essential. This is the case for modern humans: when living in conditions of adversity, they make music. Such music enables intense social bonding and facilitates mutual support. I have no doubt that the Neanderthals behaved in exactly the same way. As a consequence, communal singing and dancing would have been pervasive among the non-linguistic, non-symbolic, ice-age populations of Europe that communicated by 'Hmmmmm'.

Mimesis and hunting

One strand of 'Hmmmmm' emanating from the Neanderthals' caves would have been that of hunters either planning hunts or recently returned with game. Our best insight into this comes from the cave of Combe Grenal, located in the Dordogne region of France.[31]

This cave was used as a Neanderthal occupation site between 115,000 and 50,000 years ago, resulting in fifty-five distinct occupation layers. These were excavated during the 1950s and 1960s by the great French archaeologist François Bordes, who recovered extensive collections of stone artefacts and animal bones, along with samples of pollen and sediments that provided information about climatic changes during the last ice age. Bordes died in 1981 before his excavations were fully published, but several archaeologists have studied the materials he found. This has placed the site at the centre of our attempts to understand the Neanderthals.

Phillip Chase, an American archaeologist at the University of Pennsylvania Museum, made the most detailed study of the Combe Grenal animal bones, identifying them by species, describing which parts of the skeletons were present, and inspecting the bones for any cut-marks made by stone tools. Four species dominated the collections from each occupation layer: horse, red deer, reindeer, and bovids. Chase ascertained whether the animals had been hunted or scavenged; in the latter instance very specific types of bones are brought back to the cave and there is an absence of butchery marks associated with the removal of large fillets of meat.

His results indicated that both red deer and reindeer were hunted. Only small numbers of individual animals were represented in any one layer, and so it appears that these animals were stalked on a one-by-one basis in the vicinity of the cave. On the other hand, the collections of horse and bovid bones included a relatively large number of skulls and jaws. This suggests that these animals were scavenged from the remnants of carnivore kills, as the heads are often the only body parts left behind; they can be highly nutritious for humans who, unlike carnivores, can crack the skulls apart to extract the brains and tongues.

The picture we have, therefore, is of Neanderthal hunters departing from Combe Grenal to search for prey, in much the same manner as we imagined *Homo heidelbergensis* hunting from Schöningen. Whether the Neanderthals had similar javelin-like wooden spears is a moot point; they certainly hafted stone points to make short, thrusting-type spears. The type of animals they searched for and found depended partly upon the date at which the hunting took place. During the earlier periods of occupation, between 115,000 and 100,000 years ago, the climate was relatively warm and the cave was surrounded by woodland, so it is not surprising to find that these layers are dominated by red deer. During the later periods of occupation, when the climate had become colder and drier, woodland was replaced by open tundra and trees survived only in the sheltered valleys. As a result, reindeer bones become much more frequent.

Whether looking for animals or carcasses, hunting in woodland or on tundra, using throwing or thrusting spears, the Neanderthals who departed from Combe Grenal would have undertaken some planning and 'asked' about recent game sightings. To do so, iconic gesture, mimesis, vocal imitation, onomatopoeia and sound synaesthesia would all have been used – and were most probably more elaborate than those used by the Neanderthals' ancestors, as described in previous chapters.

The evidence from Combe Grenal suggests that animals were stalked on a one-by-one basis. Since the frequencies of reindeer, red deer, bison and horse vary in rather predictable ways with the sequence of environmental change between 115,000 and 50,000 years ago, it seems likely that the Neanderthals simply took whatever they could find in the vicinity of the cave. Elsewhere, however, there is evidence for mass killing, suggesting higher degrees of planning and cooperation in the hunt than can be inferred at Combe Grenal. This evidence comes from several 'open sites', localities where high densities of animal bones and stone artefacts have been found, which appear to have been kill and butchery sites.

The most striking example is Mauran, in the foothills of the Pyrenees.[32] This is where a massive accumulation of bison bones was excavated close to a steep riverside escarpment. It appears that the Neanderthals forced small herds of bison off the cliff edge, which then fell to their deaths. To do so, several, perhaps many, hunters would have worked together to stampede the animals. The slaughters at Mauran appear to have occurred towards the end of the Neanderthal occupation of Europe, probably around 50,000 years ago.

Considerably earlier, at around 125,000 years ago, a similar hunting technique was used at La Cotte on the Channel Island of Jersey.[33] When excavated in the 1970s, mammoth and rhino bones were found in stacks within a cave at the foot of a cliff. It appears that these animals, too, had been forced off the edge, fell to their deaths, and were then butchered by Neanderthals waiting at the base, who dragged selected parts into the cave and away from prowling carnivores.

Further examples of mass killing by either ambush or cliff-fall hunting could be cited, but my point has been made by Mauran and La Cotte: at times, the Neanderthals engaged in hunting that required extensive cooperation by groups. This should not, of course, surprise us, but it does emphasize how essential communication by mimesis must have been in order to develop and execute plans for mutual benefit.

Stone tools and mating patterns

The exchange of information would have been necessary to many aspects of Neanderthal life other than hunting. One for which we have direct evidence is the manufacture of stone tools. These can tell us far more than just how reliably technological skills must have been passed on from generation to generation; they also indicate how mating patterns had changed from the time of *Homo ergaster* and *Homo heidelbergensis*, with implications for the usage of 'Hmmmmm'.

As we examined in chapter 12, a prominent type of tool made by *Homo heidelbergensis* was the pear-shaped, ovate or sharply pointed stone hand-axe, the features of which suggest its use as a means of display to the opposite sex. My hypothesis is that males made such artefacts to advertise the quality of their genes to females; the females wanted 'good genes' but did not require male investment in terms of resource provision, owing to the existence of female kin networks that provided support to pregnant and nursing females.

Although some Neanderthals continued to make hand-axes, these had become a limited element in the technological repertoire by 250,000 years ago. Instead, flake tools were predominant. A similar shift in technology also occurred in Africa at about the same date. Both can be explained, I believe, by a significant change in male–female relationships.[34]

The increase in body and brain size of *Homo neanderthalensis* over earlier *Homo* species would have placed even greater energy demands upon the females during pregnancy, and would have made newborns even more helpless for a greater length of time. Also, we know that among the Neanderthals there were few 'grandmothers' around to support their daughters when pregnant and nursing, owing to the Neanderthals' relatively short life expectancy, with few surviving beyond the age of thirty-five. The consequence was, I believe, that females now needed to select mating partners for qualities other than intelligence and physical strength, and hence could no longer rely on assessing these simply from the stone tools they made and their physical appearance.

With the additional challenges that ice-age life presented, the females now required males to provide resources as well as their genes; they needed males to secure and share food for themselves and their joint offspring, and to provide shelter, clothing, fire and other necessities of ice-age life. They needed reliable males whom they could trust. Such attributes could no longer be demonstrated by the display of elegant stone tools, which may have been good for butchery but had limited utility for hunting. Hence the tools that became attractive were efficient hunting weapons – wooden hafts tipped with stone points. And the males that became attractive were those who could 'bring home the bacon', or at least its ice-age equivalent.

The consequence of these social and technological changes is that male Neanderthals were no longer likely to have sung and danced in order to attract females; their musical abilities were no longer subject to sexual selection, even though they would have persisted in part as a consequence of that earlier phase of human evolution. Instead, there would have been singing and dancing as a means of advertising and consolidating pair-bonding. And after that, would have come . . .

More babies

In September 1999 I had the thrilling experience of visiting Amud Cave in Israel with Erella Hovers, whose excavations during the early 1990s had uncovered the remains of several Neanderthals along with many artefacts and animal bone fragments.[35] The cave is found along a dry valley amid quite spectacular scenery. I knew we were getting near when the tall natural pillar of stone at its entrance, familiar to me from many photographs, came into view. At the cave, Erella showed me where the skeleton of a tiny Neanderthal baby had been excavated from within a niche in the rock wall. She was convinced that it had been a burial.

The skeleton was of a ten-month-old infant that had lived between 60,000 and 50,000 years ago, and is known in the academic literature as Amud 7. It was found lying on its right-hand side and was only partially preserved; the cranium was crushed, the face too badly damaged to be reconstructed, while the long bones, pelvis and shoulder bones had largely disintegrated. But the vertebrae and ribs were better preserved, and the finger and toe bones were discovered in their proper anatomical positions.

That Neanderthal infant mortality was high – this is quite evident from the skeletal record – should not be surprising. But though a great deal has been written about the mortality rates and burial practices, the manner in which Neanderthal infants were cared for has rarely been discussed. This, too, is not surprising, for it can never be more than a matter of speculation. However, such speculation can be grounded in the theoretical arguments and the data provided in my previous chapters about singing to infants. Hence we can be confident, I believe, that Neanderthal parents used music-like vocalizations, body language, gestures and dance-like movements to communicate with their infants, in a manner not dissimilar to modern humans and more highly developed than their ancestors.

With few adults surviving beyond the age of thirty-five, grandmothers would have been rare, and so mothers had to cope with their infants on their own – or at least with the support of their older offspring, their own siblings and, perhaps, the father(s) of their children. Even with such support, Neanderthal infants are bound to have been not only frequently 'put down' but also orphaned, and so they may have required a lot of singing to keep them happy.

Another factor may have placed Neanderthal parents under even greater pressure to communicate with their offspring than modern human parents: the children's rate of development. Microscopic studies of tooth growth have shown that Neanderthal children grew up at a faster rate than those of modern humans.[36] Hence they had a relatively shorter time in which to acquire their communication system, which for them meant learning a large number of holistic phrases, gestures and body language, and coming to understand the nuances of emotive expression. Consequently, there would have been a premium on 'Hmmmmm' communication between adults and Neanderthal infants, right from the moment of their birth.

Neanderthal burial

'Hmmmmm' would also have been pervasive at the other end of the Neanderthal life cycle, however soon this came. The death of, and burial and mourning for, a Neanderthal adult would have been just as emotional an event as in the case of a child, for the loss of a young, single adult could easily have spelt disaster for the group as a whole. A relatively high proportion of the burials are those of young males – the strongest members of the group, who would have undertaken the big-game hunting and whose scavenging from frozen carcasses may have been essential for group survival during the harsh winter months. Indeed, it may have been the economic value of such individuals to the group as a whole that prompted their special treatment after death.

We should envisage, therefore, the same type of grieving around the grave of the young man buried in Kebara Cave at 63,000 years ago that we imagined in Atapuerca at 350,000 years ago. In both cases, the need for the social group to bond together and express shared emotion is likely to have resulted in mournful song.

We should recall, however, John Blacking's experience among the Venda, as described in the previous chapter. He noted that it was at times of plenty that group identity was promoted via communal music-making – times when the opportunities for self-interested action were prominent. Therefore we might expect Neanderthal communal 'Hmmmmm' music-making not only when their groups were challenged by death and injury, but also during the 'good times' of Neanderthal life.

Empty spaces for performance

'Hmmmmm' communication would have involved dance-like performance, and this might explain an intriguing feature of the Neanderthal archaeological record. When either the whole or a substantial part of a Neanderthal-occupied cave is excavated, the debris they left behind is typically found in a very restricted area.

Paul Mellars, a Cambridge archaeologist with a particularly detailed and extensive knowledge of Neanderthal archaeology, has remarked upon this pattern in relation to several caves in France,[37] in particular the Grotte Vaufrey in the Dordogne, the Grotte du Lazaret, close to Nice, and Les Canalettes in the Languedoc region.

Mellars provides two possible interpretations for each case: either the 'empty' areas were used for sleeping, or the groups within the caves had been very small. There is, of course, a third: those empty areas were used for performance. One or more Neanderthals could have been gesturing, miming, singing or dancing, while other individuals squatted by the cave walls engaged in their tool-making and other activities, or perhaps joined in the song to affirm their social bonds and contribute their own information to the 'Hmmmmm' conversation piece.

When considering the site of Bilzingsleben in chapter 11, I suggested that the spaces demarcated by stone blocks and large bones might have been performance areas. A Neanderthal equivalent may have been found in the cave of Bruniquel in southern France.[38] Several hundred metres in from the cave entrance is a 5 metre by 4 metre quadrilateral structure constructed from pieces of stalactite and stalagmite. A burnt bear bone from inside this structure provided a date of 47,600 years ago, identifying the Neanderthals as having been responsible for its construction.

The Neanderthals must have used lighted torches or fires in order to build and make use of this structure, for it would otherwise have been in total darkness, being so far from the cave entrance. What they used it for is unclear; other than the structure and burnt bones there were no other signs of human presence. My guess is that it was the scene of 'Hmmmmm' singing and dancing. The surrounding cave walls are interesting because there are no traces of any paintings, engravings or marks of any kind, even though preservation conditions are ideal. Jean Clottes, the doyen of ice-age cave art, finds this significant. He has probably visited more caves, and certainly has a more extensive knowledge of cave painting, than anyone (alive or dead) and he believes that Bruniquel provided ideal walls for painting. The modern humans of Europe failed to discover the cave; the Neanderthals had evidently done so, and Clottes believes that the lack of wall painting is conclusive proof that the Neanderthals did not – indeed, could not – engage in this activity.[39]

The flute that (probably) isn't

One need not have been to Bruniquel Cave, or any cave once occupied by Neanderthals, in order to appreciate that their singing and dancing would have echoed and reverberated around the walls. I suspect that Neanderthals would have exploited this property of caves, along with the shadows thrown up against the cave walls by the firelight, to make their singing and dancing more dramatic. They may also have flicked stalagmites with their fingernails to make a high-pitched ringing sound, banged bones and chimed flint blades together, blown through horns and shells, and drummed upon mammoth skulls. None of this would have left an archaeological trace indicative of music-making. Indeed, this is true for many of the musical instruments manufactured by modern hunter-gatherers, which mainly use either skins and vegetable materials that rapidly decay, or minimally modified bones and shells that are unlikely to be interpreted as musical instruments if found on an archaeological site.

Although the evidence is lacking, it is inconceivable to me that the Neanderthals did not supplement their own voices, hand clapping, thigh slapping, and feet stamping with the use of natural objects. But whether they deliberately modified such objects to enhance their inherent musical properties is more contentious. The 'domain-specific' mentality that I believe Neanderthals possessed would have imposed considerable limits on their ability deliberately to alter natural materials. To turn, say, a piece of bone into a flute would have required a Neanderthal to take an object from the natural history realm and modify it for use in the social realm, demanding a significant degree of cognitive fluidity. Their general intelligence would, I suspect, have been sufficient to enable the use of an unmodified stick to beat out rhythms on the ground, but not to transform a bone into a flute – that would have been a cognitive step too far. So I was naturally concerned for the veracity of my theory when there was a highly publicized announcement of a Neanderthal bone flute in 1996.

Ivan Turk of the Slovene Academy of Sciences and Arts found the 'flute' in Divje Babe Cave, in Slovenia, during his 1996 excavations.[40] Objectively, the 'flute' is an 11.36 centimetre-long fragment of thigh bone from a one- to two-year-old bear. It is pierced by two complete circular holes, while both ends of the bone are fragmented in a manner that suggest additional holes may have been present. It was found in a layer of cave deposits that date to between 50,000 and 35,000 years ago, close to what Turk describes as a hearth, and could only have been made by Neanderthals. Not surprisingly, journalists leapt on the find as conclusive proof that Neanderthals made

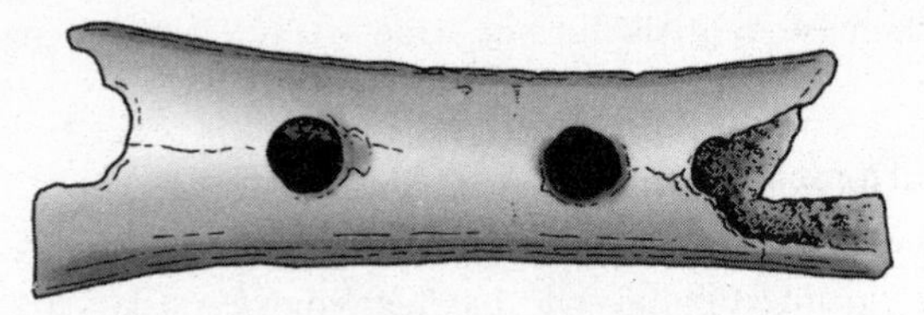

Figure 15 The Divje Babe I 'flute', around 114 millimetres long.

musical instruments; the 'flute' was featured on the cover of the *New York Times* before it had been subject to critical evaluation by academics – it became, and will for ever remain, a 'media fact'.

Is it a flute? Ivan Turk and his colleagues believe so and have published an extensive study of its musical qualities, derived from the playing of replicas. Others are not so sure. Francesco D'Errico noted that the bones within the cave belonged almost entirely to cave bears and appear to have come from a natural accumulation of bears which died in their dens. He also provided several examples of bones from similar natural accumulations that had been chewed by carnivores, leaving circular holes almost identical to those on the Divje Babe bone.[41] D'Errico submitted these and the 'flute' to a microscopic study. Telltale microscopic pittings and scorings surrounding the holes confirmed that they had been made by the canines of carnivores rather than by stone tools, which would have left quite different traces. Moreover, there were clear tooth impressions on the face of the bone immediately opposite the holes, showing how the bone had been gripped within carnivore jaws.[42] As to the type of carnivore responsible, D'Errico suggests that it was likely to have been a cave bear itself.

The 'flute' is on display in the National Museum in Slovenia. I went to see it in November 2004, and discussed with Peter Turk, the museum curator (no relation to Ivan Turk who excavated Divje Babe), whether it really is a flute. It certainly looked like a flute; and Peter challenged many of Francesco's criticisms, arguing that it could have been chewed after having been lost or discarded. But I wasn't convinced, and concluded that the bone's resemblance to a flute is simply one of chance. So we lack any evidence that the Neanderthals manufactured musical instruments. My own theoretical views suggest that they were unlikely to have been able to do so, although I suspect that unmodified sticks, shells, stones and skins may have played some role in their music-making.

Summary: walking with a Neanderthal

Trying to understand the 'Hmmmmm'-ing world of a Neanderthal is challenging, owing to the limitations of our imaginations, the inevitable speculation involved, and the restricted evidence on which these speculations must be based. Also, I believe that all modern humans are relatively limited in their musical abilities when compared with the Neanderthals. This is partly because the Neanderthals evolved neural networks for the musical features of 'Hmmmmm' that did not evolve in the *Homo sapiens* lineage, and partly because the evolution of language has inhibited the musical abilities inherited from the common ancestor that we share with *Homo neanderthalensis*.

Occasionally, however, we have an intense musical experience that may capture some of the richness that was commonplace to the Neanderthals – perhaps a song whose melodies and rhythms we find particularly moving or inspiring. One of my own experiences was when watching my first ballet, when I was suddenly made aware of how the human body can both express emotion and tell a story through its movement. We place ballerinas on a stage and very few of us watch them more than once a year; the Neanderthals watched their ice-age dancing equivalents on a daily basis by firelight within their caves.

Other experiences might also remind us how 'desensitized' we are to the music-like sounds around us. And so, to end this chapter I would like to quote for a second time how his teacher described her walk with Eddie, the musical savant we met in chapter 3:

> I found that a walk with Eddie is a journey through a panorama of sounds. He runs his hand along metal gates to hear the rattle; he bangs on every lamp post and names the pitch if it has good tone; he stops to hear a car stereo; he looks into the sky to track airplanes and helicopters; he imitates the birds chirping; he points out the trucks rumbling down the street . . . If it is aural Eddie is alert to it, and through the aural he is alert to so much more.

By the time he undertook this walk, Eddie had developed some language skills, which had been almost entirely non-existent throughout his early years. In this regard, he shared some fundamental similarities with a Neanderthal – linguistically challenged but musically advanced. So we can perhaps use this quote to help us imagine what a Neanderthal would have heard, and how he or she would have responded, when walking through the ice-age landscape. It would have been another panorama of sounds: the melodies and rhythms of nature, which have become muffled to the *Homo sapiens* ear by the evolution of language.

16 The origin of language

The origin of *Homo sapiens* and the segmentation of 'Hmmmmm'

Nina Simone singing 'Feeling Good': *Homo sapiens* at Blombos Cave, wearing shell beads, painting their bodies and feeling good

While the Neanderthals were hunting and 'Hmmmmm'-ing in Europe, another lineage of the *Homo* genus was evolving in Africa, that which produced *Homo sapiens* at around 200,000 years ago. This is the only species of *Homo* alive on the planet today; within its brief existence *Homo sapiens* has gone from living as a hunter-gatherer to living as a city banker, from colonizing Europe to exploring other planets, and from a population of less than ten thousand individuals to one of more than six billion. Most anthropologists attribute the remarkable success of *Homo sapiens* to language, a capacity they believe was absent from all other members of the *Homo* genus.

As will have been evident from my previous chapters, I think they are correct. This chapter will describe the emergence of language from 'Hmmmmm', and its profound consequences for human thought and behaviour. We must begin with the origin of *Homo sapiens*.

The evolution of *Homo sapiens*

Any statement about the origin of modern humans must be provisional, owing to the possibility that new fossils and new evidence from human genetics will change our current understanding. Many books published during the last decade have had a limited shelf-life because new discoveries have so quickly made them out of date. There is, however, a growing sense that the most recent picture of *Homo sapiens* origins may endure, because the fossil and genetic evidence have now converged on the same date.

In 2003 a key fossil discovery was revealed in the pages of the premier scientific journal *Nature*: three partial skulls discovered in Herto, Ethiopia, dating to between 160,000 and 154,000 years ago.[1] This was six years after a team of American and Ethiopian anthropologists, led by Tim White of the University of California, Berkeley, had actually found the skulls in a series of shattered fragments. It took three years for the team to clean and reassemble the skulls meticulously, and another three years to make a full analysis and

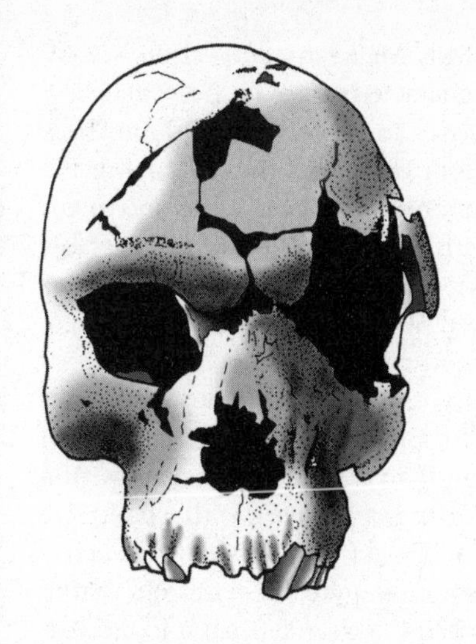

Figure 16 Adult cranium of *Homo sapiens* from Middle Awash, Ethiopia, dating to between 160,000 and 154,000 years ago.

comparison with six thousand other skulls from around the world, in preparation for the announcement – it would not be something to get wrong!

White and his colleagues had found the partial skulls of two men and a child. The most complete had the characteristic features of *Homo sapiens*, including a flat face, divided brow ridges and a cranial capacity of 1450 cubic centimetres – slightly above the modern human average. It also had some 'primitive' features linking it to more ancient African fossils – those of *Homo ergaster* – such as protruding brows. White and his colleagues decided to describe the new fossils as *Homo sapiens idaltu* – the last name being the word for 'elder' in the local Afar language. That last term seems quite unnecessary, for there has been widespread agreement among palaeoanthropologists that the skulls belong to *Homo sapiens* and provide one of the earliest fossil remains of our species.

When first published, the Herto fossils were, in fact, the oldest known specimens of our species. But early in 2005 that distinction went to two other fossil crania, also from Ethiopia, which had been excavated as long ago as 1967. These are known as the Omo I and Omo II specimens, originally discovered by Richard Leakey and previously thought to be 130,000 years old. The application of a new dating technique based on the radioactive decay of argon revised their age to 195,000 years ago.[2]

The Herto and Omo finds are part of a group of fossils from Africa that belong at between 500,000 and 100,000 years ago, some of which have been designated as *Homo helmei*,[3] the species that Robert Foley and Marta Lahr believe is the common ancestor of the Neanderthals and modern humans.

According to the palaeoanthropologists Sally McBrearty and Alison Brooks, the African *Homo helmei* specimens have characteristics of both *Homo sapiens* and earlier species, including *Homo rudolfensis* and *Homo ergaster*, dating back to 2 million years ago. The new finds from Herto and the redating of the Omo specimens suggest that some of the more recent *Homo helmei* specimens should be reclassified as *Homo sapiens* with the recognition of considerable morphological diversity. An alternative view is that several different species of *Homo* might have been alive at between 500,000 and 100,000 years ago, just one of which has survived into the modern day.

While such differences of opinion exist, there is unanimous agreement among palaeoanthropologists that the African fossil record shows a gradual evolution from the earliest *Homo* at more than 2 million years ago with a brain size of around 650 cubic centimetres, to *Homo sapiens* with a brain size of around 1500 cubic centimetres and an EQ of 5.3. By showing such a pattern of continuous change, the fossil record is quite different from that of Europe, where the striking contrasts between the Neanderthals and modern humans indicate that the latter were new arrivals in the continent.

Prior to the Herto finds and the Omo redating, the earliest fossil specimens of *Homo sapiens* in Africa came from Klasies River Mouth, South Africa, dating to around 100,000 years ago.[4] It was known, however, that the origins of *Homo sapiens* must be much further back in time. Following the publication of a seminal *Nature* article in 1987, geneticists have been comparing the DNA of modern-day humans from different populations throughout the world to estimate the date at which our species first appeared. Because of the regular frequency at which genetic mutations arise, they can do this by measuring the extent of genetic difference between individuals from different populations.[5]

The technical details are challenging to a non-specialist, but this genetic approach to human evolution has had a revolutionary impact. It has demonstrated that all modern humans had a single origin in Africa and that *Homo neanderthalensis* was an evolutionary dead-end.[6] Different teams of researchers and different methods have provided a suite of different dates for the precise point of origin of *Homo sapiens*, although since 1987 these have been consistently older than the fossil specimens from Klasies River Mouth.

In 2000, a new genetic study was published in *Nature*, conducted by Max Ingman, from the University of Uppsala, and his colleagues. This used the most advanced techniques to compare the mitochondrial DNA of fifty-three individuals from diverse populations throughout the world, and concluded that they had most recently shared a common ancestor at 170,000 years ago (plus or minus 50,000 years).[7] Three years later the same journal announced the discovery of the Herto skulls, at close to the same date, and shortly afterwards, the redating of the Omo specimens.

The evolution of a language gene

As the capacity for language is a biological attribute of *Homo sapiens*, something embedded in the genome of our species, one should conclude that it had also evolved by 170,000 years ago. Another recent genetic study provides supporting evidence for this, although its results are more contentious than those concerning *Homo sapiens* itself.

The study concerned a large, multigenerational modern family, referred to as the KEs.[8] Some members of this family, who were studied during the 1990s, had problems with some very specific aspects of grammar, including the use of inflections for marking tense. They had difficulty, for instance, in judging that 'the boys played football yesterday' is grammatically correct whereas 'the boys play football yesterday' is not.

The afflicted members of the KE family also displayed difficulties with other tasks, notably in producing fine orofacial movements of the tongue and lips, which made some of their speech unintelligible. In view of their multiple problems, linguists have debated whether the KEs' language difficulties derive from a general cognitive deficit or from something specific to their language system. The latter seems most likely, as the non-language problems appear insufficiently severe and consistent to justify the claims that the KEs suffer from, say, low intelligence or a poor memory which has then impacted on their language abilities.

In 2001 geneticists identified the specific gene the malfunctioning of which had affected approximately half of the KE family.[9] It is known as the FOXP2 gene and appears to play a crucial role in turning other genes on and off, some of which are vital for the development of neural circuits for language in the brain. Surprisingly, the FOXP2 gene is not unique to humans; it is found in an almost identical form among a great many species. Indeed, there are only three differences between the seven hundred amino acids that form the FOXP2 gene in mice and in humans. Those three differences appear to be vital, because the KE family shows us that when the human version of the FOXP2 gene malfunctions, significant language deficits arise.

While the discovery of this gene and its significance for language was a striking event, an even more remarkable finding was made in the following year. Another team of geneticists, led by Wolfgang Enard of the Max Planck Institute for Evolutionary Anthropology in Leipzig, examined the chimpanzee, gorilla and monkey versions of the FOXP2 gene, and found that it was only two amino acids different from our own.[10] They proposed that the two amino acid changes that led to the *Homo sapiens* version of the gene were critical to both speech and language. Their estimate for when the modern human form of the gene became fixed in the population (by the process of natural selection following a random mutation) is, to quote their words from *Nature*, 'during the last 200,000 years of human history, that is, concomitant with or subsequent to the emergence of anatomically modern humans'.[11]

FOXP2 is not *the* gene for grammar, let alone language. There must be a great many genes involved in providing the capacity for language, many of which are likely to play multiple roles in the development of an individual. Nevertheless, the studies of the KE family and the FOXP2 gene provide compelling evidence for a genetic basis to human language, which may have been integral to the origin of *Homo sapiens* at around 200,000 years ago. Can we, then, find evidence from the archaeological record for language-mediated behaviour in Africa at this date – evidence that is so strikingly absent from the Neanderthals in Europe?

The origin of symbolic behaviour

The earliest known representational art from Africa consists of painted slabs from the Apollo 11 Cave in Namibia.[12] These bear images of both real and imaginary animals, and may have once been part of the cave wall or were perhaps always 'portable' slabs. Their age is contentious. Radiocarbon dating of the archaeological layer in which they were found gives a date no older than 28,000 years ago. But fragments of ostrich eggshell from the layer above the slabs have been dated to 59,000 years ago, which agrees better with the types of chipped-stone artefacts found with the slabs. So the radiocarbon dates may have been contaminated with more recent material.

Even if the eggshell dates are correct, this is hardly compatible with a date of 200,000 years ago for the origin of language – if we assume that the capacity for referential symbolism required for compositional language would have also been manifest in visual form. That capacity can certainly be pushed back to 70,000 years ago in the light of incised ochre fragments coming from Blombos Cave in South Africa. This cave is located at the southern tip of South Africa, 35 metres above sea level on a rocky coastline. Christopher Henshilwood of the South African Museum initiated excavations in 1991. Within a few years he had shown Blombos Cave to be the most important currently known archaeological site for understanding the origin of modern thought and behaviour – and, by implication, language.[13]

The two most striking specimens are small rectangular pieces of what Henshilwood describes as 'shale-like ochre'; both are just a few centimetres long and marked with similar cross-hatched designs. The patterns are quite different from those on the scratched pieces of bone from European sites

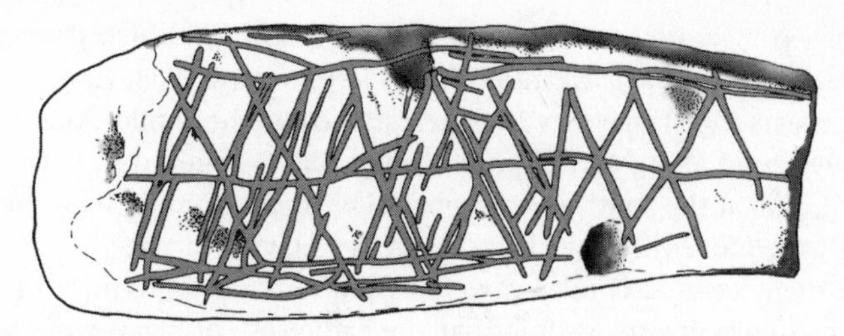

Figure 17 Incised slab of ochreous shale from Blombos Cave, dating to around 70,000 years ago. This is specimen SAM-AA 8938, measuring 75.8 millimetres long, 34.8 millimetres wide and 4.7 millimetres thick.

such as Bilzingsleben; they are sufficiently ordered to rule out any risk that they arose by chance. Moreover, as the same design is repeated on two separate artefacts, the impression is of a symbolic code.

The 70,000-year-old engravings from Blombos Cave are the earliest unequivocal symbols we have from the archaeological record, not only in Africa, but from the whole world. They were found with more than eight thousand other pieces of ochre, many of which bear signs of utilization and were probably used for body painting or for decorating items of organic material that have not been preserved. In addition, Henshilwood has found shell beads in Blombos Cave which are likely to have been attributed with symbolic meanings when worn 70,000 years ago.[14] But that date is still potentially more than one hundred thousand years after a likely origin for human language. So are there any other signs of early symbolism in Africa?

Before answering that, we must note that our knowledge of the archaeological record in Africa for the crucial period of 200,000 to 50,000 years ago is far inferior to that of Europe, where many hundreds of Neanderthal sites have been meticulously excavated. There are no more than a handful of equally well-explored sites throughout the whole of Africa. Consequently, while for Europe we can be confident that the Neanderthals did not manufacture material symbols, we must be more cautious for Early Humans in Africa; for that continent the archaeological aphorism must apply: 'the absence of evidence is not evidence of absence'. Indeed, crucial evidence may remain in the as yet unexcavated layers of Blombos Cave. Moreover, there is compelling circumstantial evidence that by 100,000 years ago, if not before, early *Homo sapiens* were making and using symbols.

This comes from the large quantities of red ochre found within archaeological deposits from at least 125,000 years ago. More than one hundred and eighty pieces have come from Klasies River Mouth Cave on the Cape Coast of South Africa.[15] The excavators have described some pieces as 'pencils', while others have clearly been scratched to remove pigment powder. Similar finds have come from Border Cave, again in South Africa, which also contains the burial of a *Homo sapiens* infant for which the most likely date is around 100,000 years ago. The bones had been stained by the application of ochre, and were found along with a pierced shell that is believed to have been a pendant. And at the South African site of Florisbad, grinding slabs that were probably used for ochre have been dated to 120,000 years ago.

Since there are no decorated cave walls or artefacts prior to those of Apollo 11 Cave, archaeologists assume that the earlier use of ochre was for body decoration – prehistoric cosmetics. Whether this should be defined as symbolic is contentious: red paint might have been used simply to draw attention to or to exaggerate selected parts of the body, such as lips, buttocks and breasts, and hence lacked a symbolic meaning in itself. Conversely, the body paintings themselves may have involved symbols, perhaps similar to those found on the two ochre fragments from Blombos Cave; and we must recall that those two unequivocally symbolic artefacts were found amid more than eight thousand pieces of ochre. Hence the absence of incised pieces from the ochre-rich Klasies River Mouth, Border Cave and other sites dating to 100,000 years ago and earlier might mean simply that they have been missed or else poorly preserved.[16]

I am more sympathetic to the idea that the red pigment used by early *Homo sapiens* was symbolic than that the black pigment used by Neanderthals in Europe was so, not only because of the vastly greater quantities in which it is found but also simply because of its colour – red. Red has been shown to have special significance for humans because of the range of physiological effects it induces, including changes in heart rate and brain activity, and because of the privileged position it bears in the colour systems of all societies. When brain lesions impair colour vision, red is the most resistant to loss and the quickest to recover.

The significance of red was both explored and explained in a brilliant short essay in 1976 by the evolutionary psychologist Nicholas Humphrey, entitled 'The Colour Currency of Nature'. Let me quote him:

> Red toadstools, red ladybirds, red poppies, are dangerous to eat, but red tomatoes, red strawberries, red apples, are good. The open mouth of an aggressive monkey is threatening, but the red bottom of a sexually receptive female is appealing. The flushed cheeks of a man or woman may indicate anger, but they may equally indicate pleasure. Thus the colour red, of itself, can do no more than alert the viewer, preparing him to receive a potentially important message; the content of the message can be interpreted only when the context of the redness is defined.[17]

Humphrey argues that the ambiguity of red as a signal colour is what makes it so disturbing, and he describes red as 'the colour currency of nature'. So it is perhaps not surprising that red ochre was selected as the first symbol to be used in human history.

How did language evolve?

Recall that 'Hmmmmm' stands for Holistic, manipulative, multi-modal, musical, and mimetic. This was the type of communication system used by the immediate ancestors of *Homo sapiens* in Africa, although in a form less highly evolved than among the Neanderthals in Europe. We need to understand how the 'Hmmmmm' used by the African *Homo sapiens* ancestors provided the evolutionary basis for language. To do so, we can start with the work of the linguist Alison Wray,[18] then elaborate her ideas in the light of my own conception of 'Hmmmmm', and then consider the findings of Simon Kirby, a linguist who has used computer models to explore the evolution of language.[19]

My 'Hmmmmm' theory draws heavily on Wray's ideas for the precursor for language, as she promoted the idea that pre-*Homo sapiens* utterances were holistic and manipulative rather than compositional and referential. Her views are diametrically opposed to those of Derek Bickerton, whose notion of proto-language as words without grammar was questioned in chapter 10. I have drawn on Wray's alternative notion when describing the communication systems of *Homo ergaster, Homo heidelbergensis* and *Homo neanderthalensis* in earlier chapters, and combined it with arguments about the significance of gesture and musicality to develop the 'Hmmmmm' theory. How then, did 'Hmmmmm' provide the evolutionary precursor to language?

Wray uses the term 'segmentation' to describe the process whereby humans began to break up holistic phrases into separate units, each of which had its own referential meaning and could then be recombined with units from other utterances to create an infinite array of new utterances. This is the emergence of compositionality, the feature that makes language so much more powerful than any other communication system.

Wray suggests that segmentation may have arisen from the recognition of chance associations between the phonetic segments of the holistic utterance and the objects or events to which they related. Once recognized, these associations might then have been used in a referential fashion to create new, compositional phrases. She provides the following hypothetical example: if, in the holistic, manipulative proto-language, there was a phrase *tebima* that meant 'give that to her', and another phrase *kumapi* that meant 'share this with her', an individual might recognize that *ma* was a common phonetic segment in both phrases, and 'her' a common aspect of their meaning. Hence that individual might conclude that *ma* could be used referentially for 'female person'. Note that in the holistic, manipulative proto-language, *ma* had no referential meaning at all, as neither *tebima* nor *kumapi* were compositional – they were arbitrary phonetic sequences that by chance had one sound in common.

Although linguists unsympathetic to Wray's approach, such as Derek Bickerton and Maggie Tallerman, have questioned the feasibility of segmentation, their criticisms have been insubstantial.[20] Tallerman claims, for instance, that the likelihood of any chance associations arising is remote, but computer simulations have shown that they could easily have happened – as will be described below – while Wray herself has explained how the process of segmentation may have been supported by 'near matches'. Tallerman also claims that holistic utterances are unlikely to have contained multiple phonetic strings with the potential for segmentation because they would have been too short. But one need do no more than consider the lengthy holistic phrases of monkeys, apes, birds, and so forth, to appreciate that those of our human ancestors may have been of considerable length, having evolved over millennia and proliferated in number to provide ever greater semantic specificity. Similarly, an argument that Wray's theory of segmentation has to assume the prior existence of discrete segments, which then invalidates her conception of holistic utterances, also has no foundation. Holistic utterances may have been multi-syllabic but they were – by definition – holistic, with none of the individual or groups of syllables mapping onto a discrete aspect of the utterance's complex meaning.

Such criticisms are further invalidated, and the feasibility of Wray's process of segmentation enhanced, when her own characterization of holistic proto-language is replaced by the rather more complex and sophisticated perspective I have developed in the form of 'Hmmmmm'.

The presence of onomatopoeia, vocal imitation and sound synaesthesia would have created non-arbitrary associations between phonetic segments of holistic utterances and certain entities in the world, notably species of animals with distinctive calls, environmental features with distinctive sounds, and bodily responses – in a previous chapter I gave the example of 'Yuk'. These non-arbitrary associations would have significantly increased the likelihood that the particular phonetic segments would eventually come to refer to the relevant entities and hence to exist as words, as compared with the entirely chance associations that Wray uses in her own examples. The likelihood would have been further increased by the use of gesture and body language, especially if a phonetic segment of an utterance regularly occurred in combination with a gesture pointing to some entity in the world. Once some words had emerged, others would have followed more readily, by means of the segmentation process that Wray describes.

The musicality of 'Hmmmmm' would also have facilitated this process, because pitch and rhythm would have emphasized particular phonetic segments and thus increased the likelihood that they would become perceived as discrete entities with their own meanings. In chapter 6, I explained that this was the case with regard to language acquisition by infants: the exaggerated prosody of IDS helps infants to split up the sound stream they hear so that individual words and phrases can be identified. In a similar way, the evolutionary process of segmentation would have been further enhanced by the repetition of phrases within 'Hmmmmm', just as words are repeated to infants and refrains are repeated within music today to achieve emotional and other effects. The musicality of 'Hmmmmm' would, moreover, have also ensured that holistic utterances were of sufficient length, so that the process of segmentation would have some raw material to work with. And such utterances would certainly have been constituted of segments without any presupposition of their comprising words. They would have derived from the oral gestures that I described in chapter 8: the physical movements of the lips, tongue tip, tongue body, tongue root, velum and larynx that create the phonetic segments of any utterance – holistic or compositional.

Further confidence in the process of segmentation derives from the use of computer models to simulate the evolution of language.

Simulating the evolution of language

Simon Kirby of Edinburgh University is one of several linguists who have begun to explore the evolution of language using computer simulation models.[21] He creates a population of agents – simulated people – within his program, which communicate with each other by using strings of symbols for language. His simulated population is multigenerational, with each new 'learning-agent' acquiring its language from the already existing 'speaking-agents'. In this way, his simulations are able to explore one of the most remarkable features of human language: how children acquire language simply by listening to their parents, siblings and other language-users.

Kirby begins his simulations by giving each speaking-agent a 'random language' – that is to say, they generate random strings of symbols for each meaning with which they are prompted. This is therefore a holistic language, because there is nothing in the structure of each string that corresponds to the specific meaning being conveyed. As the simulation runs, learning-agents are exposed to a sample of speaking-agents and by this means acquire a language of their own. Because they will only ever have heard a sample of the utterances of any single speaking-agent, their language will be unlike that of any other individual. Accordingly, Kirby describes the language system as unstable: it is highly variable within and between generations.

As the simulations proceed, Kirby finds that some parts of the language systems become stabilized and are passed on faithfully from one generation to the next. This occurs because Kirby is effectively simulating the process that Wray speculates may have happened during human evolution: a learning-agent mistakenly infers some form of non-random behaviour in a speaking-agent indicating a recurrent association between a symbol string and a meaning, and then uses this association to produce its own utterances, which are now genuinely non-random. Kirby refers to this process as 'generalization'. Other learning-agents will acquire the same association between the symbol string and its meaning, so that it spreads throughout the population, creating what Kirby calls 'pockets of stability'. Eventually, the whole language system will have been stabilized and will constitute a single, compositional language.

Through this work Kirby is turning Noam Chomsky's argument about the poverty of the stimulus on its head. Chomsky believed that it was impossible for children to learn language simply by listening to their parents and siblings; he thought the task too complicated for the time available between birth and becoming linguistically adept at around the age of three. As a consequence, he claimed that infants were born with a great deal of their language abilities already present, in the form of 'Universal Grammar'. But Kirby's simulations show that language learning may not be as challenging as Chomsky believed. Or rather, that the process of learning itself can lead to the emergence of grammatical structures. Hence, if there is such a thing as 'Universal Grammar', it may be a product of cultural transmission through a 'learning bottleneck' between generations, rather than of natural selection during biological evolution; 'poverty of the stimulus' becomes a creative force rather than a constraint on language acquisition.

Kirby's work is, of course, no more than a remarkably simplified version of real life. The unreal starting conditions – an entirely random and different language for each speaker – and unreal process of language acquisition – including the absence of a computer equivalent of infant-directed speech (IDS) – might be having an undue influence on the results. But such inevitable simplifications are most probably constraining rather than increasing the speed at which compositionality arises. Moreover, if Kirby was starting with 'Hmmmmm' rather than a random type of language, some form of grammatical structure would have already been present, most notably recursion.

As I noted in chapter 2, Chomsky and his colleagues argued that recursion is the most important element of linguistic grammar, the only element that is entirely absent from all animal communication systems.[22] Recursion is the manner in which a linguistic phrase can be embedded within itself, as in the case of a clause within a clause. It plays a key role in enabling the generation of an infinite range of expressions from a finite suite of elements. Recursion would, however, have already been present in 'Hmmmmm' – at least it would if my arguments in the previous chapters are accepted. Hence as segmentation of 'Hmmmmm' occurred, rules for recombining the newly found words were already present, and could be exploited to complete the shift to a compositional language employing grammatical rules. A further key development might have been the 'discovery' of how to combine the emergent words by the use of conjunctions such as 'but', 'and', 'or'.

Why did language only evolve after 200,000 years ago?

Together, Wray and Kirby have helped us to understand how compositional language evolved from holistic phrases. However, they have also posed us with an unexpected problem: why did this only happen in Africa after 200,000 years ago? If, as I have argued, *Homo ergaster* had a form of 'Hmmmmm' communication by 1.8 million years ago, as did *Homo heidelbergensis* at 0.5 million years ago, and *Homo neanderthalensis* at 250,000 years ago, then why did these remain intact rather than becoming segmented?

There are two possibilities, one relating to social life and one to human biology. As regards the first, we should note initially that Kirby found holistic languages remain stable in those simulations in which learning-agents hear so much of the speaking-agents' utterances that they learn every single association between symbol string and meaning. In other words, there is no learning bottleneck for language to pass through, and hence no need for generalization. It may be the case, therefore, that the social arrangements of *Homo* populations other than *Homo sapiens* were such that infants had intense and continuous exposure to a limited number of 'Hmmmmm' speakers, resulting in the acquisition of the entire 'Hmmmmm' suite of utterances, with no need for generalization.

This would indeed have been quite likely in the type of hominid and Early Human communities I have outlined in previous chapters, whether comprised of *Homo habilis* or Neanderthals. They lived in socially intimate groups with limited, if any, need for the type of novel utterances that could only be produced by compositional language, as opposed to 'Hmmmmm'. Moreoever, there would have been little need and few opportunities to communicate with members of other groups than one's own – although some contacts involving the movement of individuals would have been essential to maintain demographic and genetic viability. But little need have been said on such occasions.

It may have been only within the earliest *Homo sapiens* communities in Africa that people began to adopt specialized economic roles and social positions, that trade and exchange with other communities began, and that 'talking with strangers' became an important and pervasive aspect of social life. Such developments would have created pressures to exchange far greater amounts of information than was previously necessary in the socially intimate, undifferentiated groups of Early Humans. Only then would there have been the need for generalization, in the manner that Kirby describes within his simulation, and the need continually to generate novel utterances at a rate and of a type beyond the capability of 'Hmmmmm'.[23]

The archaeological record from Africa relating to the appearance of *Homo sapiens* certainly suggests that such social developments did occur.[24] The dilemma, of course, is whether we are dealing with cause or effect: one might argue that the development of economic specialization and exchange relations between groups was a consequence of compositional language, which enabled the necessary communication to be undertaken. My guess is that we are dealing with strong feedback between the two – they 'bootstrapped' each other to create rapid changes both in society and in communication.

The kick-start for such developments may have been a chance genetic mutation – the second possible reason why segmentation of 'Hmmmmm' only occurred in Africa with the advent of modern humans. This may have provided a new ability to identify phonetic segments in holistic utterances. Here we should recall the work of Jenny Saffran, the developmental psychologist who has shown that infants can extract statistical regularities within the continuous sound streams that they hear and hence learn about the existence of discrete words, as I described in chapter 6. What if that capacity for statistical learning, or some other related ability, was absent in species of *Homo* other than *Homo sapiens*? Well, they would not have been able to identify the phonetic segments within a holistic utterance, and thus would have been unable to take the first step in evolving a compositional language.

We have already seen that some aspects of language are dependent on the possession of the specific gene FOXP2, the modern human version of which seems to have appeared in Africa at soon after 200,000 years ago. Perhaps the process of segmentation was dependent upon this gene in some manner that has yet to be discovered. Indeed, it may be significant that those members of the KE family that were afflicted with a faulty version of the FOXP2 gene had difficulties not only with grammar but also with understanding complex sentences and with judging whether a sequence such as 'blonterstaping' is a real word.[25] These difficulties seem to reflect a problem with the segmentation of what sound to them like holistic utterances. So perhaps it was only with the chance mutation of the FOXP2 gene to create the modern human version that segmentation became possible. Or there may have been other genetic mutations at a similar date that enabled the transition from holistic phrases to compositional language, perhaps through the appearance of a general-purpose statistical learning ability.

Language: from supplementing to dominating 'Hmmmmm'

If we accept the above arguments for how compositional language evolved from holistic phrases, then the initial utterances may have suffered from the same weakness as the Bickertonian proto-language utterances: their meanings might have been highly ambiguous, if not entirely unintelligible, to other members of the community. But a key difference from Bickerton's theories is that the compositional utterances that emerged from holistic phrases by a process of segmentation would have begun as mere supplements to the 'Hmmmmm' communication system. Indeed, the transition from 'Hmmmmm' to compositional language would have taken many millennia, the holistic utterances providing a cultural scaffold for the gradual adoption of words and new utterances structured by grammatical rules. Moreover, the first words may initially have been of primary significance to the speaker as a means to facilitate their own thought and planning, rather than as a means of communication, which could have continued relying on the existing 'Hmmmmm' system.

Talking to oneself is something that we all occasionally do, especially when we are trying to undertake a complex task. Children do this more than adults, and their so-called 'private speech' has been recognized as an essential part of cognitive development. Studies in the 1990s by Laura Berk, professor of psychology at Illinois State University, demonstrated how talking oneself through a problem, especially one that involves intricate physical movements, helps to accomplish it successfully.[26] She found that children who provide their own verbal commentary when learning to tie shoelaces or solve mathematical problems learn those tasks quicker than those who remain silent. Although it never goes away entirely, self-directed speech gradually disappears as children mature, becoming replaced with the silent inner speech that we all use – thinking rather than saying words.

The basis of private speech's cognitive impact is an issue we will explore below, when the relationship between language and thought is considered. Here, we need simply note that talking to oneself can have advantageous cognitive results, and I suspect that this was especially the case in the formative days of language. Although the first segmented linguistic phrases may have been as ambiguous as the utterances of a Bickertonian proto-language, and hence of limited value for transmitting information between individuals, they must have had a significant beneficial impact on the thought of the individuals speaking them. I suspect that those who talked to themselves gained a significant advantage in problem-solving and planning. Private speech may have been crucial in the development of a compositional language to sufficiently complex a state for it to become a meaningful vehicle for information exchange.

Compositional language would then have become a supplement to 'Hmmmmm', and eventually the dominant form of communication owing to its greater effectiveness at transmitting information. With it, an infinite number of things could be said, in contrast to the limited number of phrases that 'Hmmmmm' allowed. The brains of infants and children would have developed in a new fashion, one consequence of which would have been the loss of perfect pitch in the majority of individuals and a diminution of musical abilities.

Once the process of segmentation had begun, we should expect a rapid evolution of grammatical rules, building on those that would have been inherited from 'Hmmmmm'. Such rules would have evolved by the process of cultural transmission in the manner that Kirby describes, and perhaps through natural selection leading to the appearance of genetically based neural networks enabling more complex grammatical constructions. This 'words before grammar' is the type of language evolution that Bickerton proposed – so we can see that his views are not necessarily wrong, but are simply chronologically misplaced and require a pre-existing history of holistic proto-language to be feasible. What no one will dispute is that the emergence of compositional language had a profound impact on human thought and behaviour. And so we must now return to the archaeological record of Africa.

Behavioural developments in Africa

Sally McBrearty and Alison Brooks are two leading specialists in African archaeology, based at the University of Connecticut and George Washington University respectively. In 2000 they published a comprehensive review of African archaeology dating to between 200,000 and 40,000 years ago, which had the academic distinction of appearing as a complete issue of the *Journal of Human Evolution*.[27] Their aim was to demonstrate that the archaeological record provides conclusive evidence for the origin of modern human behaviour, and by implication language, in Africa prior to 100,000 years ago.

This is certainly the case. The African record suggests a gradual change in human behaviour that is entirely compatible with a slow replacement of 'Hmmmmm' by compositional language. The key indicators are similar to those that appear in Europe at around 40,000 years ago, when *Homo sapiens* begins to overtake and replace the Neanderthals. In the latter case, the indicators of modern, linguistically mediated behaviour appear suddenly and as a single package: visual symbols, bone tools, body decoration, ritualized burial, intensified hunting practices, long-distance exchange, and structures at camping sites. The same suite of behaviours appears in Africa, but in a piecemeal and gradual fashion beginning at around 200,000 years ago.

Blombos Cave is one of the key sites that document such developments. We have already noted the appearance at this site of pigment and symbols dating to 70,000 years ago. At a contemporary date, it provides some of the earliest bone points ever discovered, showing the use of new tools for new activities. Blombos Cave also provides particularly fine stone 'Stillbay' points, which are restricted to the Cape region. These were particularly difficult to manufacture and were most likely imbued with social if not symbolic significance, in the same manner as the shell beads.

Another key site showing behavioural developments is Klasies River Mouth. Here, between 80,000 and 72,000 years ago, we see the appearance of bone artefacts, fine stone points, and delicately chipped blades of flint, known to archaeologists as microliths.[28] Strangely, however, such tools seem then to disappear from the archaeological record, with a return to more traditional plain stone flakes for another twenty thousand years, before permanent change occurs at around 50,000 years ago.[29]

This impression of temporary 'flashes' of more complex, linguistically mediated behaviour, in the middle of an archaeological record that is otherwise similar to that for the Neanderthals in its immense stability over long stretches of time, is also found elsewhere in Africa. For instance, bone harpoons have been found at Katanda in the Democratic Republic of Congo, dating to 90,000 years ago, and then nothing similar is known for another sixty thousand years.[30] Deliberately stone-lined hearths are known from Mumbwa Cave in Zambia, dating to at least 100,000 years ago.[31] These are precisely the type of hearths, so strikingly absent from Neanderthal Europe, around which hunter-gatherers would have sat talking to each other. But other than at Mumbwa and at Klasies River Mouth, such hearths are effectively quite unknown in Africa.

In summary, amid a continuation of tool-making traditions that stretch back at least two hundred and fifty thousand years, there are sporadic traces of new behaviour in Africa of the type that archaeologists associate with modern, language-using humans. The transition from a predominantly 'Hmmmmm' communication system to compositional language most likely took tens of thousands of years. Some communities may have continued primarily with 'Hmmmmm' for much longer than others; some individuals who had become proficient language-users may have died before their knowledge was passed on, while overall population sizes may have been small, with people living in widely scattered groups, which would have inhibited the cultural transmission of new ideas.[32] There is also, of course, the relative lack of archaeological exploration in many parts of Africa; vast areas effectively remain virgin territory for archaeologists when compared with the intensively explored river valleys and caves of Europe.

Passing a demographic threshold

It was not until after 50,000 years ago that many of the new behaviours became permanent features of the human repertoire. This date was once taken to be when language and modern behaviour first appeared. That was before the African archaeological evidence had become well known, before the genetic studies and fossil discoveries confirmed the appearance of modern humans by 195,000 years ago, and before the significance of the FOXP2 gene for language had been revealed. But the date of around 50,000 years ago nevertheless marks a striking change in the archaeological record.

This is now explained by the passing of a demographic threshold after *Homo sapiens* had become entirely dependent upon compositional language for communication. With sufficiently dense populations, cultural transmission would have 'fixed' new behaviours in communities so that they would no longer be lost when particular individuals died or local groups became extinct.

The significance of the 50,000–40,000 years ago date, and the likelihood of the passing of a demographic threshold, have been confirmed by a genetic study – another example of how the evidence from genetics and that from archaeology show ever-increasing convergence. In fact, this evidence comes from the same study by Max Ingman and his colleagues that proposed the genetic origin of *Homo sapiens* at around 170,000 years ago, as noted above.[33] They found a striking difference between the DNA of African people on one hand and of all non-Africans on the other, which could only have arisen from a split between the two after 50,000 years ago followed by rapid population growth in the latter group.

This date appears, therefore, to mark not only a demographic and cultural threshold in Africa but also the spread of modern humans into Europe and Asia. That was not the first dispersal of *Homo sapiens* out of Africa, as one had already occurred at around 100,000 years ago.[34] This is evident from skeletal and archaeological remains found in the caves at Skhul and Qafzeh in Israel, which include burials, beads and pigment. But both the genetic and the fossil evidence indicate that this early 'out of Africa' *Homo sapiens* population did not contribute to the modern gene pool. The Skhul and Qafzeh evidence is striking owing to the clear presence of symbolic behaviour alongside the use of stone technology very similar to that of the Neanderthals, who occupied the same region both before and after this modern human presence. It is this similarity in technology that leads Robert Foley and Marta Lahr to believe that the Neanderthals and modern humans had a relatively recent common ancestor, *Homo helmei*, at 300,000 years ago. For me, the significance of the similarity is that it indicates that the modern humans of Qafzeh and Skhul were still using 'Hmmmmm' as much as compositional language, and had not attained full cognitive fluidity.

Whether that was also the case for the earliest *Homo sapiens* in Australia, which might date to 60,000 years ago, is unclear. What is quite evident, however, is that after 50,000 years ago there was a major dispersal of *Homo sapiens* from Africa which gave rise to modern communities in all corners of the globe.

This great dispersal event was responsible for the demise and disappearance of the Neanderthals, and eventually of *Homo floresiensis* in Indonesia at 12,000 years ago. It is unlikely that the Neanderthals were actually hunted down by the incoming *Homo sapiens*; they were simply unable to compete for resources within the European landscapes owing to the superior hunting, gathering and overall survival abilities of modern humans. The reason for those superior abilities is that compositional language not only provided a superior means of information exchange compared with 'Hmmmmm' – an unlimited range of utterances – but also enabled new types of thinking. And so, to conclude this chapter, we must consider the cognitive impact of language.

The origin of cognitive fluidity

Here we can return to the central thesis of *The Prehistory of the Mind*.[35] I argued that the mind of *Homo sapiens* was fundamentally different from that of all other hominids, whether direct ancestors like *Homo helmei* or relatives like the Neanderthals. They possessed the 'domain-specific' mentality described above, which was highly adaptive because each type of 'intelligence' had been moulded by natural selection to provide ways of thinking and types of knowledge suited for solving problems within its relevant behavioural domain. I argued that there were three key intelligences: social intelligence for managing the complex social world in which hominids lived; natural history intelligence for understanding animals and plants, the weather and the seasons, and any other aspects of the natural world that were essential for a hunter-gatherer lifestyle; and a technical intelligence which enabled the complex manipulation of artefacts and especially the production of stone tools.

As I have explained, the *Homo sapiens* mind is also based on multiple intelligences, but has one additional feature: cognitive fluidity. This term describes the capacity to integrate ways of thinking and stores of knowledge from separate intelligences so as to create types of thoughts that could never have existed within a domain-specific mind. For instance, a Neanderthal or pre-*sapiens Homo* in Africa could not have taken their knowledge of a lion (from natural history intelligence) and combined it with knowledge about human minds (from social intelligence) to create the idea of a lion-like being that had human-type thoughts – a type of anthropomorphic thinking that is pervasive in all modern minds.

So how did cognitive fluidity emerge? In *The Prehistory of the Mind*, I argued that cognitive fluidity was a consequence of language: spoken and imaginary utterances acted as the conduits for ideas and information to flow from one separate intelligence to another. I was, and remain, utterly convinced that this idea is correct. But I was unable to make a sophisticated argument to that effect, lacking the necessary expertise in the philosophy of mind. Fortunately, I now have the arguments from Peter Carruthers, a distinguished professor of philosophy at the University of Maryland, to explain exactly how language did indeed bring about this change in the nature of human mentality.

While I was interpreting the archaeological evidence to conclude that language creates cognitive fluidity, Peter Carruthers was arriving at exactly the same conclusion on the basis of philosophical argument informed by linguistics, neuroscience and psychology. In his most recent contribution, a major article entitled 'The cognitive functions of language' published in the prestigious journal *Behavioral and Brain Sciences* in 2002, he proposed how language has this effect.[36] Although his terminology is slightly different from that which I use, referring to cognitive modules rather than intelligences, and inter-modular integration rather than cognitive fluidity, Carruthers argues that the 'imagined sentences' we create in our minds allow the outputs from one module/intelligence to be combined with those from one or more others, and thereby create new types of conscious thoughts. He suggests that cognitively fluid thought occurs in the unconscious through a type of linguistic representation in the mind described as 'logical form' (LF). This term was originally introduced by Noam Chomsky to describe the interface between language and cognition; some linguistic representations remain as LF, others become imagined sentences within our minds.

In an attempt to refine his explanation, Carruthers places considerable emphasis on syntax, an essential part of compositional language. Syntax allows for the multiple embedding of adjectives and phrases, the phenomenon of recursion that has frequently been referred to in this book. And so, Carruthers argues, syntax would allow one imaginary sentence generated by one type of cognitive module/intelligence to be embedded into that of another imaginary sentence coming from a different module/intelligence.

Thus a single imaginary sentence would be created, which would generate an inter-modular, or cognitively fluid, thought – one that could not have existed without compositional language.

So via segmentation, compositional language emerged from 'Hmmmmm', which in turn changed the nature of human thought and set our species on a path that led to global colonization and, ultimately, the end of the hunting and gathering way of life that had endured ever since the first species of *Homo* appeared more than 2 million years ago. Almost as soon as the last ice age finished at 10,000 years ago, agriculture was invented at several localities around the globe, which led to the first towns and the early civilizations – a remarkable history of change that I recounted in my book *After the Ice*. But we still have one question to address: what happened to 'Hmmmmm'?

17 A mystery explained, but not diminished

Modern human dispersal, communicating with the gods, and the remnants of 'Hmmmmm'

Schubert's String Quintet in C Major: an exploration of uncertain memories about a shared ancestral past

Music emerged from the remnants of 'Hmmmmm' after language evolved. Compositional, referential language took over the role of information exchange so completely that 'Hmmmmm' became a communication system almost entirely concerned with the expression of emotion and the forging of group identities, tasks at which language is relatively ineffective. Indeed, having been relieved of the need to transmit and manipulate information, 'Hmmmmm' could specialize in these roles and was free to evolve into the communication system that we now call music. As the language-using modern humans were able to invent complex instruments, the capabilities of the human body became extended and elaborated, providing a host of new possibilities for musical sound. With the emergence of religious belief, music became the principal means of communicating with the gods. And what I have learned from writing this book is that throughout history we have been using music to explore our evolutionary past – the lost world of 'Hmmmmm' communication – whether this is Franz Schubert writing his compositions, Miles Davis improvising or children clapping in the playground.

Modern human dispersals and the archaeology of music

The modern humans that dispersed throughout the Old World soon after 50,000 years ago, and who entered the New World at some time after 20,000 years ago, had language, music and cognitive fluidity. In these respects they would have contrasted with those first *Homo sapiens* who ventured into the Near East at around 100,000 years ago and are represented in the caves of Skhul and Qafzeh. Those 'modern humans' were still using a communication system partially dependent upon 'Hmmmmm', one in which language and music had not become fully differentiated and in which only limited cognitive fluidity had been achieved. Hence some of their thought and behaviour was similar to that of the fully 'Hmmmmm'-ing and cognitively domain-specific Neanderthals, as is evident from the stone tools they made.

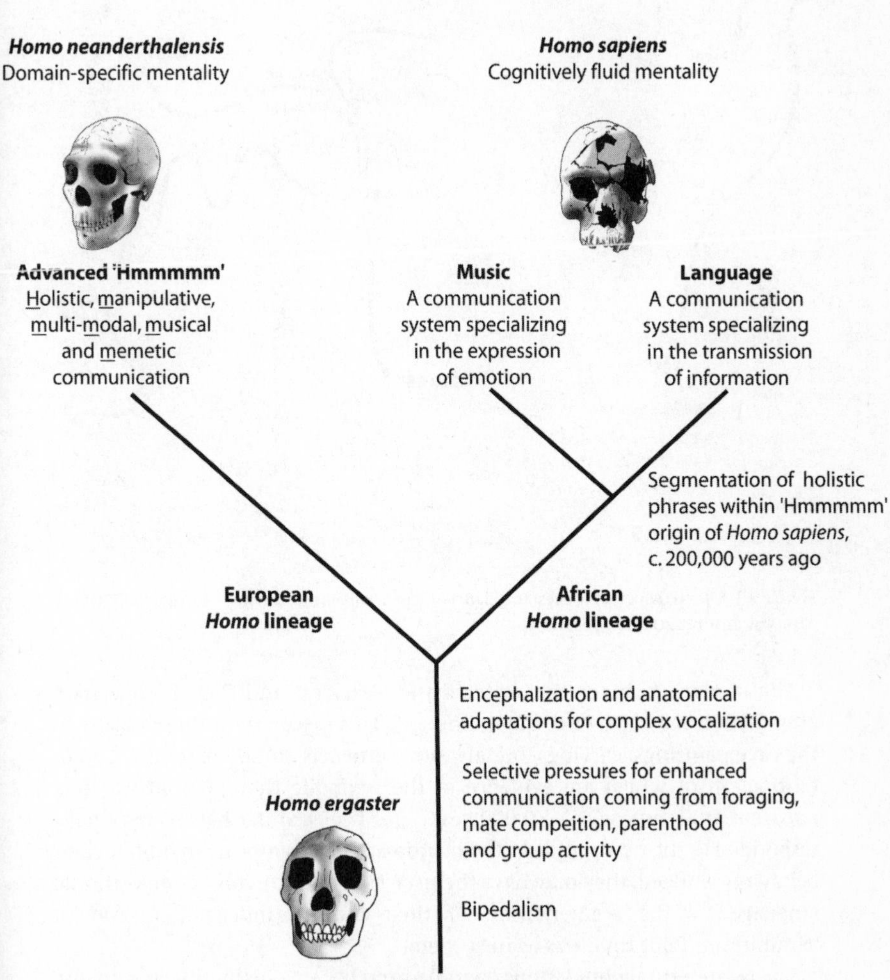

Figure 18 The evolution of music and language.

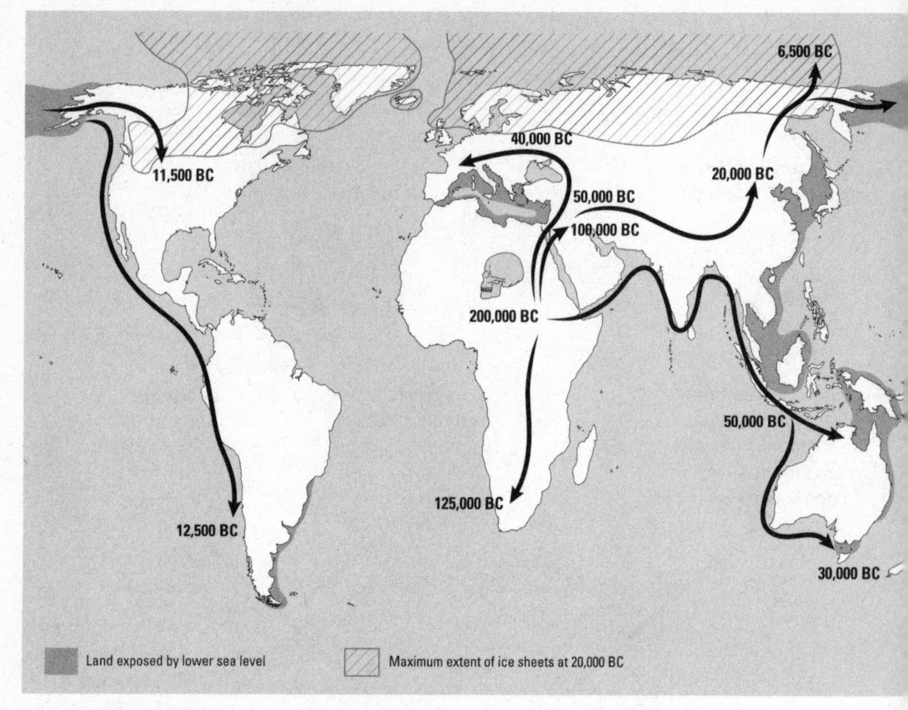

Figure 19 Likely dispersal routes and dates of *Homo sapiens* following its origin in East Africa at around 200,000 years ago.

The modern humans that entered the Near East and then Europe soon after 50,000 years ago were quite different. This is seen most dramatically in the cave paintings, carvings, burials, stone artefacts and bone tools found in Europe, all of which are evidence of the symbolic thought that was first apparent at Blombos Cave 70,000 years ago. The Neanderthals seem to have responded to the new people in their landscape by attempting to imitate their behaviour. Indeed, they may have themselves begun to evolve compositional language by the segmentation of their own Neanderthal version of 'Hmmmmm'. But time was against them.

The Neanderthal populations dwindled and became susceptible to climatic fluctuations, which were often severe during the latter stages of the ice age. When the Neanderthals became extinct at 28,000 years ago, and *Homo floresiensis* at 12,000 years ago, so did the 'Hmmmmm' communication system that had sustained early hominid and Early Human communities for more than two million years.

One of the most striking features of the modern human archaeological record in Europe is the presence of musical instruments. While I am sympathetic to the idea that the Neanderthals and other types of Early Humans may have used natural or minimally modified materials to make sounds – banging sticks together or on the ground, blowing though reeds, swirling strings of animal or plant fibres in the air – I do not think they could have manufactured instruments. To do so requires a cognitively fluid mind, which is precisely what enabled the occupants of Geissenklösterle to make the earliest known instruments in the world – bone flutes dating to 36,000 years ago.[1]

Geissenklösterle is a cave in southern Germany and one of the earliest known sites of modern humans in Europe. Although fragmentary, its flutes are entirely different to the claimed Neanderthal flute from Divje Babe. They were made from the wing bones of large birds, identified as a swan in the case of the best-preserved specimen, and each has at least three well-formed finger holes, arranged with a larger gap between holes two and three than between holes one and two. The Geissenklösterle flutes have linear incisions on the surface, which, one assumes, were either ownership marks or else carried symbolic meanings.

A much larger sample of bird-bone instruments has come from the cave of Isturitz in the foothills of the Pyrenees, which has also produced the greatest number of art objects of any ice-age site in Europe. More than twenty specimens of pipes have been found since the original discovery in 1921. The oldest may date back to 35,000 years ago and, like the flutes at Geissenklösterle, they display sophisticated workmanship.

Francesco D'Errico and Graham Lawson have recently subjected several of the specimens that date to between 27,000 and 20,000 years ago to microscopic and acoustic examination. Although they may originally have been made several thousand years apart, these specimens are strikingly similar, having been made from the wing bones of vultures, with four finger holes arranged in two pairs. The mouthpieces lacked a hard edge to blow against and so must have been inserted into the mouth and blown, probably with the insertion of a reed; they were pipes rather than flutes. The finger holes had been very carefully worked to lie within a shallow concave surface, which ensured a complete seal, and were aligned at a slightly oblique angle to the pipe's axis. These features are common in historically documented pipe traditions and are usually associated with blowing the pipe from the

Figure 20 Bone pipe (*c.* 120 mm) from Geissenklösterle, southern Germany, dating to around 36,000 years ago and currently the oldest known musical instrument in the world.

corner rather than centre of the mouth, as this helps the lips to create a secure pressure seal.

We should assume that the ice-age modern humans who made these pipes also made a variety of other instruments. We certainly know that they made use of natural formations in some of their caves, as stalagmites have been found marked with paint and showing clear traces of having been struck.[2] In some caves, paintings appear to have been made in chambers that have very marked acoustic properties. So whereas we now visit painted caves in a hushed reverence, they probably once reverberated with the sounds of pipes, stalagmite xylophones, singing and dancing.

Technology, expertise and elitism

The bone pipes of Geissenklösterle and Isturitz were produced by the use of stone tools. As technology advanced through the ages, new types of instruments became possible. All of these were made to extend the musical abilities of the human voice and body with material culture, reflecting both an evolved passion for music-making that originated from 'Hmmmmm' and the cognitive fluidity that is required to collapse the barriers between the human and technical worlds.

The ethnomusicologist Bruno Nettl reflected on the impact of technological advance in one of his 1983 essays:

> the invention of pottery adds kettle drums and clay percussion instruments, and the development of metal work makes possible the construction of brass gongs. Modern industrial technology provides the opportunity for highly standardized instruments. Beyond instruments, development of complex architecture gives us theater and concert halls, electronics, the possibility of precise reproduction and unchanged repetition of performances. And so on to hundreds of other examples.[3]

Today we must, of course, add the impact of the Internet to the manner in which technology continues to change the nature and dissemination of music.

Such technological developments have served both to democratize the availability of music and to create a musical elite. They have created a world in which music is all-pervasive and all around us – the world to which I referred in chapter 2, in which we hear music at airports, when shopping, when our mobile phone rings, and whenever we turn on the radio. But, as John Blacking explained, technological development also leads to musical complexity and then exclusion.[4] When the technical level of what is defined as musicality is raised, some people will become branded as unmusical, and the very nature of music will become defined to serve the needs of an emergent musical elite. In the West, this elite has been formed by those who have the resources to learn complex instruments such as the piano and violin, and to attend classical concerts, and who have mistakenly come to believe that this type of activity *is* music, rather than just one type of musical activity. The 'appalling popularity of music' may not be to their liking. That phrase was a chapter title in a 1934 book by the composer Constant Lambert, in which he (in)famously wrote that 'the present age is one of overproduction . . . never has there been so much music-making and so little musical experience of a vital order'.[5]

John Blacking's lifetime of ethnomusicological study led him to despise such elitist conceptions of music, which set arbitrary standards of musical competence and inhibited popular participation in music-making. The world would be a better place, he thought, if we were all able to communicate unselfconsciously through music.[6] I agree – bring back 'Hmmmmm'!

The religious and other functions of music

Although I lack any evidence and doubt if any could be found, I am confident that the music played through the Geissenklösterle pipes and sung within ice-age painted caves had a religious function. As I noted in chapter 2, the use of music to communicate with the supernatural appears to be a universal feature of all historically documented and present-day human societies.[7] Why should this be the case?

To answer this, we must first appreciate that ideas about supernatural beings are a 'natural' consequence of the cognitive fluidity that language delivered to the human mind, as I argued at length in *The Prehistory of the Mind*. By combining, say, what one knows about people with what one knows about animals one can imagine an entity that is part human and part animal – such as the lion/man figurine from Hohlenstein Stadel in Germany dating to 33,000 years ago. Similarly, by combining what one knows about people with what one knows about physical objects, one can create ideas of beings that can live for ever (like rocks), walk on water (like a floating branch), or are invisible (like the air we breathe).

Ideas about supernatural beings are the essence of religion.[8] But when such beings cannot be seen, except perhaps during hallucinations, how should one communicate with them? The problem is not only that they can be neither seen nor heard, but that they remain as vaguely defined entities because they do not relate to any evolved part of the human mind, such as the categories of 'human', 'animal', or even simply 'living being'. On the other hand, supernatural beings are usually believed to have mind-reading powers – they know what we and everyone else think; they have been described by the anthropologist Pascal Boyer as 'complete strategic agents'.[9] Accordingly, there is no need to communicate with them by language, the primary role of which is the transmission of information. Music, however, is ideal.

After language had evolved as the principal communication system of modern humans, people were left with the question of who to communicate with through music. Music is, after all, a derivative of 'Hmmmmm', which itself evolved as a means of communication, so the communicative function could not easily be dropped; there remained a compulsion among modern humans to communicate with music, as there still is today. How could this be fulfilled? Communication with other humans was now far better achieved by language than by music, other than for prelinguistic infants. But in the minds of modern humans there was now another type of entity with whom they could and should communicate: supernatural beings. So the human propensity to communicate through music came to focus on the supernatural – whether by beating the shaman's drum or through the compositions of Bach.

Music also plays a secondary role in relation to religion. Ideas about supernatural beings are unnatural in the sense that they conflict with our deeply evolved, domain-specific understanding of the world.[10] As a consequence they are difficult to hold within our minds and to transmit to others – try, for instance, explaining to someone the concept of the Holy Trinity, or try understanding it when someone explains it to you. Matthew Day, a professor of religious studies, has recently written, 'one of the bedevilling problems about dealing with gods is that . . . they are never really *there*';[11] hence we have difficulty in knowing not only how to communicate with them, but also how to think about them.

In my previous work, I have argued that modern humans compensate for this by the use of material symbols that provide 'cognitive anchors'.[12] Whether supernatural beings are made tangible in a representational manner, as we suppose is the case with the lion/man from Hohlenstein Stadel, or in abstract form as in the Christian cross, such material symbols function to help conceptualize and share the religious entities and ideas in which one believes. Such objects constitute an extension of the human mind, just as musical instruments extend the human vocal tract and body.

My previous arguments regarding this issue focused on material objects alone. But music can also be a cognitive anchor and an extension of the mind. Just as a community can share a non-discursive religious idea by all wearing the same material symbol, they can do so by all making the same music. In this manner, music plays an active role in maintaining and manipulating ideas about supernatural beings in the human mind; without such support from music and material symbols, such ideas would simply be too difficult to think about and to share, other than for a fleeting instance. Matthew Day agrees: 'the broad spectrum of rituals, music, relics, scriptures, statues and buildings typically associated with religious traditions are no longer seen as mere ethnographic icing on the cognitive cake. Rather than thin cultural "wrap arounds" that dress-up the real cognitive processes going on underneath, they begin to look like central components of the relevant machinery of religious thought.'[13]

Entertainment and the importance of song

Communication with supernatural beings and enabling religious thought are not, of course, the only functions of music; neither of these can be isolated from music's other roles since religious activities frequently also serve as occasions for social bonding, emotional expression and healing, all of which are supported by music. Indeed, music has evolved to meet many diverse functions and to show immense cultural diversity. Since making music is an evolved propensity that is no longer required for one of its precursor's original functions, because information transmission is much better achieved by language, societies have come to use music in many different ways. Hence we find immense cultural diversity in how music is made and the role it plays, both within and between different human societies.

One of these roles is simply entertainment. The evolutionary history of 'Hmmmmm' ensured that the human mind evolved to enjoy melody and rhythm, which were critical features of communication before becoming usurped by language. In this regard, music is very similar to food and sex. Those of us living in the affluent West eat more food and have more sex than we biologically require. We have evolved to enjoy both types of behaviour and often engage in them for entertainment value alone, although their roles in social bonding continue. Music is the same: in certain circumstances it still provides some of the adaptive value that was central to 'Hmmmmm', especially in forging group identities; but we also enjoy making music and pursue it at will.

Here we must note the importance of song – the combination of music and language. Song can be considered as the recombination of the two products of 'Hmmmmm' into a single communication system once again. But the two products, music and language, are only being recombined after a period of independent evolution into their fully evolved forms. Consequently, song benefits from a superior means of information transmission, compositional language in the form of lyrics, than ever existed in 'Hmmmmm', combined with a degree of emotional expression, from music, that cannot be found in compositional language alone. Moreover, that music is often produced by instruments which, as an extension of the human body in material form, are themselves a product of cognitive fluidity. And that is a further consequence of the segmentation of 'Hmmmmm'.

Back inside the brain

Here we should return to look inside the brain, as it is in the firing of neurons and the swilling of chemicals that our enjoyment of music lies. As described in chapter 5, Isabelle Peretz used the case histories of people such as NS, the man who lost his ability to recognize words but could recognize melodies, HJ, the man for whom music sounded 'like awful noise', and Monica, who suffered from congenital amusia, to argue that the music and language systems within the brain are constituted by a series of discrete modules. Her argument was represented in Figure 5 (p. 63) and is entirely compatible with the evolutionary history that I have proposed within this book.

In general, an evolutionary approach to the mind leads to an expectation that the mind will have a modular structure. In accordance with the specific evolutionary history I have proposed, we should expect pitch and temporal organization to have the degree of independence that Peretz suggests, because the latter appears to have evolved at a later date, being associated with the neurological and physiological changes that surrounded bipedalism. Similarly, we should not be surprised that Peretz found that the root of congenital amusia lies in the inability to detect variations in pitch, because from an evolutionary perspective that appears to be the most ancient element of the music system within the brain.

The fact that the music and language systems in the brain share some modules is also to be expected given the evolutionary history I have proposed, because we now know that both originate from a single system. Conversely, the fact that they also have their own independent modules is a reflection of up to two hundred thousand years of independent evolution. The modules relating to pitch organization would once have been central to 'Hmmmmm' but are now recruited only for music (with a possible exception in those who speak tonal languages); while other 'Hmmmmm' modules might now be recruited for the language system alone – perhaps, for example, those relating to grammar. This evolutionary history explains why brain injuries can affect either music alone (chapter 4), language alone (chapter 3), or both systems if some of the shared modules are damaged.

It is, of course, profoundly difficult, if not impossible, to reconstruct evolutionary history from present-day mental structures, for the neural networks within any one person's brain are as much a product of developmental as evolutionary history (chapter 6). Indeed, some would argue that the type of environment within which the brain develops is the principal determinant of its neural circuitry. Babies are born into and develop within cultures that have language as the dominant form of aural communication, and this influences the neural networks that are formed within their brains. Nevertheless, the genes we inherit from our parents derive from our evolutionary history and must channel that development; it is the extent of this channelling that remains highly debated among psychologists. My own view is one that gives equal weight to evolution and culture as regards the manner in which neural networks develop. All I expect is a broad compatibility between evolutionary history and brain structure – and this is indeed what appears to be present.

Remnants of 'Hmmmmm' within music and language

Music evidently maintains many features of 'Hmmmmm', some quite evident, such as its emotional impact and holistic nature, others requiring a moment's reflection. It is now apparent, for instance, why even when listening to music made by instruments rather than the human voice, we treat music as a virtual person and attribute to it an emotional state and sometimes a personality and intention. It is also now clear why so much of music is structured as if a conversation is taking place within the music itself, and why we often intuitively feel that a piece of music should have a meaning attached to it, even though we cannot grasp what that might be.

Remnants and reflections of 'Hmmmmm' can also be found in language. Perhaps the most evident is in the acquisition of language by infants.

Ever since the late nineteenth century, scientists have debated whether there is a correspondence between the stages through which we passed during human evolution and the stages through which a foetus or infant passes during development. The strongest form of this argument is referred to as recapitulation, and was proposed by Ernst Haeckel in 1866 in the form: 'ontogeny is the short and rapid recapitulation of phylogeny'. Whether or not that theory has any value, my concern is with a far more general argument concerning the development of language.[14] It is, of course, whether the earliest stages of infant-directed speech (IDS, as described in chapter 6) are likely to be similar to the type of 'Hmmmmm' used by Early Humans to their infants, since in both cases the infants lack a language capability but need support for their emotional development.

If there is a similarity – and I strongly suspect there is – then when we hear mothers, fathers, siblings and others 'talking' to babies, are we perhaps hearing the closest thing to 'Hmmmmm' that we can find in the world today? Modern-day IDS uses words along with 'ooohs' and 'aaahs', but those words are quite meaningless to the infants and simply provide further acoustically interesting sound strings for the infant to hear. As well as the content of IDS, the contexts in which it is used may also be strongly similar to those of our evolutionary past. The *Homo sapiens* mother in her suburban home will have the same propensity to sing to her infant, to make it laugh, jiggle and dance, as did the *Homo ergaster* mother under the African sun and the *Homo neanderthalensis* mother in an ice-age cave. In each case, they will be comforting their babies when physically separate from them, and enabling then to learn about communication and emotional expression.

Any similarity between IDS and ancient 'Hmmmmm' starts to diminish by the time the babies reach one or two, the age when they begin segmenting sound strings and acquiring a lexicon and a grammar for compositional language. This in itself mirrors the evolutionary process of the differentiation of 'Hmmmmm' into two separate systems of music and language. With modern-day infants, this is also the time when a bias towards absolute pitch is replaced by a bias towards relative pitch, a development that facilitates language learning but inhibits music learning – except in those cases where there are either intense musical experiences and/or a genetic predisposition to maintain perfect pitch.

If IDS is one remnant of 'Hmmmmm', then another is the use of spontaneous gestures when speaking. As I explained in chapter 2, hand signals and body movements often convey information to supplement what is being said. The listener/watcher may be quite unaware that some of the information he/she is receiving is coming from the gesture rather than the words being heard. Spontaneous gestures maintain the key features of 'Hmmmmm' – they are holistic and often both manipulative and mimetic. Had we not evolved/developed language we might be far more effective at inferring information from such gestures, and would have grown up in a culture where such gesturing was recognized as a key means of communication, rather than as a curious hangover from our evolutionary past.

The most significant survival of 'Hmmmmm' is within language itself. One aspect of this is the presence of onomatopoeia, vocal imitation and sound synaesthesia, which are probably most readily apparent in the languages of present-day people who still live traditional lifestyles and are 'close to nature'. Another is the use of rhythm, which enables fluent conversations to take place.

Perhaps of most significance, however, is our propensity to use holistic utterances whenever the possibility arises. Although the creative power of language certainly derives from its compositional nature – the combination of words with grammar – a great deal of day-to-day communication takes place by holistic utterances, or what are more frequently called 'formulaic phrases'. This is the principal argument of Alison Wray's 2002 book entitled

Formulaic Language and the Lexicon, which has been so influential on my own thinking. She describes formulaic phrases as 'prestored in multiword units for quick retrieval, with no need to apply grammatical rules'.[15] In my chapter 2, I gave the example of idioms, such as 'straight from the horse's mouth' and 'a pig in a poke', while Wray provides many more examples which are often phrases used as greetings or commands: 'hello, how are you?', 'watch where you're going', 'keep off the grass', 'I'm sorry', 'how dare you!'[16]

Critics of Wray's views about the prevalence and nature of formulaic phrases have noted that the majority do actually conform to grammatical rules and are evidently constructed from words.[17] They rely, therefore, on the prior existence of compositional language. This is true, but misses the point. Even though we have compositional language, we have a propensity to slip into the use of formulaic phrases/holistic utterances whenever appropriate occasions arise. These are frequently the oft-repeated social situations, such as greeting friends ('hello, how are you?') and sitting down to meals ('bon appétit'), especially in company with people with whom we already share a great deal of knowledge and experience, such as the members of our close family. One might argue that we use such formulaic phrases simply to reduce the mental effort of having to put together words with grammatical rules whenever we wish to say something. But to my mind, their frequency in our everyday speech reflects an evolutionary history of language that for millions of years was based on holistic phrases alone: we simply can't rid ourselves of that habit.

Finally, we should recall that form of vocal expression I referred to in chapter 2, that can be defined neither as music nor as language, while it exhibits aspects of both: Indian mantras. The philosopher Franz Staal has undertaken extensive studies of mantras and concluded that these lengthy speech acts lack any meaning or grammatical structure, and are further distinguished from language by their musical nature. As relatively fixed expressions passed from generation to generation, they are, perhaps, even closer than IDS to the type of 'Hmmmmm' utterances of our human ancestors.[18]

Finis

In 1973 John Blacking published *How Musical Is Man?* It was based on the John Danz lectures given at the University of Washington and was, to Blacking, an unexpected success. It was published in London by Faber and Faber and translated into several languages. According to a later colleague of Blacking, the book 'made bold and sweeping assertions on sometimes rather slender evidence, and occasionally none at all, about the innate musical capacities of humankind'.[19]

John Blacking had a profound knowledge of how music was made and used in societies around the world. He did not, however, have the types of evidence that I have been able to marshal and interpret within this book – that from the archaeological and fossil record, that about monkeys and apes, about child development, and about what goes on inside the brain. My conclusion is the same as John Blacking's in *How Musical Is Man?*: 'it seems to be that what is ultimately of most importance in music can't be learned like other cultural skills: it is there in the body, waiting to be brought out and developed, like the basic principles of language formation'.[20]

The evolutionary history I have proposed explains why there are on the one hand such profound similarities between music and language, and why on the other there are such differences, as described in chapter 2. I have also explained how that history accounts for how music and language are constituted in the brain (chapters 3–5), how we communicate with infants (chapter 6), and the impact of music on our emotions (chapter 7), and I have shown how an evolutionary approach explains why we so often make music together (chapter 14).

In spite of all this, words remain quite inadequate to describe the nature of music, and can never diminish its mysterious hold upon our minds and bodies.[21] Hence my final words take the form of a request: listen to music. When doing so, think about your own evolutionary past; think about how the genes you possess have passed down from generation to generation and provide an unbroken line to the earliest hominid ancestor that we share. That evolutionary inheritance is why you like music – whatever your particular taste.

So listen to J. S. Bach's Prelude in C Major and think of australopithecines waking in their treetop nests, or Dave Brubeck's 'Unsquare Dance' and think of *Homo ergaster* stamping, clapping, jumping and twirling. Listen to Vivaldi's Concerto in B flat Major for trumpet and imagine a member of *Homo heidelbergensis* showing off a hand-axe, then let Herbie Hancock's 'Watermelon Man' help you picture a hunting group celebrating a kill, and with Miles Davis's 'Kind of Blue' imagine them satiated with food and settling to sleep amid the security of trees. Or simply listen to a mother singing to her baby and imagine *Homo ergaster* doing the same. When you next hear a choir perform, close your eyes, ignore the words, and let an image of the past come to mind: perhaps the inhabitants of Atapuerca disposing of their dead, or the Neanderthals of Combe Grenal watching the river ice melt as a new spring arrives.

Once you have listened, make your own music and liberate all of these hominids that still reside within you.

Notes

1. The mystery of music

[1] The slow movement from Schubert's Quintet in C was described as 'the most perfect piece of music' by a listener to BBC Radio 3's programme '3 For All', introduced by Brian Kay on 26 December 2004 at approximately 16.30. I am most grateful to that anonymous listener and to Brian Kay for then playing the piece while I was driving home from a party on Boxing Day and wondering how I should rewrite the introduction to my book.

[2] A history and list of ongoing activities of the Société de Linguistique de Paris is provided on its website ‹http://www.slp-paris.com›.

[3] See, for instance, Arbib (in press), Bickerton (1990, 1995), Hurford et al. (1998), Knight et al. (2000), Wray (2002b), Christiansen and Kirby (2003).

[4] Darwin (1871), Blacking (1973). It is useful to cite Blacking's insight: 'There is so much music in the world that it is reasonable to suppose that music, like language and possibly religion, is a species specific trait of man. Essential physiological and cognitive processes that generate musical composition and performance may even be genetically inherited, and therefore present in almost every human being' (Blacking 1973, p. 7). Another exception to this neglect is William Benzon's *Beethoven's Anvil* (2001) and the volume edited by Wallin et al. (2000). Key texts by archaeologists that have addressed the evolution of the mind with minimal, if any, mention of the musical and emotional capacities of human ancestors include Gibson and Ingold (1993), Renfrew and Zubrow (1994), Mellars and Gibson (1996), Noble and Davidson (1996), and Nowell (2001).

[5] Thomas (1995) provides a detailed discussion of the relationship between music and language in the writing of the French Enlightenment, in a book appropriately entitled *Music and the Origins of Language*.

[6] Jespersen (1983 [1895], p. 365). I am grateful to Alison Wray for drawing my attention to Jespersen's views and this specific quote.

[7] For instance, Morten Christiansen and Simon Kirby are two of the most

impressive linguistic scholars writing today. In 2003 they coedited a book entitled *Language Evolution,* which they conceived as 'a definitive book on the subject'. But it failed to mention music in any of the seventeen chapters written by the 'big names in every discipline' (Chistiansen and Kirby 2003, p. vii). I apologize to any contributors to that volume if they did indeed mention music; if so, this must have been only in passing, as my meticulous reading of all chapters failed to note any such reference.

[8] This is, of course, a gross simplification of a more complex situation. Language is extremely good at transmitting some types of information, such as that concerning social relationships, but less good at other types, such as how to tie knots. Some linguists believe that transmitting information may not be the main function of language, preferring to emphasize its role in concept formation and creative thought. Carruthers (2002) discusses the cognitive functions of language.

[9] I am asking this for rhetorical reasons, as it is quite clear that music has numerous different functions, which will be variously referred to throughout my book. Merriam (1964) listed ten major functions of music: emotional expression, aesthetic enjoyment, entertainment, communication, symbolic representation, physical response, enforcing community and social norms, validation of social institutions and religious rituals, contribution to the continuity and stability of culture, contribution to the integration of society (cited in Nettl 1983, p. 149).

[10] Pinker and Bloom (1995) made the most persuasive argument for the gradual evolution of language capacity by the process of natural selection, while Jackendoff (1999) has proposed possible stages in the evolution of language, based around the gradual appearance of simple and then grammatical rules. Bickerton (1990), however, has argued for no more than two stages in the evolution of language, the first creating a proto-language and the second involving a catastrophic change to fully modern language.

[11] The two views of proto-language have also been termed 'synthetic' (for 'compositional') and 'analytic' (for 'holistic'). Tallerman (in press) discusses their respective virtues, strongly supporting synthetic rather than analytic approaches.

[12] Bickerton has published prolifically on the evolution of language, making any simple characterization of his views difficult. A selection of his key works includes Bickerton (1990, 1995, 1998, 2000, 2003) and Calvin and Bickerton (2000).

[13] Key works by Jackendoff include *Foundations of Language* (2000) and *A Generative Theory of Tonal Music* (Lerdahl and Jackendoff 1983). Jackendoff (1999) proposes that simple rules might have been present in proto-language.

[14] Alison Wray's work will be discussed in later chapters. Her key publications on

this issue are Wray (1998, 2000, 2002a). The other key proponent of holistic proto-language is Michael Arbib (2002, 2003, in press). Like Wray (1998, 2000), he argues that hominid utterances consisted of arbitrary strings of sounds rather than words, and he uses the term 'fractionation' for the manner in which these were transformed into words. Unlike Wray, however, whose scenario implies vocalization throughout, Arbib believes that holistic proto-language developed via the use of gesture and was related to the presence of mirror neurons (Rizzolatti and Arbib 1998).

[15] Although not strictly an evolutionary psychologist, one might also mention Donald's (1991) limited concern with musical abilities in his study of cognitive evolution.

[16] I do not mean to dismiss Pinker's (1997, pp. 529–39) ideas out of hand, as he has developed a cogent position that I believe is partially correct, in the sense that our musical ability draws on a range of mental mechanisms that originally evolved for other purposes. But, unlike Pinker, I believe that underlying these is a musical ability that arose as the result of either natural and/or sexual selection. Pinker specifies five areas of human thought that may contribute to musical ability: language, auditory, scene analysis, emotional calls, habitat selection and motor control. The last provides our love of rhythm, since motor actions such as walking, running, chopping and digging are performed most efficiently in a rhythmic mode.

[17] Pinker (1997, p. 528).

[18] Cross (1999; see also Cross 2001). He draws on my own previous work (Mithen 1996) regarding the significance of cognitive fluidity and proposes that music may have been the means by which this emerged: 'I am suggesting that proto-musical behaviours may play a functional role in general development – and by implication, in cognitive evolution – by virtue of their multi-domain properties and their floating intentionalities' (Cross 1999, p. 25).

[19] A key foundation for the recent resurgence of interest in the evolutionary basis of music and its relationship to language was Nils Wallin's 1991 book entitled *Biomusicology*. This was a synthesis of his research into the biological basis of music in the human brain, physiology, and auditory and vocal systems. In 1997 Wallin and two colleagues organized a conference in Florence entitled 'The Origins of Music', the proceedings of which were published in 2000 (Wallin 2000).

With over thirty contributors from the fields of linguistics, anthropology, psychology and zoology, as well as musicology itself, this volume provided a suitable riposte not only to Pinker, but also to those like myself who have neglected music in our studies of the human mind.

However, while *The Origins of Music* may have exposed the neglect of music in accounts of human evolution, it was quite unable to fill the gap. As is the nature of edited, interdisciplinary volumes, each author seemed to view the origins of music from their own highly specialized and rather idiosyncratic perspective. This would often conflict with the views of another specialist writing in the same book. The resulting volume was quite unable to provide a coherent account of how music evolved and its relationship to language and other features of the human mind. In fact, human evolution had a relatively minor part in *The Origins of Music*, as it was dominated by studies of animal vocalizations and speculative theories with limited grounding in the evidence from the fossil and archaeological records. Moreover, a great deal of material relevant to the origins of music was quite lacking from that volume, and more has appeared since its publication, especially as concerns the biological basis of music within the brain. None of its contributors – nor anyone else writing about music or evolution – seems to have realized what has become quite obvious to me: that Neanderthal big brains were used for music-like, rather than language-like, communications.

[20] Tolbert (2001) draws heavily on the work of Donald (1991) regarding mimesis.

[21] Bannan (1997). Although not (as far as I know) musicologists, Vaneechoutte and Skoyles (1998) have also argued that singing has an evolutionary priority over language.

[22] Dunbar (2004).

[23] Blacking wrote that '"Musical" discourse is essentially nonverbal, though, obviously, words influence its structures in many cases; and to analyse non-verbal languages with verbal language runs the risk of distorting the evidence. "Music" is therefore, strictly speaking, an unknowable truth, and discourse about music belongs to the realm of metaphysics.' This is taken from Blacking's 1984 essay entitled 'Music, culture and experience' originally published in the *South African Journal of Musicology*; I have drawn on the edited version in Byron (1995) who describes it as providing one of 'Blacking's most concise, cogent, and comprehensive statements of his later theoretical views'.

[24] The process by which linguistic and musical diversity arises is, of course, partly conditioned by the evolved nature of the human mind, as this conditions what choices, conscious and unconscious, are made regarding the adoption of linguistic and musical elements by individuals. It is also claimed that the diversity may provide some clues about origins. Bruno Nettl, a musicologist from the University of Illinois and arguably the most distinguished ethno-musicologist working today, has written that 'the only even remotely reliable guide to the origins of music is the plethora of contemporary and recent musics known to us' (Nettl 1983, p. 171). I disagree: while acknowledging the nature and especially the diversity of such musics is important, they are nevertheless the product of more than a hundred thousand years of cultural evolution since the first music arose – or at least that will be my claim. In this regard I am following Blacking (1973, p. 55), who described the study of the musical practices of living people as a means of understanding the origin of music as a

'futile exercise'. Just as we should not expect to find the origin of language by examining the diversity of languages spoken today, neither should we expect to do this for music, although Blacking's comment on this issue appears rather extreme: 'Speculative histories of world music are a complete waste of effort' (Blacking 1973, p. 56). Indeed, one of my arguments will be that music is relatively unconstrained; once the capacity for making music had evolved, it could be expressed in a great many ways for a great number of different reasons. Although I will frequently draw upon Bruno Nettl's understanding of musics throughout the world, my own 'reliable guides' to its origin are quite different: the human brain, infant development, the psychology of emotion, and the communication systems of non-human primates. And also, of course, the archaeological and fossil record, which is covered in the second part of my book.

[25] Birdsong and whale song share at least one feature with human music that is not found in the calls/songs of any other animals, even those of our closest living relatives, the African apes. This is what the biologist Peter Marler (2000) has described as 'learnability'. Before we address this, however, we must consider another feature that links human music and animal song: 'phonocoding'.

Phonocoding is the combination of meaningless acoustic elements into a sequence (Marler 2000). This is the nature of the majority of animal calls; the individual acoustic elements that constitute a chimpanzee's pant-hoot or a nightingale's song have no meaning. As far as we understand, the combination of such elements into an extended call or song also lacks a symbolic reference. Even the most complex and exquisite birdsong appears to be communicating no more than the bird's presence, and hence its availability for mating or preparedness to defend its territory. There is a close similarity here with music, as neither individual notes nor their sequence within a musical phrase are symbolically meaningful. But both music and animal calls are expressive of emotions and arouse them in other individuals.

Not all animal calls lack meaning. Some non-human primate and bird species are known to have alarm and food calls that should be characterized as symbolically meaningful. A much-cited example is the vervet monkey (Cheney and Seyfarth 1990), as described in chapter 8. This has distinctive alarm calls referring to snakes, birds of prey and leopards. When they are made, the monkeys respond in an appropriate way – for example, by climbing trees following a leopard call, or by looking around on the ground if the call refers to a snake. Such calls appear to be similar to the words that we use in language. But, as Marler has explained, they come as indivisible packages. They can neither be broken down into meaningful elements, nor be combined to create emergent meanings in the way that humans are able to do with words. Nor are they learnt in the manner as a child learns the words of its language.

While the calls/songs of all animals might be characterized as examples of phonocoding, and hence have a strong link to human music, 'learnability' is far more restricted in the animal world. This is the ability to learn new acoustic elements and phrases from other individuals, and spontaneously to create new phrases. Learnability is a key feature of both human music and language but is surprisingly rare in the animal world. It is quite absent in our closest living relatives, the African apes. As will be further examined in chapter 8, chimpanzees and gorillas have their calls innately specified; each individual may have its own variant of a specific type of call, but these remain fixed throughout its lifetime (see Mitani 1996 for a review of ape calls). The same is found among the most 'musical' of primates, the gibbons. These engage in extended singing, often involving duets by males and females, which appear to be about claiming territory and social bonding. But the young gibbons do not have to learn such songs, and neither are gibbons able to modify their songs during their lifetime – a gibbon's song is biologically fixed at birth (Geissmann 2000).

Songbirds are quite different, as learning plays a critical role in their vocal development (Slater 2000). This is also the case for parrots and hummingbirds, indicating that 'learnability' has evolved independently at least three times among the birds. It is principally the males that sing, and they do so simply to attract mates and to deter others from their territory. A young male will learn a set of songs from the adults that he hears; these will be broken down into shorter phrases and then rearranged to create a repertoire of new songs. Some species acquire a limited number of songs; others, such as the winter wren, acquire a seemingly endless set of variations throughout their lifetime – many thousands of songs, all of which mean the same thing: 'I am a young male'.

Such vocal learning and creative output are similar to those found in human language and music, in which a finite set of words/notes can be used to create a infinite set of utterances/musical phrases. Another similarity is that songbirds have a sensitivity period during which they are particularly susceptible to learning song, this often being a few months after hatching. Such sensitivity periods also exist for young children when learning language; if they have not acquired language before adolescence, perhaps due to social deprivation or illness, then it can only be learnt with considerable effort and they never achieve high levels of fluency. Whether there is an equivalent sensitivity period for acquiring musical skills is unknown.

We must be cautious not to read more into this analogy between birdsong and human language/music than is appropriate, for there are significant differences. Songbirds are confined to learning the songs of their own species, even if surrounded by the songs of many different species which, to the human ear, sound remarkably similar. In contrast, children acquire any language to which they are exposed during their sensitivity period – a child of English-speaking parents will learn Chinese if adopted into a Chinese-speaking family just as easily as will a Chinese child learn English, even though such languages may sound very different to us.

The composer and musicologist François-Bernard Mâche (2000) believes that there are far deeper links between birdsong and human music than merely those of phonocoding and learnability. He thinks it significant that some common features of human music occur within birdsong. For an example, he cites the aksak family of rhythms, which are found in musical systems of eastern Europe and western Asia. These oppose an irregular number of basic units, very often grouped by three or two, and this is found too, Mâche claims, in the songs of the red-legged partridge and the red-billed hornbill.

Other birds use stable and precise sets of pitches in their signals, which is a feature of human music but not of language, although it is perhaps similar to the use of pitch in tonal languages such as Chinese. Mâche cites the chocolate-backed kingfisher, which moves up and down its own scale, which is characterized by very small intervals; while the appropriately named music wren sings with a near-perfect scale. Other birds not only enumerate the tones of their scales but also build 'melodic motives as elaborate as many human achievements, and even sounding so close to them that one might be mistaken' (Mâche 2000, p. 477). More generally, birdsong includes many of the processes of repetition that are so critical to human music, such as refrains, rhymes, symmetry and reprises.

The similarities between human music and birdsong are far greater than those between human music and the calls/songs of non-human primates or, indeed, any other type of animal – with one possible exception: whales. Katharine Payne (2000) of Cornell University has spent over fifteen years listening to and analysing the songs of humpback whales; when combined with the work of her colleagues, more than thirty years' worth of songs have been recorded from populations in the North Pacific and North Atlantic. The similarities to human music are striking. Whale song consists of long, highly structured sequences which can repeat for many hours, often without pause, even when the singer comes to the surface. The songs have a hierarchical structure: notes are combined into phrases; these are combined into themes, up to ten of which are found in any one song; and the songs themselves are linked together into cycles.

Katharine Payne and her colleagues have discovered that while the songs in different whale populations are similar in structure, they differ in content. She compares them to the dialects of a single human language. Of more interest, and showing a greater resemblance to human music than to language, is the manner in which the whale songs of a single population change over time.

Each individual whale is constantly altering its song by modifying the phrases and themes it uses, and the manner in which these are combined. Nevertheless, all whales within a single group appear to agree as to which themes are stable and which are changing. No single whale appears to be driving this change; each seems to change its song about as much as any other whale in the population. But listening and learning must be essential to the evolution of their songs; Payne suggests the process is best understood as being analogous to improvisation in human music. It certainly occurs more quickly than change in human language; within one decade a population's song may have undergone so much change that one can no longer recognize its relation to the earlier version.

How can we explain the similarities between human music, birdsong and whale song? Had such similarities been found between chimpanzee calls and human music we would readily attribute them to a common ancestor of these species – an ape-like creature that lived at around 6 million years ago. But the common ancestor of birds, whales and humans lived so long ago, and has given rise to so many species that lack any song/musical abilities at all, that this is a quite unfeasible explanation for such similarities. Moreover, the physiological means by which birds, human and whales actually generate their songs are quite different. Rather than a larynx located in the throat as in mammals, birds have a syrinx at the point where the ducts from the two lungs join to form the trachea.

The similarities between birdsong and whale song are more readily explained by convergent evolution: natural and/or sexual selection has driven the evolving vocal apparatus and neuronal networks in the same direction to create communication systems with both phonocoding and learnability. Marc Hauser, a leading authority on animal communication systems, argues that such convergent evolution suggests 'important constraints on how vertebrate brains can acquire large vocabularies of complex learned sounds. Such constraints may essentially force natural selection to come up with the same solution repeatedly when confronted with similar problems' (Hauser et al. 2002, p. 1572).

Pinker (1997) would have us believe that the presence of phonocoding and learnability in human music is for quite different reasons – a chance spin-off from quite different evolved capacities which coincidentally match the key features of whale song and birdsong. While the learnability of music may be derivative from that necessary for language, it seems unlikely that this spin-off could have so completely lost the grammatical and symbolic characteristics of language. A more persuasive and parsimonious idea is that human musical ability has also been shaped by evolution as a means to express and induce emotions, and has converged to share properties with birdsong and whale song. The extent to which that shaping occurred independently from that of language can be further explored by delving inside the human brain.

2. More than cheesecake

[1] Hauser et al. (2002, p. 1569) argue that many of the 'acrimonious debates' in the field of language evolution have arisen from a failure to distinguish between language as a communication system and the computations underlying the system.

[2] In being equally concerned with listening as much with the production of music, I am following the lead of John Blacking in his 1973 book, *How Musical is Man?*

[3] Nettl (1983, p. 24). Readers who wish to tackle this question should consult this introductory essay of Bruno Nettl's 1983 collection, *The Study of Ethnomusicology*, in which the problem of definition is discussed.

[4] Nettl (1983, chapter 1) discusses varying concepts of music. The Hausa of

Nigeria, for instance, have no term for music, but have words for music-like activities related to different cultural contexts with which they are verbally associated. The Blackfoot Indians have a word that can be roughly translated as 'dance', which includes music and ceremony but does not cover 'musical activities' that do not include dance. They have a word for song but not for instrumental music. The very idea of a 'piece' of music might be restricted to Western concepts; Nettl (1983, chapter 8) describes how ethnomusicologists have to work with the notion of 'tune families' in other musical traditions.

[5] John Cage's *4'33"* consists of four minutes and thirty-three seconds of uninterrupted silence for any instrument or combination of instruments. I had the pleasure of listening to its first-ever radio broadcast, made by BBC Radio 3 in 2003.

[6] Many linguists dismiss such formulaic utterances as being a trivial element of language, but Wray's book on this subject – by far the best book on linguistics I have ever read – shows that they are quite wrong (Wray 2002a).

[7] See Wray and Grace (in press). In making their argument, they cite the following: Thurston writing in 1989 that 'since diachronic linguistics has developed within literate societies using primarily written materials for data, perhaps the discipline has been unduly biased'; Olson writing in 1977 that 'Chomsky's theory is not a theory of language generally but a theory of a particular specialized form of language ... it is a model for the structure of autonomous written prose'; and Tomasello, writing in 2003 that 'there is very little in [spontaneous spoken speech] that corresponds to a "sentence"' and that 'spontaneous spoken speech ... has properties of its own that are different, in some cases very different, from the intuitive model of language that literate, educated people carry around in their heads'. Chomsky did, of course, draw an important distinction between 'competence' and 'performance', which is sometimes equated with that drawn by Saussure between 'langue' and 'parole'. Jackendoff (2000, pp. 29–37) provides a very useful review, and part-interpretation, of Chomsky's distinction.

[8] For an argument that mantras lack meaning, see Staal (1988). I am most grateful to Björn Merker for referring me to Staal's book and commenting on its possible relevance to the theory I develop.

[9] Blacking (1973) claims that all human societies of which we have knowledge appear to have music. The similarities, and the following differences, between music and language that I cite in this chapter are described at greater length by John Sloboda in his 1985 book, *The Musical Mind*.

[10] Nettl (1983, chapter 3) discusses different types of universals in music, and also features that are very widespread.

[11] Merriam's (1964) ten functions of music are discussed by Nettl (1983), who

promotes the idea that a religious function is a universal feature of music throughout the world.

[12] Nettl (1983, chapter 3 – the quote is on p. 40) stresses the role of cultural transmission in generating universal features of music, whereas my emphasis is on the shared physical and psychological propensities of *Homo sapiens*.

[13] Blacking (1973, p. 8).

[14] Blacking (1973, p. 47).

[15] Nettl (1983, chapter 3) describes how the history of ethnomusicology during the twentieth century has been one of fluctuation between a concern with universals and with diversity. Bickerton (2003) has noted that very few linguists have been concerned with the evolution of the language capacity, implying that their interests principally rest with explaining linguistic diversity. The most interesting study of the latter has been undertaken by Nettle (1999).

[16] See Nettl (1983, chapter 4) for a discussion about the blurred boundaries of languages and musics. A further aspect of diversity is that between the structures of the language and of the music of the same people. Nettl (1983, p. 239) notes that the typical pattern of linguistic tone sequences in some west African languages seems to affect how their melodies are composed, and that Czechs and Hungarians have languages in which utterances begin with stressed syllables and folk songs that typically begin with stressed beats. But he states that 'it is difficult to establish, in even these cases, that music was moulded to fit the characteristics of language, that the structure of language determined the musical style'. This appears to quash any notion that music is somehow derivative of language.

[17] There can, of course, be transpositions of one musical style into another, such as Bollywood versions of Western music and the 1980s' 'pop' versions of classics.

[18] Blacking (1973, pp. 108–9).

[19] The gestural theory of language origins was pioneered by Gordon Hewes (1973) and substantially developed by Michael Corballis (2002). Most recently, it has been championed by Arbib (in press).

[20] Unless otherwise specified, the following text about gesture and its relationship to spoken utterances draws on Beattie (2003).

[21] See Pinker (1994).

[22] By 'informal' I am attempting to distinguish such singing and dancing from that which conforms to set of cultural rules and requires extensive training, such as ballet and opera in modern Western society.

[23] Blacking (1973, p. 100).

[24] Pinker (1997, p. 529) contrasts the ability of all adults to produce language with the scarcity of those who can produce music as a means to support his argument that music is a 'technology, not an adaptation'.

[25] Hauser et al. (2002).

[26] Sloboda (1998, p. 23) states that representation in music is hierarchical. He suggests that musical grammars 'determine elements within a musical sequence which control or subsume neighbouring elements, and this process is recursive so that lead elements at one level become subsumed at the next higher level (metre is probably the simplest element of this)'.

[27] Cross (in press).

[28] Auer et al. (1999, p. 202).

[29] This assumes that we take words and notes to be equivalent units; it might be more appropriate to equate notes with speech sounds, in which case they would both lack symbolic meanings (Wray, pers. comm.). It is also the case that some notes or combinations of notes might be given a symbolic meaning that becomes shared within a community.

[30] For sound synaesthesia I will be drawing on the work of Berlin (1992, 2005).

[31] Cross (in press).

[32] Blacking (1973, p. 68). The ethnomusicologist Bruno Nettl (1983, p. 148) gives the example of an outsider watching the sun dance of the Blackfoot Indians. For the outsider, this would primarily be a musical rather than a religious event, whereas the Indians themselves accord religion the primary role. He also makes the observation that for many concertgoers in the West, attending a concert is primarily a social rather than a musical event, as evidenced by the obligatory forms of dress, the standard length, intermissions and programmes of such events, and the ritualistic nature of applause.

[33] Charles Pierce's distinction between symbols, signs and icons is widely cited and used in many academic fields, although the distinctions are not as clear as is often implied. An icon has a representational relationship with its referent, as in a portrait. A sign is non-representational but has a non-arbitrary relationship with its referent – smoke is a sign of fire. Signs are likely to be universally understood; one does not have to belong to a particular culture to understand their use. A symbol, on the other hand, has an arbitrary relationship to its referent which is agreed upon by a particular culture or subculture within a wider community.

[34] I am grateful to Alison Wray for giving me the example of *mae'n flin 'da fi* and making me aware of the significance of formulaic utterances and their similarity with musical phrases.

[35] The classic case is Bernstein (1976). Sloboda (1985) reviews other attempts to

find musical grammars, while Nettl (1983, pp. 212–15) discusses how linguistic approaches have been applied to music with limited success.

[36] Lerdahl and Jackendoff (1983).

[37] Dempster (1998).

[38] Dempster (1998). It is the case, however, that linguistic grammars evolve through time. If this were not the case, we would not have the six thousand or so languages that exist in the world today – assuming that these derive from a single 'proto-world language'. And this figure is likely to be a small fraction of the languages present less than a millennium ago, in view of the rate at which languages are becoming extinct. Moreover, some linguists question the notion that there is such a thing as 'universal grammar', meaning an innate set of grammatical principles that are common to all languages. Grammars may simply be emergent properties of communication by the use of symbols, and some languages may have more complex grammars than others, although this is generally denied by linguists who invoke the so-called 'uniformitarian hypothesis' (Kenny Smith, pers. comm.).

Linguists such as James Hurford of Edinburgh University use the term 'grammaticalization' to describe the way in which a word can acquire a specialized grammatical role simply by virtue of particularly frequent use. The key idea of grammaticalization theory is that 'syntactic organization, and the overt markers associated with it, emerges from non-syntactic, principally lexical and discourse, organization ... through frequent use of a particular word, that word acquires a specialized grammatical role that it did not have before' (Hurford 2003, pp. 51–2). Hurford and other linguists have used computer models to demonstrate how complex syntax can emerge from simple communication systems that have small numbers of initial assumptions and are prone to make minor errors. Such computer models start with simulated populations that have no language at all but that develop communicative codes which, though still extremely simple, share some characteristics with human language (Hurford 2003, pp. 53–4). Key examples are Batali (1998, 2002), Kirby (2000, 2002), Hurford (2000). Tomasello (2003), Davidson (2003) and Dunbar (2003) have stressed the significance of such models for understanding the evolution of language – in essence, they remove the problem of how to explain the evolution of syntax by characterizing this simply as an emergent property. Bickerton (2003, p. 88) is dismissive of such arguments: 'It will just not do to dismiss it [i.e. syntax] as due to self-organization, or the grammaticization of discourse, or analogies with motor systems or syllabic structure, or any of the other one-paragraph explanations that writers on language evolution typically hope to get away with.'

[39] Pinker (2003, pp. 16–17) states that 'language can communicate anything from soap opera plots to theories of the origin of the Universe, from lectures to threats to promises, to questions'.

[40] Cross (in press).

[41] Sloboda (1985, p. 12).

[42] The following draws on Sloboda (1985).

[43] When Alison Wray read this paragraph for me she made the following important and entertaining comment: 'However, ask musicologists rather than naïve outsiders and they will say "the transition from the major to the minor via the augmented chord was a masterstroke of a cheeky reference to Schoenberg and the chromatic descent just before the cadence was worthy of Ives". Between the two you will get those who will say "that second phrase would make a great ring tone for my mobile" or "that bit is just what I want for the film that I'm making, for where it gets really creepy". Others will say "the performance was poor – too little sensitivity in the quiet passages and not enough energy later on".'

[44] The cognitive function of language has recently been considered by Carruthers (2002).

[45] As Ian Cross (in press) has recognized, 'without the certainty of mutual knowledge there will always be a degree of ambiguity in linguistic communication'.

[46] See Dunbar (1998, 2004) and Mithen (2000) for a discussion of the relationship between theory of mind and language.

[47] Brown (2000) places substantial emphasis on the presence of expressive phrasing in both music and language, arguing that this was one of the key features of their shared ancestral state, which he calls 'musilanguage'.

[48] Watt and Ash (1998).

[49] Sloboda (1998, p. 28). While a degree of emotional response to music occurs quite instinctively, there is also an element of acquired skill. John Sloboda gives the example of an inexperienced listener finding Mozart 'pale, insipid, and all vaguely "jolly"'. Once a listener has gained closer knowledge of Mozart, and perhaps of the emotional world, Sloboda suggests that music becomes 'richly and sublimely expressive' (Sloboda 1985, p. 63).

[50] Meyer (1956), Cooke (1959).

[51] For the activation of motor areas of the brain when listening, see Janata and Grafton (2003).

[52] The following alternatives have been clearly outlined and discussed by Brown (2000), who also suggests the possibility that music and language may have evolved in parallel and then become 'bound together' by relatively recently evolved neural circuits. This also seems unlikely; it is difficult to imagine how either a 'proto-music' or 'proto-language' could have evolved without the characteristics that language and music share being in place very early on in

their evolutionary history, as these characteristics appear to be so fundamental to their nature.

[53] Brown (2000).

[54] For a discussion of J.-J. Rousseau's *Essai sur l'origine des langues* (1781), see Thomas (1995).

3. Music without language

[1] Changeux (1985). These figures use the UK definition of a trillion.

[2] The cerebral cortex is linked to the brainstem by four anatomical structures, two of which are relatively small – the epithalamus and subthalamus – and two of which are particularly important parts of the brain – the thalamus and hypothalamus. The cerebral cortex lies upon the thalamus, which is appropriate since this name derives from the Greek word for bed. The cerebral cortex receives most of its inputs via the thalamus. The hypothalamus is a part of the brain that specializes in the control of internal organs and the automatic nervous system; it regulates hunger, thirst, sexual functions and body temperature. The cerebral cortex and these four anatomical structures are known as the forebrain; they are, in essence, the products of a massively swollen end of the brainstem. A rather less swollen part forms what is known as the midbrain. This has several distinct anatomical structures, one of which is the substantia nigra. This contains neurons that release dopamine, which is one of the most important neurotransmitters. If it fails to function correctly, an insufficient amount of dopamine is produced, muscles lose the ability to move freely, and body movements become jerky. Behind the midbrain one finds (not surprisingly) the hindbrain, which has a further set of anatomical structures. The most notable of these is the cerebellum. This looks like a smaller version of the cerebrum, as it has a highly convoluted surface with an outer 'bark' of grey matter. The cerebellum is involved in fine motor movements, and when it is damaged the individual becomes very clumsy. Although it constitutes a relatively small fraction of the human brain, owing to the massive size of the cerebral cortex, the cerebellum constitutes as much as 90 per cent of total brain size in certain fish species, and almost half in certain bird species. The hindbrain is directly connected to the brainstem, which in turn leads into the spinal cord and the rest of the central nervous system.

[3] Non-human brains can also be studied by methods such as single-cell recordings, which may then permit inferences as to how human brains function.

[4] Damasio and Damasio's 1989 volume *Lesion Analysis in Neuropsychology* provides the classic work describing how lesion analysis is integrated with brain imaging to provide insights into the brain.

[5] The case of Shebalin is described by Luria et al. (1965).

[6] Schlaug et al. (1995).

[7] Chan et al. (1998).

[8] An account of NS is provided by Mendez (2001).

[9] Although, as covered in this and the following chapter, there is a considerable amount of work comparing the perception of musical and environmental sounds, I have not been able to find a formal means of distinguishing between these.

[10] The following account draws on Metz-Lutz and Dahl (1984).

[11] There are several possible explanations as to how language abilities are able to return following their temporary loss. The neural networks may have become inhibited – for example, by swelling in the brain, and then return to normal working. Alternatively, new neural networks might be formed to replace those that have been damaged, or existing networks might take on new functions.

[12] Her doctors suspected that this was due to a lowered auditory concentration span, rather than inhibition of her ability for rhythm.

[13] Yaqub et al. (1988).

[14] This case is described by Takahashi et al. (1992).

[15] This was a small sample, and without the use of controls (that is, individuals without any impairments undertaking the same tests), so it is difficult to know whether there had been a selective impairment of his ability to process rhythm.

[16] Godefroy et al. (1995).

[17] It should be noted here that the way in which a music 'faculty' develops in the brains of 'savants' might be quite different from the 'normal' development of the musical faculty – I am grateful to Ilona Roth for this note of caution.

[18] The following text draws on Miller (1989).

[19] Another is LL, a man in his twenties who had been blind since birth. LL had been born with cerebral palsy and various other congenital anomalies. He was fortunate in having very caring foster parents, but his mental development was extremely slow. Like Eddie and 'Blind Tom', his early language was characterized by echolalia; now, as an adult, he has minimal ability at verbal communication. In spite of these difficulties, LL is a brilliant pianist, especially of classical and country and western music. It was once claimed that LL's musical talent was only discovered when, at the age of sixteen, he woke his foster family in the middle of the night by playing Tchaikovsky's First Piano Concerto. But Miller found evidence that LL had shown an interest in music since at least the age of three; on one occasion he was discovered strumming the bedsprings under his bed while humming a tune. He was given toy instruments as a child and when aged eight his family purchased a piano. His foster mother provided informal music lessons, encouraging the development of his

talent – although she actively discouraged LL's particular liking for country and western music.

[20] The case studies of Eddie and CA rely to some extent on anecdotal accounts of their musical abilities. But with Noel, another musical savant, rigorous scientific tests have been undertaken to evaluate and understand his musical abilities (Hermelin 2001). These have demonstrated a quite astonishing musical memory. At the age of five, Noel had been placed in a school for children with severe learning difficulties. He was autistic – unable to make contact with other children or initiate speech – and had patterns of repetitive and obsessive behaviour associated with this condition. One of these was listening to music on the radio and then playing it by ear the following day on the school piano. The psychologists Beate Hermelin and Neil O'Connor heard about Noel when making a study of autistic savants, including those with outstanding abilities at drawing, arithmetic and poetry, all of which are described in Hermelin's fantastic 2001 book *Bright Splinters of the Mind*.

They examined Noel when he was nineteen years old, when he had an IQ of 61 and an almost total absence of spontaneous speech. Hermelin and O'Connor played Greig's 'Melody', Op. 47 No. 3, to Noel and to a professional musician, neither of whom were familiar with the piece. It was played from beginning to end, and then again in short sections. After each section, Noel and the professional musician had to play it back, along with the previous sections, until finally they played the whole piece from memory. Noel gave an almost perfect rendering of all sixty-four bars of 'Melody', retaining both the melody and harmonic components and making mistakes on only 8 per cent of the 798 notes. The professional musician was only able to play 354 notes, 80 per cent of which were incorrect. Twenty-four hours later Noel gave a second near-perfect performance.

By analysing Noel's mistakes, and undertaking tests with further pieces of music, Hermelin and O'Connor concluded that Noel's musical memory was based on an intuitive grasp of musical structure. Through his obsessive listening to music, he had acquired a profound understanding of the diatonic scales in which the majority of Western music from between 1600 and 1900 is composed, including Greig's 'Melody'. This is what I referred to in chapter 2 as 'tonal knowledge', the closest musical equivalent to a linguistic grammar. Noel combined his tonal knowledge with a tendency to focus on discrete musical phrases, rather than attending to the piece as a whole as was the inclination of the professional musician. This interpretation of Noel's ability was confirmed when he was asked to repeat *Mikrokosmos* by Bartók under the same conditions. This piece was composed in the 1930s and eschews the diatonic scale, falling into the category of atonal music. Noel was now markedly less successful, making errors in 63 per cent of the 277 notes he played, in contrast to the 14 per cent of errors in 153 notes played by the professional musician.

[21] That is, if Eddie had been asked whether one melody was the 'same as or

different to' another melody, his response would have been simply to repeat the phrase 'same or different'.

[22] Hermelin's studies on perfect pitch were undertaken with Pamela Heaton and Linda Pring and are described in Hermelin (2001, pp. 168–78). They were made on children suffering from autism, some of whom had exceptional musical abilities. By definition, none of them had normal linguistic abilities, but all were found to have perfect pitch. The figure of one in ten adults having perfect pitch is given in Takeuchi and Hulse (1993).

[23] Ilona Roth (pers. comm.) has informed me that echolalia is found among autistics as a matter of degree and may diminish with development, which appears, from Miller's (1989) description, to have been the case with Eddie.

[24] Mottron et al. (1999) consider the association between perfect pitch and autism. They studied a girl known as QC, a low-functioning adolescent with autism and with perfect pitch. They suggested that her autism resulted from a lack of cognitive flexibility in a person with a marked interest in auditory stimuli, which had arisen at the critical age for the appearance of perfect pitch.

[25] Simon Baron-Cohen (2003) argued 'the female brain is predominantly hardwired for empathy. The male brain is predominantly hardwired for understanding and building systems.'

[26] Miller (1989, p. 183).

[27] In Hermelin's (2001) studies of autistic children, she found a tendency to focus on segments rather than wholes – which is recognized as a widespread characteristic and is often described as a lack of 'central coherence'. So, for instance, an autistic child would be interested in the individual elements of a picture rather than in what it appeared to look like as a whole, and would interpret each element quite independently of the others. The only exception to this was in the domain of music, where autistic children showed some ability to respond to the characteristics of a piece of music as a whole, rather than to focus entirely on its individual elements. These autistic children showed no particular musical, or any other, ability. But the results might suggest that savant musical ability is not to be explained in the same manner as, say, savant artistic or savant arithmetical abilities.

[28] Miller (1989, p. 241).

4. Language without music

[1] The following account draws on Henson (1988) and Sergent (1993).

[2] Alajouanine (1948).

[3] Wilson and Pressing (1999, p. 53).

[4] The following account draws on Piccirilli et al. (2000).

[5] Wilson and Pressing (1999, p. 53).

[6] Peretz (1993, p. 23).

[7] Wilson and Pressing (1999).

[8] The case of KB is described by Steinke et al. (2001).

[9] The case of CL is described in Peretz et al. (1994), where she is compared and contrasted with GL.

[10] Wilson & Pressing (1999).

[11] Patel et al. (1998).

[12] Allen (1878).

[13] Fry (1948) evaluated a 1200-subject sample on tests requiring the subjects to compare two notes or two musical phrases in order to detect a change in pitch, leading him to conclude that 5 per cent of the British population was amusical, and that congenital amusia arose from deficiencies in both musical memory and pitch discrimination. Ayotte et al. (2002) argue that such claims are not supported by Fry's data analysis. Kalmus and Fry (1980) examined the ability of a sample of 604 adults to detect anomalous pitches inserted into melodies, concluding that 4.2 per cent of the British population were amusical. Ayotte et al. (2002) also describe this finding as problematic and criticize the methodology employed.

[14] Kazez (1985).

[15] The case of Monica is described by Peretz et al. (2002).

[16] Although speech excerpts from well-known people were used, these were selected so as not to include any words that might have given a clue as to the speakers' identities (Peretz et al. 2002).

[17] Ayotte et al. (2002).

[18] The studies by Isabelle Peretz and her colleagues demonstrate that congenital amusia exists – tone-deafness is not a myth. Between 3 and 6 per cent of the population suffer from a language-learning deficiency in the absence of any other cognitive impairment, and a similar percentage can be suggested for congenital amusia.

An obvious question is whether congenital amusia is a biologically heritable condition – can those who are unable to hold a tune blame it on their parents? Monica's mother and her brother were described as being musically impaired, but were not formally tested. Her father and sister were not reported as suffering any such deficits. Of the ten other subjects tested by Peretz, six reported that one of their parents (often their mother), and at least one sibling, were musically impaired, while every family also had members who were musically competent.

There has, to my knowledge, been just one study designed to explore the heritability of musical ability, aiming to assess the relative significance of genetic and environmental factors (Drayna et al. 2001). This appeared in the distinguished journal *Science*, and is of particular interest because it focuses on the same musical deficit that Peretz had identified as the essence of congenital amusia: the recognition of pitch. It was conducted by a joint team of scientists from the National Institute on Deafness, in Maryland, and the Twin Research Unit at St Thomas' Hospital, London. The latter was crucial to the study as twin studies provide the means to evaluate the relative contributions of inherited genes and developmental environment to individual thought and behaviour.

Identical twins – those originating from a single egg – have exactly the same genes as each other; while fraternal, or dizygotic, twins are of no greater similarity to one another than to any other sibling from the same parents. Except in very unusual cases, twins grow up in environments that are as identical as is possible for two individuals, undergoing the same education and life experiences. Very occasionally, twins are separated at birth and grow up with different families, in quite different environments. By examining the similarities and differences between identical and fraternal twins when they are adults, and comparing those who remained together and those separated at birth, it is possible to tease out the relative significance of genes and of environment on their looks, thoughts and behaviour.

Dr Dennis Drayna, from the National Institute on Deafness, led the study of pitch discrimination abilities. He was able to test 284 twin pairs, 136 of which were identical. All twins were asked to undertake a version of the 'distorted tunes test', which was used by Peretz and others when examining the musical abilities of brain-damaged and other subjects. This is the test in which familiar melodies are played, in some of which notes with an incorrect pitch have replaced one or more correct notes, and the subjects have to identify whether the melodies are correct or not. Each individual listened to twenty-six melodies and was scored according to how many they correctly identified as being correct. The test was carried out on a very rigorous basis, designed to ensure that factors such as hearing deficits and gender did not influence the results.

Drayna found that the pairs of identical twins showed greater similarity to one another in their ability to detect distorted tunes than did the fraternal twins. By a sophisticated statistical analysis, he calculated that approximately 80 per cent of pitch recognition ability derives from inherited genes and 20 per cent from the specific environment within which one matures and the musical experiences to which one is exposed.

5. The modularity of music and language

[1] Ilona Roth (pers. comm.) has noted that there is another interpretation of the case studies I have presented. It is possible that the complementary mechanisms for music in the absence of language, or vice versa, are not the same as those that exist when both language and music are present and function normally. Although we might know that the 'ouput' behaviour for music or language is the same as in a normal individual, we do not know that the neural mechanisms that generate this output are necessarily the same.

[2] Peretz (2003, p. 200).

[3] Peretz and Zatorre (2003) provide a collection of papers describing the latest research in the cognitive neuroscience of music.

[4] Patel (2003) proposes how this apparent contradiction can be resolved, by arguing that there is one specific point of convergence between syntactic processing in language and music.

[5] Maess et al. (2001). Koelsch et al. reached a similar conclusion from an fMRI study, concluding that 'the cortical network known to support language processing is less domain-specific than previously believed' (Koelsch et al. 2002, p. 956).

[6] The following draws on the report by Parsons (2003).

6. Talking and singing to baby

[1] Fernald (1991) reviews evidence that compares the prosody in mothers', fathers' and siblings' speech to infants. The higher pitch of siblings, however, might be a reflection of their own heightened levels of arousal, rather than a means of influencing the infant.

[2] Fernald (1991) cites evidence that infants prefer to listen to IDS rather than adult-directed speech.

[3] Malloch (1999) has made one of the most detailed studies of speech behaviour between mother/father and infant, characterizing it as 'communicative musicality'. Robb (1999) has explored the impact upon turn-taking when mothers fail to exaggerate prosody in their IDS owing to postnatal depression.

[4] Monnot (1999).

[5] Fernald and Mazzie (1991) examine how prosody is used to facilitate word learning by children.

[6] Nelson et al. (1989) undertook experiments that demonstrated how IDS contributes to the learning of syntax.

[7] The following text draws on Fernald (1991), who provides a four-stage, age-related model for the developmental functions of IDS, which she describes as: (1) intrinsic perceptual and affective salience; (2) modulation of attention, arousal and affect; (3) communication of intention and emotion; and (4) acoustic highlighting of words.

[8] Fernald (1989, 1992).

[9] Fernald (1989, 1992).

[10] This correlation has been identified by Monnot (1999). It depends, however, on rather subjective measures of the quantity and quality of IDS. Monnot

argues that the correlation she finds suggests that IDS is instrumental in language evolution.

[11] Fernald et al. (1989). This study also contains a critique of studies that claim to have found human languages in which no specific prosodic register is used in speech to infants.

[12] The origin, spread and diversification of Indo-European languages are matters of considerable dispute, but the problems have become potentially soluble owing to the recent integration of evidence from genetics, linguistics and archaeology. This is exemplified in the volume edited by Renfrew and Bellwood (2002), which contains several articles about the possible origins of Indo-European languages. The date I have given is a broad time horizon for the spread of farmers across Europe, which some equate with the spread of proto-Indo-European.

[13] Studies of Xhosa IDS are cited in Fernald (1991). Papousek et al.'s (1991) results confirmed the findings of Grieser and Kuhl (1988) regarding similarities between Mandarin and English IDS.

[14] Saffran et al. (1996, 1999).

[15] Cited in Saffran et al. (1999).

[16] Saffran and Griepentrog (2001). For a further development of Saffran's work on perfect pitch in infancy, see Saffran (2003).

[17] Heaton et al. (1998); Mottron et al. (1999).

[18] Saffran and Griepentrog (2001, p. 82).

[19] Zatorre (2003) argues that the higher incidence of perfect pitch in Asian people is found irrespective of any cultural factors such as type of language or childhood exposure to music. He suggests that it derives from a genetic predisposition for the possession and/or maintenance of perfect pitch.

[20] This appears to be a critical period in cognitive development for acquiring/maintaining perfect pitch, just as there is a critical period for language learning (Takeuchi and Hulse 1993). The presence of perfect pitch in adults appears to derive partly from such experiential factors and partly from one's genetic make-up (Zatorre 2003).

[21] I am grateful to Alison Wray (pers. comm.) for this suggestion as to how the possession of perfect pitch might enhance musical ability.

[22] Hermelin (2001, pp. 168–78) describes experiments that compared the abilities of musically naive, autistic children and 'normal' children (who were matched for levels of mental development and intelligence) at identifying and recalling musical pitches. She also found that such autistic children were more able at segmenting chords into their constituent parts and at relating notes to one

another. A detailed case study of a low-functioning adolescent with autism is provided by Mottron et al. (1999).

[23] D'Amato (1988) argues that cebus monkeys have perfect pitch, while Wright et al. (2000) explore the musical sensitivity of monkeys. There is, of course, an immense methodological challenge with such work, and these results should be used with caution. Hauser and McDermott (2003) note that the value of Wright et al.'s results depends critically on whether the monkeys they tested had had prior exposure to Western music. Other species have perfect pitch abilities, most notably songbirds. Weisman et al. (2004) compared the perfect pitch abilities of songbirds, humans and rats. Experiments to test for perfect pitch in the African apes would be of immense interest.

[24] For studies on lullabies see Trehub and Schellenberg (1995), Trehub and Trainor (1998). Cross-cultural similarities and the abilities of adults to identify lullabies are examined in Trehub et al. (1993) and Unyk et al. (1992).

[25] Trehub and Schellenberg (1995). See also Trehub et al. (1997) and Trehub and Trainor (1998).

[26] Trehub and Nakata (2001).

[27] This study was undertaken by Shenfield, Trehub and Nakata, and is reported in Street et al. (2003).

[28] These studies have been reported in Standley (1998, 2000) and Standley and Moore (1995), and are summarized in Street et al. (2003).

[29] Street et al. (2003, p. 630).

[30] Trehub (2003, p. 13).

[31] Trevarthen (1999, p. 173).

[32] I have deliberately avoided the issue of infant movement in this chapter, even though rhythmic kicking of legs, waving of arms and rocking may have significant 'musical' qualities. Thelen (1981) has described such movements as rhythmic stereotypes and considered their adaptive significance and origin.

[33] Trevarthen (1999, p. 174). He goes on to stress that fathers also interact with their infants in the same manner.

[34] Knutson et al. (1998).

[35] For brain development see Changeux (1985).

[36] For neural Darwinism, see Changeux (1985), Edelman (1987, 1992).

[37] For neural constructivism, see Greenough et al. (1987) and Rosenzweig et al. (1999).

7. Music hath charms and can heal

[1] See Nettl (1983, chapter 11), which coincidentally uses the same title as my chapter but has more interesting reflections on Congreve's poem.

[2] Meyer (1956), Cooke (1959). Meyer (2001) provides an updated review of his ideas.

[3] Juslin and Sloboda (2001) review the history of the study of the relationship between music and emotion.

[4] Juslin and Sloboda (eds, 2001).

[5] Oatley and Jenkins (1996) describe the history of the study of emotions.

[6] Evans (2001) provides a short and very readable introduction to the view of emotions as central to human rationality and intelligence. Key works in the development of this viewpoint are De Sousa (1987), Damasio (1994), Oatley and Jenkins (1996).

[7] See, for instance, Damasio (1994, 2003), Le Doux (1996).

[8] Both Oatley and Jenkins (1996) and Evans (2001) describe the debates between those who believe that all emotions are cultural constructions and those who believe that some at least are biologically based human universals. Ekman's work has been essential in establishing the latter viewpoint (Ekman 1992, Ekman and Davidson 1995).

[9] Goodall (1986). See also De Waal (1982) for the complicated social and emotional lives of chimpanzees.

[10] Oatley and Johnson-Laird (1987).

[11] Frank (1988).

[12] Ekman (1985, 2003) provides brilliant studies of human emotions, offering evidence for cross-cultural universals of expression and exploring the limits of lying about one's emotions.

[13] Baron-Cohen (2003) provides an entertaining 'reading the mind in the eyes test' as an appendix to his book. Similarly, Ekman (2003) provides a suite of 'photos of people with particularly subtle expressions that may be a mixture of two emotions for a reader to test themselves against'.

[14] Niedenthal and Setterlund (1994). More specifically, they played Vivaldi's Concerto in C Major and Mozart's *Eine kleine Nachtmusik* to induce positive moods; when sad subjects were required they played the Adagietto by Mahler or the adagio from the Piano Concerto No. 2 in C Minor by Rachmaninov.

[15] Blacking (1973, pp. 68–9). Blacking was concerned that his own evidence from the Venda did not show the same association between musical intervals and emotions as was found by Cooke. He suggested that 'it is possible that Venda

musical conventions have suppressed an innate desire in Venda people to express their emotions in a specific, universal way'. He also stressed how Cooke's theory is not general enough to apply to any culture and hence is inadequate for European music.

[16] Oelman and Lœng (2003, p. 244). Her work was presented at the 5th Triennial ESCOM conference held in September 2003 at Hanover University.

[17] Cooke (1959, p. 60).

[18] Cooke (1959, p. 105).

[19] The following draws on Juslin (1997). This paper involves far more than the experiments described in my text, as Juslin seeks to develop a generalized model for a functionalist perspective on emotional communication in music.

[20] The female advantage in inferring the intended emotion from the musical performance was not, however, statistically significant, so we must be cautious in interpreting this result (Juslin 1997).

[21] The following draws on Scherer (1995).

[22] Scherer and Zentner (2001) describe the problems with subjective reporting of emotional states. In their studies they have explicitly requested their subjects to differentiate between the emotions they feel themselves and those they recognize in the music to which they are listening.

[23] Krumhansl (1997).

[24] Scherer and Zentner (2001) provide a critical review of such experiments.

[25] Scherer and Zentner (2001).

[26] Dibben (2003) provides experimental evidence that indicates that 'arousal level influences the intensity of emotion experienced with music, and therefore that people use their bodily sensations as information as to the emotion felt and expressed by music'. Viellard et al. (2003) provide some insights into the influence of musical expertise on the ability to extract emotional information from a piece of music.

[27] Bunt and Pavlicevic (2001) explore the relationship between music and emotion from the perspective of music therapy.

[28] This is the volume edited by Schullian and Schoen (1948), which is analysed and discussed by Gouk (2000).

[29] Standley (1995) provides a review of the use of music therapy in medical and dental treatment. Robb et al. (1995) provide a particularly interesting study of how music therapy has been used to reduce preoperative anxiety in paediatric burns patients, while Davila et al. (1986) consider the use of music to relax mentally handicapped patients while undergoing dental treatment, and Robb (2003) provides an overview of music therapy in paediatric healthcare.

[30] This is a quote from Paul Robertson describing work at the Spingte Clinic and is cited by Horden (2000, p. 12). Robertson presented a series on Channel 4 television in the UK in May 1996 entitled 'Music and Mind'.

[31] Mandel (1996).

[32] Mandel (1996).

[33] Unkefer (1990) provides an overview of the use of music therapy for adults with mental disorders.

[34] Mandel (1996, p. 42).

[35] This point has been made by Horden (2000, p. 17).

[36] Evans-Pritchard (1937), and see remarks regarding this and other anthropological accounts in Horden (2000, pp. 17–18).

[37] Roseman (1991), Janzen (1992), Friedson (1996).

[38] Our knowledge of how music has therapeutic and healing properties continues to develop. The 2003 triennial ESCOM conference held in Hanover – probably the world's largest gathering of musicologists – had a whole session devoted to music therapy. Aiyer and Kuppuswamy (2003) provided the results of their use of Indian classical music for therapy for patients suffering from stress, depression, high blood pressure, sleeplessness, muscle pain and mental retardation. Bailey and Davidson (2003) claimed that participation in group singing is more beneficial than less active types of music involvement.

[39] Isen (1970). Oatley and Jenkins (1996) provide a review of Isen's experiments.

[40] Isen et al. (1987).

[41] Baron (1987).

[42] Forgas and Moylan (1987).

[43] Fried and Berkowitz (1979).

[44] Fried and Berkowitz (1979) note that sitting in silence is not a very good control for such experiments as this might itself induce negative emotions arising from boredom.

8. Grunts, barks and gestures

[1] Byrne (pers. comm.) suggests that ape language experiments have simply come to a halt at 250 or less symbols, and so it remains unclear whether apes could potentially learn a much larger number of symbols. It is evident, however, that their rate of learning is a great deal slower than that of children.

[2] There is a large literature on chimpanzee language-learning experiments, which began in earnest during the 1960s. Key points in this literature are Beatrice and Alan Gardner's teaching of sign language to Washoe (Gardner et al. 1989),

and David Premack's language experiments with Sarah using plastic chips of different colours and signs (Premack and Premack 1972). A long-term research programme was begun in the 1970s at the Yerkes Language Research Centre by Sue Savage-Rumbaugh and Duane Rumbaugh (1993) who used symbols on a computer keyboard to represent words. Some of this work came under serious criticism from Terrace (1979, Terrace et al. 1979), who argued that the chimpanzee language abilities had been inadvertently exaggerated by poor methodology. Such criticisms were one factor that led to a switch from a teaching regime in which young apes were specifically trained to use symbols, to one in which they were simply exposed to language. In 1990 Sue Savage-Rumbaugh began work with Kanzi, a male bonobo, and soon claimed that he had acquired not only a large vocabulary but also a simple grammar (Greenfield and Savage-Rumbaugh 1990; Savage-Rumbaugh and Rumbaugh 1993). Both chimpanzees and bonobos have been shown to have an extensive understanding of spoken English. Even if this is the case, a vast gulf exists between chimpanzee 'language' abilities and those of humans – by the age of six children typically have a vocabulary of more than ten thousand words, can construct long sentences, and are constant commentators on the world around them. Steven Pinker's (1994) description of chimpanzee linguists as 'highly trained circus acts' is a bit harsh, but my assessment is that these experiments have very little to tell us about how human language evolved.

[3] Some scholars such as Bickerton (1990), Arbib (in press) and Tallerman (in press) argue against any direct evolutionary continuity between human language and ape vocalizations, claiming, for instance, that all ape vocalizations are involuntary and rely on quite different parts of the primate brain from those used in human language. While there are undeniable and significant differences, to argue that these invalidate evolutionary continuity strikes me as bizarre. Seyfarth (in press) has succinctly summarized the case for continuity, citing studies that demonstrate continuities between the behaviour, perception, cognition and neurophysiology of human speech and primate vocal communication.

[4] There are some disagreements among palaeoanthropologists regarding the dates of lineage divergence. The dates are calculated by examining the extent of difference in the DNA of different species, and then estimating the rate at which those differences would have arisen. Boyd and Silk (2000) provide a succinct explanation of how such differences are measured and the divergence dates calculated. The dates I cite are towards the younger end of the spectrum within the literature, with others pushing the gorilla/human divergence back to 10 million years ago and the chimpanzee/human divergence to 6 million years ago.

[5] Mitani (1996) provides a comparative study of African ape vocal behaviour, explaining the reasons for both similarities and differences.

[6] Mitani and Gros-Louis (1995) studied the species and sex differences in the

screams of chimpanzees and bonobos, concluding that body size was an influential factor. The relatively larger males of both species emit calls of lower frequency than the smaller females.

[7] For a study of vervet calls that used playbacks to identify their meaning, see Cheney and Seyfarth (1990). Richman (1976, 1978, 1987) describes the vocalizations of gelada baboons, while a summary of gibbon song, and its possible relevance for the evolution of human music, is provided by Geissmann (2000).

[8] Struhsaker (1967). The following section draws primarily on material in Cheney and Seyfarth 1990.

[9] Cheney and Seyfarth (1990) describe how a vervet gave a leopard alarm call in the context of an inter-group encounter. Byrne (pers. comm.) suggests that such 'tricks' are used without any deep understanding of how they work.

[10] Zuberbühler (2003, p. 270) notes that the small vocal repertoires of non-human primates make it difficult to support the case that their vocalizations are referential. He also notes that it is unclear whether the small number of acoustic utterances reflects a similarly small number of mental representations, or whether these matters are quite unrelated.

[11] Wray (1998).

[12] The term 'functionally referential' means that they serve to refer to something in the outside world without the necessary intention of doing so. Zuberbühler (2003, p. 269) summarizes the debate regarding whether or not the calls of non-human primates are referential.

[13] Zuberbuhler (2003) provides detailed descriptions of the alarm calls of Diana monkeys and the experiments undertaken to ascertain whether they are referential. He is sympathetic to this view, concluding that their calls are processed on a 'conceptual-semantic' rather than a 'perceptual-acoustic' level.

[14] Studies on the calls of rhesus monkeys are described by Hauser (2000).

[15] Richman (1987, p. 199). See also Richman (1976, 1978) for his studies of gelada monkeys, and Dunbar and Dunbar (1975) for a study of the social dynamics of geladas.

[16] Richman (1987, p. 199).

[17] Leinonen et al. (1991) and Linnankoski et al. (1994) tested the ability of human children and adults to identify the emotional content of monkey calls.

[18] Leinonen et al. (2003) explored whether there are similarities in the waveforms of utterances made by humans and macaques.

[19] Gibbons may not be as monogamous as once thought. Oka and Takenaka (2001) did parentage tests on a group of wild *Hylobates muelleri* and found

evidence for extra-pair copulations. A similar finding had been reported by Reichard (1995) for *Hylobates lar.*

[20] Geissmann (1984) demonstrated genetic predisposition for specific songs by showing that gibbon hybrid siblings (from a *Hylobates pileatus* father and a *Hylobates lar* mother) could not have learnt their vocalizations from either of their parents.

[21] Cowlishaw (1992) tested the hypothesis that females sing to attract males. If this was the case then females should stop singing after mating; but they do not. Ellefson (1974) and Tenaza (1976) proposed the hypothesis that females sing to defend their territory. It was supported by observation and singing playback experiments, which showed that females respond differently to play-backs made at the centre and at the edge of their territory. Cowlishaw (1992) also used playback experiments to argue that female singing has a territorial defence function.

[22] Mitani and Marler (1989) explain that male gibbon songs are highly versatile with songs rarely repeated but governed by distinct rules.

[23] Cowlishaw (1992) reviews the evidence for male gibbon songs and discounts the territorial defence hypothesis, finding that there is no relationship between the number of territorial conflicts and singing intensity. Unmated males sing more frequently and for longer bouts than those who have mated, and some males cease to sing after mating. Those who continue to sing may be attempting to repel rival males. Cowlishaw supports this hypothesis by showing that there is a positive correlation between the amount of singing and the amount of opposition from other males.

[24] Geissmann (2000, p. 107).

[25] Brockleman (1984) hypothesized that the function of gibbon duets was to form and maintain pair-bonds. Geissmann (2000) summarizes his tests of this hypothesis with siamangs.

[26] Schaik and Dunbar (1990) suggested that males sing with females to advertise their paired status. Cowlishaw (1992) argues that duetting occurs to reduce the costs of territorial conflicts.

[27] Mitani (1985).

[28] Such grading of one call into another was demonstrated by Marler and Tenaza (1977).

[29] Unless otherwise specified, material in the following section draws on Mitani (1996).

[30] Byrne (pers. comm.).

[31] Mitani and Brandt (1994) have explored the factors that influence the long-

distance pant-hoots of chimpanzees, finding that calls varied more within than between individuals.

[32] Mitani et al. (1992) is the first study to identify geographical differences in the pant-hoots of chimpanzees. They discuss various interpretations, before concluding that vocal learning is the most likely. The other likely possibility is that the call differences could have arisen from an artefact of provisioning – different sounds were unintentionally rewarded by feeding (Byrne pers. comm.).

[33] Crockford et al. (2004).

[34] Seyfarth et al. (1994) provide a study of gorilla double grunts, identifying the 'spontaneous' and 'answering' double grunt, and showing that double grunts are individually distinctive.

[35] 'Bigger' here refers to absolute brain size. If brain size is measured relative to body size there is some overlap between apes and monkeys (Byrne pers. comm.).

[36] Zuberbühler (2003, p. 292) notes that while vocal repertoires often appear small, close examination may reveal subtle acoustic variations. For instance, a study of baboons in the Okavango Delta of Botswana showed that their barks varied continuously in their acoustic structure across a number of contexts, such as when given in response to carnivores or crocodiles, during contests between males, or when callers were separated from the group. Zuberbühler suggests that subtle acoustic variations may provide recipients with rich information about the external event perceived by the signaller.

[37] See Crockford et al. (2004) for arguments that chimpanzees engage in vocal learning and Zuberbühler (2003, p. 292) for a summary of evidence, from tamarins, rhesus monkeys and other chimpanzee studies, for the ability of non-human primates to modify their call structures.

[38] Duchin (1990) provides a comparative study of the oral cavity of the chimpanzee and human with regard to the evolution of speech.

[39] See Zuberbühler (2003, p. 293) for a summary of the relationship between the vocal tract in non-human primates and their vocalizations.

[40] Chimpanzee hunting is described by Stanford (1999), with useful discussion about its implications for the origins of human behaviour. He cites success rates for lone hunters of about 30 per cent, increasing to as much as 100 per cent for parties of ten or more.

[41] Boesch (1991) records two instances of mothers apparently 'teaching' their infants how to crack nuts, neither of which involved any vocal instruction. These instances emphasize the extraordinary rarity of such teaching, in view of the nutritional value of nuts and the difficulty that infants have in learning how to open them. Cheney and Seyfarth (1990) have commented on the surprising absence of further functionally referent calls among vervets to

supplement the predator alarm calls. They describe how a call that means 'follow me', made by a mother to her infant, would be of considerable benefit to the protection of the young, and hence to her evolutionary advantage.

[42] This is most likely to be a very significant constraint on monkeys, which are most unlikely to have a theory of mind. Indeed, it seems unlikely that they perceive of other individuals as mental agents whose actions are driven by beliefs, desires and knowledge (Zuberbühler 2003, p. 298). The extent to which this is also the case for the African apes is more contentious.

[43] There is a very extensive literature discussing theory of mind in general, and more particularly whether chimpanzees possess it. Key publications include the edited volumes by Byrne and Whiten (1988, 1992), which concentrated on the evidence from tactical deception, and the edited volumes by Russon et al. (1996) and Carruthers and Smith (1996). Povinelli (1993, 1999) has undertaken extensive laboratory experiments, which, he argues, show that chimpanzees cannot understand mental states. This has now been challenged by the latest findings of Tomasello et al. (2003). Dunbar (1998) discusses the relationship between the evolution of theory of mind and language.

[44] Tomasello et al. (2003).

[45] Dunbar (2004) argues that the orders of intentionality possessed by a living primate or extinct hominid can be inferred from absolute brain size. He maintains that increasing orders of intentionality are the critical aspect of cognitive evolution.

[46] The following draws on Tanner and Byrne (1996).

[47] Zuberbühler (2003, p. 297) explains that there is no evidence for referential gestures among non-human primates.

[48] Kohler (1925, pp. 319–20).

[49] Savage-Rumbaugh et al. (1977). See also the studies of gestural communication undertaken by Tomasello and his colleagues (Tomasello et al. 1985, 1989, 1994).

[50] See Cheney and Seyfarth (1990, pp. 98–102) for a brief summary of changing views about primate vocalizations.

[51] My conclusion that non-human primates lack referential calls will not be agreeable to many other writers, some with far more specialized knowledge of primate communication than my own. See Zuberbühler (2003) for a comprehensive review of the existing literature and an interpretation that attributes referential signalling to primates.

[52] Byrne (pers. comm.).

9. Songs on the savannah

[1] The Sahel discovery has been designated as a new species from a new genus, *Sahelanthropus tchadensis*. It is described by Brunet et al. (2002) while Wood (2002) provides a commentary regarding its potential significance in the evolutionary history of humans and apes.

[2] Unless otherwise specified, the following text draws on material in Boyd and Silk (2000) and Johanson and Edgar (1996).

[3] Collard and Aiello (2000) discuss the implications of recent identification of knuckle-walking traits in *Australopithecus afarensis* and *Australopithecus anamensis*, commenting on the mix of terrestrial and arboreal adaptations in *Australopithecus afarensis*.

[4] See Richmond and Stait (2000) for an analysis of the wrist bones of Lucy and similar specimens that suggest knuckle-walking. Collard and Aiello (2000) comment on their conclusions.

[5] For the difficulties involved in classifying these fossils, see Wood (1992), Wood and Collard (1999).

[6] For reviews of the archaeology within these two localities see Potts (1988) and Isaac (1989).

[7] Key articles in this debate include Isaac (1978), for the proposal of home bases, substantial meat-eating and food sharing; Binford (1981) for a critique proposing that hominids were 'marginal scavengers'; Bunn (1981) and Potts and Shipman (1981), for the presence of cut-marks. Isaac (1983), Bunn and Kroll (1986), and Binford (1986) continued the debate.

[8] Dunbar (1988) provides evidence for a correlation between group size and within-group aggression, along with a review of the factors that influence group size in non-human primates.

[9] The home base hypothesis, later rebranded as central place foraging, was originally proposed by Isaac (1978).

[10] For the relationship between group size and brain size see Aiello and Dunbar (1993) and Dunbar (1992). Aiello and Dunbar suggest that the cranial capacity of fossils classified to early *Homo* indicate a group size between seventy-one and ninety-two individuals.

[11] For the costs of a large brain and the need for a high quality diet, most likely involving meat, see Aiello and Wheeler (1995).

[12] For a summary of early hominid technology, known as the Oldowan industry, see Toth (1985).

[13] Toth et al. (1993) describe tool-making experiments involving Kanzi.

[14] Dunbar (2004) argues for a strong relationship between brain size and theory

of mind abilities as measured by the orders of intentionality one can handle.

[15] The following text regarding oral gestures draws on Studdert-Kennedy (1998, 2000) and Studdert-Kennedy and Goldstein (2003).

[16] Moggi-Cecchi and Collard (2002) describe the study of the fossil stapes from Stw 15. Their conclusion of sensitivity to high pitch is supported by the studies undertaken by Spoor and Zonneveld (1998) on the bony labyrinth of the inner ear. By using CT imaging techniques, Spoor was able to record the height and width of the cochlea in a sample of hominid fossils coming from several different species. He found that the bony labyrinth of early hominids was relatively short, which is known to correlate with the perception of higher frequencies among mammals as a whole.

[17] Tobias (1981, 1987), Falk (1983). I am sceptical of the value of endocasts. Deacon (1997) notes that other palaeoneurologists have suggested that the presence of additional folding in the brain of *Homo habilis* may simply have been a consequence of increasing size rather than of the evolution of any functionally distinctive brain regions.

[18] Rizzolatti and Arbib (1998), Arbib (in press).

[19] Rizzolatti and Arbib (1998, p. 189).

[20] Studdert-Kennedy and Goldstein (2003). Also see Wray (2002a) regarding the acquisition of formulaic language by children.

[21] Studdert-Kennedy and Goldstein (2003, p. 246).

[22] Rizzolatti and Arbib (1998) and Arbib (in press) give mirror neurons a more fundamental role in the evolution of language than I am suggesting here, in what is, to me, an unnecessarily complicated scenario which rejects the idea of a direct evolutionary continuity between primate vocalization and human speech.

[23] Rizzolatti and Arbib (1998, p. 192).

[24] Brain (1981) studied hominid remains from South African sites and felt it was appropriate to characterize these as the 'hunted' rather than the 'hunter'.

[25] Berger and Clarke (1995) have identified the eagle as the accumulator of the Taung bone assemblage.

[26] The HAS site is described and interpreted by Isaac (1978) and within Isaac (1989), and played a key role in the development of his home base/food sharing hypothesis.

[27] Dunbar (1991) considers the functional significance of grooming, while the 'vocal grooming' hypothesis is described by Aiello and Dunbar (1993) and Dunbar (1993, 1996).

[28] Aiello and Dunbar (1993, p. 187).

[29] Knight (1998) and Power (1998) argue that because vocalizations are easy and cheap to make, they are unreliable and can be used to deceive. Consequently, they cannot be used to express commitment to another individual in the same manner as the time-consuming physical grooming.

[30] Power (1998) suggests that vocal grooming needs to be song-like if it is to provide the reinforcement mechanisms of opiate stimulation that make it pleasurable to be groomed.

[31] The presence of mirror neurons alone, which may have enabled manual gestures to have been imitated, would not have been sufficient for the development and cultural transmission of Oldowan technology. To master the art of making stone tools such as these one must appreciate the intention that lies behind an action, and this requires a theory of mind.

[32] The site of FxJj50 is described is described by Bunn et al. (1980) and within Isaac (1989).

10. Getting into rhythm

[1] Wood and Collard (1999).

[2] Blacking (1973, pp. vi–vii).

[3] For an excellent description of the anatomical differences between humans and apes, see Lewin (1999). This work is drawn on in the following text.

[4] Lewin (1999, p. 93).

[5] For a description of this specimen, and others referred to in this text, see Johanson and Edgar (1996).

[6] Spoor et al. (1994).

[7] Hunt (1994).

[8] East Africa experienced a major increase in aridity around 2 million years ago. DeMenocal (1995) provides a succinct overview of climate change in East Africa and how it appears to relate to human evolution.

[9] 'Stand tall and stay cool' was the title of the *New Scientist* article within which Wheeler (1988) summarized his theory. Longer and more academic articles addressing specific elements of the theory are provided by Wheeler (1984, 1991, 1994).

[10] Aiello (1996).

[11] Clegg (2001).

[12] The following text draws on Wray (1998, 2002a, c).

[13] Wray (1998, 2002c) also suggests that, paradoxically, a Bickertonian protolanguage would have been too powerful for what hominids and Early Humans

required. The availability of words allows novel messages to be constructed, but the majority of social communication among non-human primates, and a great deal among modern humans (for example, 'Good morning', 'how are you today'), is holistic in character – a set of routines that are understood and responded to with minimal processing.

[14] Wray and Grace (in press) have persuasively argued that there is a significant relationship between the nature of language and social organization. To quote them: 'The modern world fosters huge communities of people who are surrounded by strangers, and often separated geographically from their family and from those who most closely share their interests and concerns. Modern complex society favours specialized professions and pastimes that furnish the individual with particular expertise in some areas to the exclusion of others. In contrast, the first language-users presumably lived in fairly small groups of familiar, mostly closely genetically related, individuals, engaged in common purposes and activities as a "society of intimates ... where all generic information was shared".'

[15] Jackendoff (1999).

[16] Bickerton (1995) associated proto-language with *Homo erectus* and a long period of 'semi-stagnation' in its development. Bickerton (2000) suggests there were 2000 years of proto-language.

[17] Bickerton's criticisms of holistic proto-language are found in Bickerton (2003). Tallerman (in press) provides a much longer critique.

[18] Tallerman (in press) refers to five such discontinuities: human speech and primate vocalizations are handled by different parts of the brain; primate calls are involuntary; human speech and primate calls have a different physiological basis; primate calls are genetically transmitted; human speech has a dissociation between sound and meaning that is absent from primate calls. While there is undeniably some truth in each of these claims, the differences are shades of grey rather than black and white. Evolution is a process of change – it is the process that transformed the particular neural, physiological and behavioural aspects of primate calls used by the human/chimpanzee common ancestor into those of human language.

[19] Seyfarth (in press). I appreciate that some linguists may wish to discriminate between Seyfarth's use of the word 'speech' and their use of the word 'language'.

[20] Berger (2002, p. 113).

[21] The following text draws on Thaut et al. (1997, 2001) and Hurt et al. (1998).

[22] For this association, see Friberg and Sundberg (1999) and Shove and Repp (1995). Friberg and Sundberg (1999, p. 1469) cite a 1938 study by Truslit that argued: 'Every *crescendo* and *decrescendo*, every *accelerando* and *decelerando*, is nothing but the manifestation of changing motion energies,

regardless of whether they are intended as pure movement or as expression of movement.'

[23] For Sundberg's studies on the relationship between music and motion, see Sundberg et al. (1992), which shows how information on locomotion characteristics could be transferred to listeners by using the force patterns of the foot as sound-level envelopes; and Friberg and Sundberg (1999), which examined the association between deceleration in running and the final ritardandi of musical performances.

[24] Blacking (1973, p. 111).

[25] Bramble and Lieberman (2004).

[26] I am grateful to my daughter, Heather, for the terms 'arabesque' and 'échappé', which I have drawn from her book about ballet. An arabesque is a pose on one straight leg with the working leg extended straight behind the body. An échappé is a jump in which the feet begin in a closed position and separate in the air before landing apart.

[27] Modern humans also have two other classes of gesture, neither of which should be attributed to pre-modern humans. Emblematic gestures are intentional and come in highly standardized forms, which are often specific to particular cultures. These are the V for victory, thumbs-up or OK-type gestures. These are also often iconic in form, sometimes being derivative of the spoken word itself. A third type of hand movement is those used in sign language. These are symbolic in form and constitute a 'natural language' because they have a grammar and are equivalent in all respects to human spoken language.

[28] This was demonstrated by experimental studies using English- and Arabic-speakers by Aboudan and Beattie (1996).

[29] Beattie (2003) summarizes his work on the relationship between gesture and spoken utterance. A sample of his key experiments is found in Beattie and Shovelton (1999, 2000, 2002).

[30] McNeill (1992).

[31] McNeill (2000, p. 139).

[32] For a popular but informed and informative book on body language see Pease (1984), which is drawn upon in the following text.

[33] It is claimed that the size of the intimate zone varies between cultures, that for Americans being significantly larger, at 46–112 centimetres, than for Europeans and Japanese, at 20–30 centimetres. This can lead to misinterpretations of behaviour – an American might feel that a Japanese is being inappropriately intimate during a conversation (Beattie 2003).

[34] I have found Marshall (2001) a useful insight into how actors use their bodies for performance and expression.

[35] Laban (1988 [1950], p. 19).

[36] Pease (1984, p. 9).

[37] Trevarthen (1999, p. 171).

[38] Laban (1988 [1950], p. 1).

[39] Laban (1988 [1950], p. 94).

[40] Laban (1988 [1950], pp. 82–3).

[41] This was demonstrated by MacLarnon and Hewitt (1999), who measured the size of the canals of the thoracic vertebrae – through which the nerves for controlling the diaphragm, and hence breathing, pass – on a variety of fossil specimens. The thoracic canal of KNM-WT 15000 is similar in size to those of the australopithecines and the African apes, while those of the *Homo neanderthalensis* specimens are equivalent to that of *Homo sapiens*.

[42] Ruff et al. (1997).

11. Imitating nature

[1] For a description of the Schöningen spears, see Thieme (1997). Two other spear fragments were known from Europe prior to the Schöningen discoveries, coming from Clacton (England) and Lehringen (Germany). In both cases only tips were preserved, which were open to various interpretations regarding their effectiveness as spears or, indeed, whether they were from spears at all.

[2] In the late 1960s there was a consensus view that all pre-modern humans had been big-game hunters; indeed, big-game hunting was implicitly linked to the definition of humanity. Publications by Brain (1981) and Binford (1981) challenged the consensus with regard to Plio–Pleistocene hominids, arguing that they were the 'hunted rather than the hunters' and were 'marginal scavengers'. Binford (1984, 1985, 1987) questioned whether the Early Humans in Europe were big-game hunters, while Gamble (1987) reinterpreted the Clacton and Lehringen 'spears' as snow-probes for finding frozen carcasses.

[3] Dennell (1997, pp. 767–8) provided an important comment regarding the Schöningen spears in the issue of *Nature* in which their discovery was announced.

[4] For good reviews of changing attitudes to hominid dispersal, see Straus and Bar-Yosef (2001) and Rolland (1998). This is, however, a fast, fast-moving field and new discoveries that change our understanding should not be unexpected.

[5] These new dates remain controversial because they derived from sediment attached to the bone rather than the bone itself, and the skulls might simply have become buried in sediment much older than the bones themselves. For redating of the Java *Homo erectus* specimens see Swisher et al. (1994). For an evaluation of the claim, see Sémah et al. (2000). Many archaeologists were,

however, very sympathetic to the new dates, as they seemed to fit with a proposed date of 2 million years ago for stone artefacts from the Riwat Plateau in Pakistan. Robin Dennell and his colleagues had found these 'artefacts' in 1987 (Dennell et al. 1988). They look just like Oldowan chopper tools – cobbles of stone from which one or more flakes have been detached. Dennell's problem is that the detachments could have been made by collisions with other cobbles when being carried by river water. Consequently, the Riwat 'artefacts' might be a product of natural forces rather than hominid knapping. And yet, had they been found within Olduvai Gorge no one would have questioned their cultural authenticity.

[6] Wanpo et al. (1995) describe the artefacts and claimed *Homo* remains from Longuppo Cave, China.

[7] Zhu et al. (2004).

[8] Tchernov (1995) provides a review of the earliest fossil and archaeological evidence from the southern Levant, including those from 'Ubeidiya.

[9] Gabunia et al. (2000) describe the *Homo ergaster* skulls from Dmanisi, while Lordkipanidze et al. (2000) provide information about the site and context of discovery.

[10] Lordkipanidze et al. (2000).

[11] Carbonell et al. (1995, 1999) describe the hominid remains from TD6, Atapuerca, their dating and archaeological context. Important recent publications about the colonization of Europe are by Palmqvist (1997) and Arribas and Palmqvist (1999).

[12] Artefacts, dating evidence and palaeontological remains from Orce are described by Gibert et al. (1998), while Roe (1995) provides a commentary about the site. Moya-Sola and Kohler (1997) and Gibert et al. (1998) discuss the status of the hominid remains.

[13] Magnetic reversal dating is based on the fact that the earth's magnetic field has periodically reversed during the course of the planet's history, so that what is now the South Pole was the North Pole, and vice versa. It is not well understood why such reversals occur, or how long it takes for a reversal to becomes complete, but their existence is well documented from sedimentary and igneous rocks from throughout the world. Iron-bearing minerals within sediments act like tiny magnets and record the magnetic field at the time the sediment becomes consolidated into rock. Similarly, when an igneous rock cools below a certain temperature, its iron-bearing minerals also become fixed in accordance with the direction of the magnetic field. The dates of the magnetic field reversals have been established by radioisotopic dating of the reversal boundaries; some reversals have been shown to have lasted for several million years, while others lasted for a few tens of thousands of years and show a pattern of relatively rapid fluctuation between normal and reversed polarities. At sites such as Orce,

a pattern of reversals has been established for a sequence of rocks, which then has to be correlated with the known and dated global pattern in order to establish their date.

[14] For Boxgrove, see Roberts and Parfitt (1999).

[15] As reported by Simon Parfitt at the Quaternary Research Association meeting in London, 5 January 2005.

[16] For the description of *Homo floresiensis*, see Brown et al. (2004), and for its associated archaeology, Morwood et al. (2004). Lahr and Foley (2004) provide a useful commentary on the significance of this find.

[17] These were described by Morwood et al. (1998).

[18] For a description of hand-axe manufacture and how this contrasts with Oldowan tools, see Pelegrin (1993).

[19] Boxgrove is described by Roberts et al (1994).

[20] Turner (1992) has treated hominids from this perspective.

[21] This is the principle underlying the simulation work of Mithen and Reed (2002).

[22] My impression from meetings with academic colleagues is that the weight of opinion suggests not: we should treat hominids at 2 million years ago like any other large Pleistocene mammal. I disagree, believing that the spirit of adventure was born early within humankind.

[23] Periods of lower sea level are likely to have been crucial for hominid disperals (Mithen and Reed 2002). Coastal landscapes would have provided rich foraging grounds for parties of *Homo ergaster.* Indeed, it may have been only by following the coast, and hence avoiding thick forests and mountain chains, that hominids could have spread to south-east Asia in the short time frame suggested by the early dates on the Java skulls. Hominids may also have been able to make short water crossings, as stone artefacts claimed to date to 800,000 years ago have been found on Flores Island, which could only have been reached by a water crossing (Morwood et al. 1998).

[24] For detailed descriptions of hunting by the !Kung see Lee (1979). Those hunters spent many hours discussing game movements and the tracks and trails they had seen, often drawing on the observations made by female plant gatherers. This use of language has been neglected in theories of the origin of language that focus on the role of language in social interaction, such as the vocal grooming theory of Dunbar (1996).

[25] Donald (1991, p. 168).

[26] Donald (1991, p. 169).

[27] I first read Donald's ideas about mimetic culture as soon as they were published

in 1991, at a time when I was just formulating my own ideas about the evolution of the human mind, and was entirely unconvinced. I was unable to imagine non-linguistic hominids engaged in mime as a means of communicating either things they had seen in the landscape or events that might occur. A key problem for me was that mimesis of this form involved 'displacement' – reference to either the past or the future – which is a key feature of modern human language. Indeed, the capacity for displaced thought is often argued to be dependent on language, and in 1991 I found that idea persuasive. But as the evidence for more complex behaviour by *Homo ergaster* and its immediate descendants has accumulated, especially that concerning dispersal out of Africa, the idea that such species lacked an ability to plan for the future and reflect on the past has come to seem untenable, if not quite ridiculous. How, for instance, could a spruce tree have been cut down and shaped into a throwing spear without some anticipation, however vague, of a future hunting event? Now that I have come to appreciate the significance of non-linguistic vocal communication and body language, I have also recognized the seminal contribution that Donald made when he proposed mimesis as a key means of Early Human communication. Indeed, I now find it almost impossible *not* to think of *Homo ergaster, Homo erectus* and *Homo heidelbergensis* as being engaged in various forms of mime to supplement their other means of communication about the natural world.

[28] The following text draws on Marshall (1976).

[29] Boys also used mimicry in games of encounter and attack: one would imitate a lion growling and springing, or pretend to be a hyena slipping into the camp at night to bite the other boys, who pretended to be asleep and would jump up shouting and waving their arms to scare the hyena away. Another game involved one boy imitating a gemsbok, using sticks in his hair to represent horns, while the others imitated the hunters who stalked and killed the animal. Marshall once saw some men play the game of porcupine: they wore porcupine quills in their hair and made claw marks in the sand with their fingers while snarling ferociously. One porcupine-man pretended to be killed; he rolled into a ball and imitated the porcupine's twitching death throes, while the onlookers guffawed with laughter.

[30] There is a lengthy history of debate among psychologists as to how imitation should be defined. To an outsider, much of this debate appears to dwell obsessively on fine terminological distinctions of no practical significance. But one issue is of considerable importance when considering how Early Humans may have imitated animals and birds as part of their communication system. Some psychologists believe that 'true' imitation requires not only imitating the body movements of another person or animal, but also understanding the intention behind those movements. This requires that the imitator 'gets into the mind' of the person/animal that is being copied. Whatever Merlin Donald believes, my view of Early Human mimesis is that it did not involve 'thinking

animal' in the sense of imagining or believing that they were becoming the animals they were imitating. In this respect, it might have been quite different from the imitation of animals undertaken by modern hunter-gatherers. A great deal of the animal mimesis undertaken by the !Kung, the Australian aborigines, or, indeed, any modern hunter-gatherers, relates to a mythological world in which people and animals can transform into each other. The mimesis often takes place as part of religious ceremonies in which the animals represented are ancestral beings rather than merely entities of the natural world. Those undertaking the performance may take on the powers of the animals involved, with those animals often being ascribed human-like thoughts and intentions – just as we do when we talk about the 'sly fox', 'brave lion' and 'wise owl'. When shamans enter their trances they believe that they are transformed into entities with animal-like qualities, especially the ability to fly. We can refer to this as 'mental' mimesis, and it must surely help with the capture of the precise movements of the animal in what we can term 'physical' mimesis. I suspect that Early Humans were largely constrained to the latter; they did not 'think animal' in the same manner as modern hunter-gatherers or, indeed, any member of *Homo sapiens*. My reason for this lies in the structure of their minds. As I remarked in the introduction to this book, I have previously argued that Early Humans had a domain-specific mentality, one in which their thoughts relating to 'people', 'animals' and 'artefacts' were quite isolated from each other – 'social', 'natural history' and 'technical' intelligences. This domain-specific mentality explains their lack of art, religion and cultural innovation, and I will return to how it both enabled and constrained behaviour when discussing the Neanderthals in chapter 15. When modern humans engage in 'mental' mimesis they are evidently integrating their social and natural history intelligences – imagining, indeed creating, an entity that is part human and part animal. This type of 'cognitively fluid' thought finds its first material manifestation in some of the earliest images of ice-age art, such as the lion-man carving from Hohlenstein Stadel and the 'bison-man' painting from Chauvet, both dating to around 30,000 years ago. Mimesis by Early Humans was, I believe, more prosaic in nature: it was concerned with physical imitation alone, as a means to communicate information about the natural world. There was no transformation into human-animal entities or thoughts of 'what it is like to be' a lion, horse, eagle or bear. This may have inhibited the elegance of their displays; it would, for instance, be difficult for us today to mime a bird without imagining what it would be like to fly. On the other hand, we are quite able to imitate trees blowing in the wind or racing cars without attributing to those entities thoughts and feelings. In summary, we might characterize Early Humans as having a capacity for simile – they could be 'like' an animal – but not for metaphor – they could not 'become' an animal.

[31] Berlin (1992, p. 249).

[32] In the following I draw on Berlin (1992, chapter 6) and Berlin (2005).

[33] Jespersen cited in Berlin (2005).

[34] I should note that Berlin (2005) undertook a very rigorous statistical study to reach this conclusion.

[35] In the following text I draw on Wray (2002b).

[36] One possible example of a fireplace comes from Beeches Pit in Norfolk, dating to around 0.5 million years ago, but this consists of no more than an area of heavily burnt soil where the fire appears to have been too intense and localized to have had natural causes. What is lacking is evidence for stone constructions to indicate that the fireplaces had been used to structure social activity, as occurs with modern humans. Bellamo (1993) discusses the uses of fire and the problems of its identification in the archaeological record.

[37] Gamble (1999, pp. 153–72) provides a detailed description of the Bilzingsleben excavations and the interpretations of Dietrich Mania (the excavator). He also provides his own interpretation, which I draw on in my text.

[38] Gamble (1999, p. 171).

12. Singing for sex

[1] Darwin (1871, p. 880). This quote was also cited by Miller (2000, p. 329) in his polemical article regarding the evolution of music through sexual selection. The following text about sexual selection draws on Miller (1997, 2000).

[2] Zahavi (1975), Zahavi and Zahavi (1997).

[3] This example is given by Miller (2000).

[4] Darwin (1871, p. 705).

[5] Darwin (1871, p. 878).

[6] Darwin (1871, p. 880).

[7] Miller suggests that making music would have been a handicap for our hominid ancestors because its noise would have attracted predators and competitors, and because all music requires time and energy to learn and perform – time that early hominids could have otherwise spent on tasks with some utilitarian function.

[8] Miller (2000, p. 349).

[9] Miller (2000, p. 340).

[10] Miller (2000, p. 343).

[11] Miller (2000, p. 337).

[12] The fact that for modern humans so much of music-making is a group activity need not be a problem for Miller's theory, as pointed out to me by Anna Machin (pers. comm.). She notes that the mating dance of the Wodaabe occurs in a

group and is definitely a mate choice activity. Group activity allows the comparison of individuals while still allowing individuals to display.

[13] The following text about primate social and mating behaviour draws on chapters in McGrew et al. (1996).

[14] When mating systems from all societies are taken into account, humans as a whole are probably most accurately described as a weakly polygynous species. Human mating patterns are, of course, highly influenced by social, economic and cultural contexts, and hence we find a very substantial degree of variability, which has been a keen concern of social anthropology. Among primates, polygyny can take the form either of multi-male, multi-female groups, or of groups with multiple females and a single male, who then becomes liable to attacks from the other males, who form all-male groups.

[15] As their supporters may be related (as brothers or sons) such rewards simply help to maintain their own inclusive fitness.

[16] Plavcan and van Schaik (1992) explored the correlation between the intensity of male competition and canine sexual dimorphism.

[17] McHenry (1996) provides figures of 1.0 for *Hylobates syndactylus* and 1.1 for *Hylobates Lar*, and 1.2 for *Homo sapiens*.

[18] The following text draws on McHenry (1996).

[19] We must, however, be cautious, as sexual dimorphism can also be influenced by factors such as predator pressure and diet, and hence its reduction in *Homo ergaster* may be quite unrelated to mating behaviour. Rowell and Chism (1986, p. 11) do not think it possible to infer the social systems or mating patterns of extinct species from the degree of sexual dimorphism shown by their fossils.

[20] The following text draws on Key and Aiello (1999). They cite a figure of a 66–188 per cent increase in the mean calorific intake of lactating females, and note that many females lose weight during lactation.

[21] Key and Aiello (1999) explain that up to the age of eighteen months, the energy requirements of human infants are 9 per cent greater than those of chimpanzees.

[22] McHenry (1996).

[23] Key and Aiello (1999) provide numerous examples of such female cooperation among a wide variety of species.

[24] Key and Aiello (1999) illustrate this by the example of the brown hyena, which has a similar level of female cooperation as found among many primates but which also engages in food sharing and provisioning because of the importance of meat within the diet. Meat is difficult to acquire and the supply unreliable, quite unlike the supply of plant foods for the non-food-sharing apes.

[25] For the grandmothering hypothesis, see Hawkes et al. (1997); and for its application to *Homo ergaster* and the evolution of the human life history, see O'Connell et al. (1999).

[26] Key and Aiello (1999) argue that when men focus on small game, hunting can be an effective means of provisioning females and their infants.

[27] Key (1998) and Key and Aiello (2000) used the prisoner's dilemma model to explore the conditions in which males would begin to cooperate with females in terms of provisioning and protection.

[28] This has often been claimed by archaeologists without any formal evaluation. Recently, however, Anna Machin, one of my graduate students at the University of Reading, has undertaken controlled experiments to demonstrate that symmetry has no influence on butchery efficiency.

[29] Wynn (1995) discussed many of these issues in an article appropriately entitled 'Handaxe enigmas'.

[30] For arguments about the significance of raw material characteristics for handaxe shape, size and symmetry, see Ashton and NcNabb (1994) and White (1998).

[31] The following draws on Kohn and Mithen (1999).

[32] For swallow tails, see Møller (1990); for peacock tails, see Manning and Hartley (1991).

[33] For the significance of bodily and facial symmetry in human sexual relationships, see Thornhill and Gangstead (1996).

[34] Roe (1994, p. 207).

13. The demands of parenthood

[1] Boxgrove is described by Roberts and Parfitt (1999). Their report provides the geological background to the site, mammalian fauna, dating and a summary of the archaeology.

[2] For the flaking technology at Boxgrove and refits, see Bergman and Roberts (1988) and Wenban-Smith (1999).

[3] For patterns of butchery, see Parfitt and Roberts (1999).

[4] For the hominid tibia, see Roberts et al. (1994).

[5] For Gamble's view of Boxgrove and Middle Pleistocene hominid behaviour in general, see Gamble (1987, 1999).

[6] The Schöningen spears are described by Thieme (1997).

[7] The following text draws on Dissanayake (2000). Cordes (2003) provides evidence that is claimed to support the derivation of music from IDS-type

communication. She found that the melodic contours of IDS correspond closely to the melodic contours of songs made by humans in traditional societies. She also claims that there is a significant similarity to the melodic contours of primate calls, although from the data she provides this is far more contentious.

[8] Dissanayake (2000, p. 390).

[9] The following text draws on Falk (in press).

[10] Wheeler (1984, 1992) argued that the loss of body hair was a physiological adaptation for thermoregulation, evolving at the same time as bipedalism.

[11] Pagel and Bodmer (2003).

[12] Bellamo (1993) not only describes the evidence from FxJj50 but also provides a thorough review of the methodological issues involved in identifying traces of fire in the archaeological record.

[13] Kittler et al. (2003).

[14] Falk (in press).

[15] Darwin (1872) counted disgust as one of the six universal emotions. Rozin et al. (1993) review studies of disgust, describing the characteristic facial expressions, neurological signs and actions that are found cross-culturally.

[16] Curtis and Biran (2001) provide an excellent overview of theories about disgust, making a strong argument that it is an evolved response to avoid infection by disease. I am most grateful to Val for saying 'Yuk!' to me immediately after I had presented a paper on the origin of language at a conference at the Institute of Child Health, London, on 16 December 2004.

14. Making music together

[1] The following text draws on Merker (1999, 2000).

[2] Cited in Merker (2000, p. 319).

[3] Merker (2000, p. 320).

[4] McNeill (1995).

[5] McNeill (1995, p. 2). He coins the term 'muscular bonding' for the 'euphoric feeling that prolonged and rhythmic muscular movement arouses among nearly all participants'.

[6] Dunbar (2004). I am grateful for conversations with Robin Dunbar on this issue and for his presentation at the 2004 'Music, Language and Human Evolution' workshop at the University of Reading, where he stressed the importance of endorphins.

[7] John Blacking published a great many studies concerning the Venda – see

Byron (1995) for a complete bibliography – the most important of which may have been his study of Venda children's songs (1967). I have drawn on his comments about Venda communal music-making in Blacking (1973, p. 101).

[8] The following text about the prisoner's dilemma model draws on Axelrod and Hamilton (1981) and Axelrod (1984).

[9] The prisoner's dilemma model is also a powerful research tool. The anthropologists Leslie Aiello and Cathy Key made excellent use of it to examine the evolution of social cooperation in early hominid society (Key and Aiello 2000).

[10] Caporael et al. (1989, p. 696).

[11] Caporael et al. (1989).

[12] Benzon (2001, p. 23).

[13] Freeman (2000).

[14] Cited in Benzon (2001, p. 81).

[15] The impact of blocking oxytocin on female prairie rats is referred to in Damasio (2003, p. 95).

[16] Blacking (1973, p. 44).

[17] The following text concerning Atapuerca draws on Arsuaga et al. (1997) for a general description of the site. The dating of the specimens remains unclear; some may be younger than 350,000 years old, but a sample at least appear to be sealed by sediments with this date.

[18] The arguments for a catastrophic accumulation of bodies in the Sima de los Huesos is made by Bocquet-Appel and Arsuaga (1999).

[19] See Arsuaga (2001, pp. 221–32) for discussion of the various explanations for the accumulation of bodies at Atapuerca.

15. Neanderthals in love

[1] The final Neanderthals lived in southern Spain and Gibraltar before becoming extinct at around 28,000 years ago. Barton et al. (1999) describe recent excavations on Gibraltar that have recovered important new information on Neanderthal lifestyles, notably their use of coastal foods. Further recent work about Neanderthal extinction with a focus on southern Iberia is found in the volume edited by Stringer et al. (2000), suitably called *Neanderthals on the Edge*.

[2] Although the specific identity of the common ancestor is unclear, there is wide consensus on the general evolutionary relationship between *Homo sapiens* and *Homo neanderthalensis*. Ever since the first Neanderthal specimens were discovered in 1856, this relationship has been debated not only within the pages of scientific journals but also in literature and art, as the Neanderthals became the subject of novels and paintings. This was far more than a mere

academic disagreement, because the Neanderthals were employed as the 'other' against which our own identity as a species was defined. Those anthropologists who wished to exclude the Neanderthals from the human family depicted them as uncouth ape-men with slouching shoulders and a shambling gait; conversely, those who believed that *Homo neanderthalensis* was closely related to *Homo sapiens* generated 'noble savage' depictions of the Neanderthals – described by Clive Gamble as looking merely a haircut away from civilization. Such opposing views were present at the start of the twentieth century and continued in one form or another until less than a decade ago; see Stringer and Gamble (1993) and Trinkaus and Shipman (1995) for the history of Neanderthal studies. In recent years, debate became focused on the origins of *Homo sapiens*: was there a single origin in Africa or multiple origins in different regions of the Old World? If the latter, then *Homo neanderthalensis* would have been the direct ancestor of *Homo sapiens* in Europe. The Neanderthal debate was sustained for so long because the fossil record is fragmentary and difficult to interpret. In 1997, a new source of data became available and provided a resolution to what had become a tedious debating point at conferences and in journals, when Krings et al. (1997) announced the first extraction of DNA from a Neanderthal skeleton. Quite remarkably, this skeleton was the original find, coming from a cave in the Neander Valley, in Germany, in 1856. A comparison of this ancient DNA sample with DNA from modern humans indicated that *Homo neanderthalensis* and *Homo sapiens* shared a common ancestor at around 500,000 years ago. This means that *Homo neanderthalensis* could not have been a direct ancestor of *Homo sapiens*. The DNA also indicated that the Neanderthals had made no contribution to the modern gene pool, and hence that there had been no interbreeding. The vast majority of anthropologists now agree that *Homo sapiens* evolved in Africa and *Homo neanderthalensis* in Europe, both deriving ultimately from *Homo ergaster* and possibly sharing *Homo heidelbergensis* as their most recent common ancestor. For commentary on the significance of the Neanderthal DNA, see Kahn and Gibbons (1997) and Ward and Stringer (1997).

[3] The fossil specimens from Atapuerca are described by Bermúdez de Castro et al. (1997). Their use for identifying a new species of *Homo* is controversial as the claimed type-specimen is a juvenile in which the adult facial characteristics may not have fully developed. Gibbons's (1997) commentary on the finds provides an insight into the differences that exist among palaeoanthropologists regarding the interpretation of the fossil record. By saying that these are the earliest known *Homo* remains from Europe, I mean they are the earliest unambiguously identified or dated specimens.

[4] Foley and Lahr (1997) argue that palaeoanthropologists should pay as much attention to the archaeological as to the fossil record when reconstructing the evolutionary history of the Neanderthals. They stress the similarities between the chipped-stone technology used by Neanderthals and by modern humans in the Near East prior to 50,000 years ago. This is Levallois-based and was once classified as a 'mode 3' technology – this term now being very rarely used by

archaeologists themselves but thought to be significant by Foley and Lahr. They argue that the technological similarity implies a recent common ancestor for *Homo sapiens* and *Homo neanderthalensis*, which they designate as *Homo helmei*. This term was originally introduced in the 1930s to refer to a specimen from Florisbad, South Africa. McBrearty and Brooks (2000) also refer to *Homo helmei* as an immediate ancestor of *Homo sapiens*, but not of *Homo neanderthalensis*.

[5] Unless otherwise stated, this section draws on the descriptions of Neanderthal and other fossils in Stringer and Gamble (1993) and Johanson and Edgar (1996).

[6] For the relationship between body size and brain size see Kappelman (1996). A key feature of his argument is that the increase in relative brain size of modern humans after 100,000 years ago arose from selection for a smaller body mass rather than a larger brain.

[7] Churchill (1998) provides a review of Neanderthal anatomy and its physiological adaptations for cold climates.

[8] See Mellars (1996) for the most extensive review and interpretation of Neanderthal archaeology from Western Europe.

[9] See Mellars (1996) for a review and interpretation of the relevant archaeology. All leading archaeological authorities have agreed with this view of Neanderthal social life. Lewis Binford (1989, p. 33) described their groups as 'uniformly small'; Paul Mellars (1989, p. 358) concluded that Neanderthal 'communities ... were generally small ... and largely lacking in any clear social structure or definition of individual social or economic roles'; and Randall White (1993, p. 352) stated that Neanderthals lived in 'internally un- or weakly differentiated' societies. The only argument for an equivalent degree of group size and social complexity to modern humans has come from Dunbar (1993, 1996, 2004) and is based on Neanderthal brain size, which is similar to that of modern humans. This argument relies, however, on a quite inappropriate extrapolation from a relationship between group size and brain size that exists for relatively small-brained modern-day primates.

[10] For a momentous study of Levallois technology, see Dibble and Bar-Yosef (1995); Van Peer (1992) provides a very detailed study of the Levallois reduction strategy. Both volumes indicate the high degree of technical skill that had been mastered by the Neanderthals and apparently employed on a routine basis.

[11] The Kebara I burial is described by Bar-Yosef et al. (1992). There has been a vigorous debate among archaeologists over the existence of Neanderthal burial; see, for example, Gargett (1989, and commentary). With the discovery of the Kebara specimen and re-evaluation of those from sites such as La Ferrassie, very few archaeologists now deny that Neanderthals buried some of their dead. But there remains disagreement as to why this was undertaken.

[12] The Kebara I hyoid bone is described by Arensburg et al. (1989). See Arensburg

et al. (1990) and Lieberman (1992) for discussions of its significance.

[13] Lieberman and Crelin (1971). Their influential reconstruction was based on the idea that the cranial base of a skull can indicate the laryngeal position. They found that the cranial base of the Chappelle-aux-Saints specimen was closer to that of an adult chimpanzee or a human neonate than of an adult modern human, and concluded that the larynx was in a similarly heightened position. Burr (1976) made one of the earliest critiques of their work, which is now widely repeated in works on human evolution.

[14] Kay et al. (1998). The specimens examined were three from Sterkfontein, South Africa (Sts 19, Stw 187 and Stw 53); the Kabwe and Swanscombe fossils, which are ill defined as to species but reasonably described as *Homo heidelbergensis;* two Neanderthal specimens, La Ferrassie 1 and La Chapelle-aux-Saints; and one early *Homo sapiens,* Skhul 5. As the Kabwe and Swanscombe specimens also fell within the modern human range, motor control of the tongue had most likely attained its modern character by 400,000 years ago. See DaGusta et al. (1999) for a critique of their conclusions.

[15] MacLarnon and Hewitt (1999).

[16] Martinez et al. (2004).

[17] Gamble (1999) provides an excellent summary of the climatic changes during the Middle and Late Pleistocene of Europe and their implications for human behaviour. Particularly important recent work has been that by D'Errico and Sanchez Goni (2003) regarding the significance of climatic variability for Neanderthal extinction. Recent research by Felis et al. (2004) indicates that the Neanderthals may have survived through periods of greater seasonal variation during the last (OIS stage 5) interglacial than is the case for modern humans during the Holocene.

[18] See Marshack (1997) for a description of the Berekhat 'figurine' and impressive colour photographs, and then see D'Errico and Nowell (2000) for a microscopic analysis of this artefact, followed by discussion from various commentators.

[19] For the Bilzingsleben bones, see Mania and Mania (1988). In addition to the Berekhat Ram 'figurine' and the Bilzingsleben bone, writers such as Robert Bednarik (1995) cite a small number of other supposedly marked or shaped objects to support their claim that Neanderthals and previous hominids were capable of symbolic thought. But in every case an alternative explanation can be provided, and the failure of critical evaluation by Bednarik and others is striking to everyone except themselves. Mithen (1996) provides a critique of Bednarik's arguments.

[20] Bednarik (1994) argues that the rarity of symbolic artefacts in the archaeological record of pre-modern humans arises for taphonomic reasons. I find this a very weak and in some places entirely misconceived argument.

[21] The following text draws on a personal communication from Francesco D'Errico in May 2004 and a visit to his laboratory to see several pigment specimens from Neanderthal sites.

[22] Humphrey's (1984) chapter has been curiously neglected by archaeologists writing about the origin of art. He makes a very interesting argument about the ambiguity of red as a colour, which is used as a sign of both anger/danger and sexual attraction/love.

[23] See Kuhn (1995) for further aspects of Neanderthal lithic technology, and the general review provided in Mellars (1996).

[24] For a description of the Chatelperronian industry and the debates regarding its interpretation see D'Errico et al. (1998) and the commentary following that article.

[25] Mithen (1996/1998).

[26] Wynn and Coolidge (2004).

[27] Zubrow (1989) demonstrated by the use of demographic modelling that seemingly minor differences in reproduction rates might have led to the rapid replacement of the Neanderthals by modern humans.

[28] Trinkaus (1995) infers Neanderthal mortality patterns from their skeletal remains. There are many methodological problems in doing so, including the difficulty of ageing adults. One of the key difficulties faced by archaeologists is that the Neanderthals may have treated different age classes in different fashions. 'Old' people (that is, over thirty-five years) might simply not have been buried, leading to very rare preservation of their remains.

[29] Trinkaus (1985) describes the pathologies of the La Chapelle-aux-Saints specimen, which are summarized in Stringer and Gamble (1993), along with further examples of diseased or injured specimens.

[30] Trinkaus (1983) describes the Neanderthals from Shanidar Cave.

[31] The animal bones from Combe Grenal have been analysed and interpreted by Chase (1986) and further discussed by Mellars (1996).

[32] Mauran is described by Farizy et al. (1994). Mellars (1996) provides further descriptions of Neanderthal mass kill sites.

[33] Callow and Cornford (1986) describe the excavations at La Cotte, while Scott (1980) interprets the mammoth remains as the consequence of big-game hunting.

[34] See Kohn and Mithen (1999) for the original argument that development of flake technology, sometimes referred to by archaeologists as 'mode 3' technology, reflects a change in mating patterns.

[35] The infant burial in Amud Cave is described by Rak et al. (1994), while further burials in the cave are described by Hovers et al. (1995).

[36] Ramirez Rozzi and Bermúdez de Castro (2004) identified the rate of growth of Neanderthals and compared this with those of Upper Palaeolithic and Mesolithic modern humans, and *Homo antecessor* and *Homo heidelbergensis*, by comparing the speeds of enamel formation on teeth. Life-history traits correlate closely with dental growth. The Neanderthals had the shortest period of dental growth, indicating that they were the fastest developing of all species of *Homo*. They use this to stress the clear evolutionary separation between *Homo neanderthalensis* and *Homo sapiens*. Prior to their work there had been other indicators of a relatively rapid developmental rate for the Neanderthals, as summarized in Churchill (1998). Mithen (1996) suggested that the rapid rate of growth may have had significant implications for the development of Neanderthal cognition.

[37] The following text regarding 'empty spaces' in caves draws on Mellars (1996).

[38] The discovery in Bruniquel Cave was made by François Rouzard. A useful commentary about the discovery and its implications is provided by Balter (1996).

[39] Clottes (pers. comm., May 2004).

[40] Turk (1997) described the discovery of the bone 'flute' from Divje Babe while Kunej and Turk (2000) undertook a musical analysis of this artefact and defended its flute status against the criticisms that had been raised by D'Errico et al. (1998) and others. Whether or not a Neanderthal flute was found, Divje Babe is an important site that has been meticulously excavated (Turk 1997).

[41] D'Errico et al. (1998) undertook a microscopic analysis of the Divje Babe 'flute' and demonstrated that it is most likely a product of carnivore gnawing that fortuitously looks like a flute.

[42] Further evidence against the flute interpretation sounds so obvious that one wonders how it could ever have been ignored by Turk: the ends were blocked with bone tissue, preventing the passage of air – the 'flute' could simply never have been played.

16. The origin of language

[1] White et al. (2003). A useful commentary on the significance of these finds is provided by Stringer (2003).

[2] McDougall et al. (2005) redated feldspar crystals from pumice clasts in the sediments immediately below the Omo I and Omo II specimens, giving a date of 196,000±2000 years ago. They estimated the age of the fossils as being close to this and proposed 195,000±5000 years ago.

[3] McBrearty and Brooks (2000) provide a succinct overview of the fossil evidence from Africa and discuss various interpretations.

[4] Johanson and Edgar (1996) provide descriptions of these particular fossils and excellent colour illustrations.

[5] The original paper that marked a new phase in the study of human evolutionary genetics was Cann et al. (1987). A measure of the astounding progress that has been made can be appreciated by consulting the recent textbook on human evolutionary genetics by Jobling et al. (2003).

[6] This has been demonstrated by studies of three Neanderthal specimens, one being the type-specimen from the Neander Valley (Krings et al. 1997), the others being from Vindjia (42,000 years ago) and Mezmaiskaya (29,000 years ago). Jobling et al. (2003, pp. 260–3) provide an excellent summary and interpretation of these studies.

[7] Ingman et al. (2000) studied the complete mitochondrial DNA sequence of fifty-three modern humans, rather than just concentrating on the control region which constitutes less than 7 per cent of the mitochondrial genome. Their estimate for the most recent common ancestor of their sample of modern humans was 171,000±50,000 years ago.

[8] Bishop (2002) provides a useful and succinct summary of the studies of the KE family and the results of the genetic studies, discussing various interpretations of the so-called 'language gene'. Pinker (2003) also provides a useful summary of this work, arguing that it supports his own ideas about the natural selection of the language capacity.

[9] Lai et al. (2001).

[10] Enard et al. (2002). There is an inconsistency between Enard et al. (2002) and Pinker (2003) as to the number of amino acids that are different between the FOXP2 gene in different mammals. I have followed Enard et al.

[11] Enard et al. (2002, p. 871).

[12] Unless otherwise specified, the following section draws on the excellent, detailed and comprehensive review of African archaeology provided by McBrearty and Brooks (2000).

[13] There are several recent reports of the work at Blombos Cave that provide details about the discoveries and their context. Of particular value are Henshilwood and Sealy (1997), describing bone artefacts, Henshilwood et al. (2001), describing the 1992–9 excavations, and Henshilwood et al. (2002), summarizing the ochre engravings and arguing that these provide crucial evidence for the emergence of modern human behaviour.

[14] Forty-one perforated shell beads made from *Nassarius kraussianus* from Blombos Cave are described by Henshilwood et al. (2004).

[15] Knight et al (1995) provide quantitative data on the ochre from Klasies River Mouth, showing a very substantial increase in quantity at around 100,000 years ago. But be careful of their interpretation!

[16] Even those sites do not provide the earliest traces of ochre use. Archaeological deposits at the site of GnJh-15 in Kenya have been dated to 280,000 years ago and are described as having 'red-stained earth' with numerous friable and poorly preserved ochre fragments. Three pieces of ochre have come from artefact-rich deposits in Twin Rivers Cave, Zambia, which are below a layer dated to 230,000 years ago (McBrearty and Brooks 2000).

[17] Humphrey (1984, p. 152).

[18] My references to Wray's work draw on Wray (1998, 2000, 2002a, b, c).

[19] My references to Kirby's work draw on Kirby (2001, 2002) and Kirby and Christiansen (2003).

[20] See Bickerton (2003) and Tallerman (in press) for criticisms of Wray's notion of holistic proto-language in general, and especially of the process of segmentation.

[21] Simon Kirby is just one of several scholars who have been exploring the evolution of language by computer simulation. Important work of this type has been undertaken by Martin Nowak (Nowak et al. 1999; Nowak and Komarova 2001a,b; Komarova and Nowak 2003). This has focused on examining the evolution of 'Universal Grammar' by the use of game theory. Also see Brighton (2002) and Tonkes and Wiles (2002).

[22] Hauser et al. (2002).

[23] See Wray and Grace (in press) for an important discussion of how social organization impacts upon language, in an article appropriately entitled 'The consequences of talking to strangers'.

[24] See McBrearty and Brooks (2000).

[25] Bishop (2002).

[26] Berk (1994), and see the edited volume by Diaz and Berk (1992).

[27] McBrearty and Brooks (2000). They frame their article around a challenge to those who have argued that modern behaviour originated in the Upper Palaeolithic of Europe around 40,000 years ago – setting up, I believe, something of a straw man against which to argue. The period 200,000 to 40,000 years ago in Africa is referred to as the Middle Stone Age and has often been taken to be equivalent to the Middle Palaeolithic of Europe. Although there are strong similarities in the stone technology, McBrearty and Brooks argue that these two periods are fundamentally different, with traces of modern behaviour in the Middle Stone Age archaeology.

[28] These microliths are part of the Howisons Poort industry, which is interstratified between layers with Middle Stone Age artefacts.

[29] This permanent change is the start of the Later Stone Age. The dating of this is, in fact, rather contentious, and might be highly variable across the continent. In some regions it may not have become established until the last glacial maximum at 20,000 years ago.

[30] Katanda harpoons are described by Yellen et al. (1995).

[31] McBrearty and Brooks (2000).

[32] See Shennan (2000) for computational models that examine the relationship between cultural transmission and population density.

[33] Ingman et al. (2000) estimate that the population expansion began at about 1925 generations ago. They assume a generation time of twenty years and conclude that this equates to 38,500 years ago.

[34] Lahr and Foley (1994) explain that there are likely to have been multiple dispersals out of Africa prior to the permanent establishment of *Homo sapiens* in Eurasia. On the basis of the genetic evidence from Ingman et al. (2000), and of morphological characteristics, the Qafzeh and Skhul populations are evidently not ancestral to modern *Homo sapiens*.

[35] Mithen (1996).

[36] Carruthers (2002) is a long article that explores various views of the relationship between thought and language, ranging from those who believe that all thought is dependent upon language, to those who see the two as effectively unrelated. He takes a specific type of middle ground, seeing language as the vehicle for intermodular integration. There are numerous comments on his article from a host of linguists, psychologists and philosophers, followed by his response.

17. A mystery explained, but not diminished

[1] The following text about the Geissenklösterle and Isturitz flutes/pipes draws on material in D'Errico et al. (2003).

[2] The use of natural 'lithophones' in caves is described by Dams (1985). It is argued that some painted walls in caves also have particularly good acoustic properties.

[3] Nettl (1983, p. 138).

[4] Blacking (1973, p. 34).

[5] In fact, he wrote much worse: 'We live in an age of tonal debauch ... It is obvious that second-rate mechanical music is the most suitable fare for those to whom musical experience is no more than a mere aural tickling, just as the prostitute provides the most suitable outlet for those to whom sexual experience

is no more than the periodic removal of a recurring itch. The loud speaker is the street walker of music' (Lambert 1934, cited in Horden 2000, p. 4).

[6] See Byron (1995, pp. 17–18) for Blacking's views about elitism in music.

[7] Nettl (1983, chapter 3) argues that a universally found feature of music is that it is used to communicate with the supernatural.

[8] For the role of the supernatural in religion, and discussion about the evolution and cognitive basis of religion, see Boyer (2001).

[9] Boyer (2001).

[10] The following text draws on my own ideas in Mithen (1996, 1998).

[11] Day (2004, p. 116).

[12] For example, Mithen (1998).

[13] Day (2004, p. 116).

[14] I note here that the psychologist Kathleen Gibson has written that 'ontogenetic perspectives have become the rule, rather than the exception, among serious scholars of cognitive and linguistic evolution' (Gibson and Ingold 1993, p. 26).

[15] Wray (2000, p. 286).

[16] Wray (2002a, p. 114).

[17] Tallerman (in press).

[18] Staal (1988). Staal (1989, p. 71) himself speculated that mantras actually 'predate language in the development of man in a chronological sense'.

[19] Byron (1995, pp. 17–18).

[20] Blacking (1973, p. 100).

[21] For a discussion of this issue, see Nettl (1983, chapter 7).

Bibliography

Aboudan, R., and Beattie, G. 1996. Cross cultural similarities in gestures: the deep relationship between gestures and speech which transcend language barriers. *Semiotica* 111, 269–94.

Aiello, L. C. 1996. Terrestriality, bipedalism and the origin of language. In *Evolution of Social Behaviour Patterns in Primates and Man* (ed. W. G. Runciman, J. Maynard-Smith and R. I. M. Dunbar), pp. 269–90. Oxford: Oxford University Press.

Aiello, L. C., and Dunbar, R. I. M. 1993. Neocortex size, group size, and the evolution of language. *Current Anthropology* 34, 184–93.

Aiello, L. C., and Wheeler, P. 1995. The expensive-tissue hypothesis. *Current Anthropology* 36, 199–220.

Aiyer, M. H., and Kuppuswamy, G. 2003. Impact of Indian music therapy on the patients – a case study. *Proceedings of the 5th Triennial ESCOM conference*, pp. 224–6. Hanover: Hanover University of Music and Drama.

Alajouanine, T. 1948. *Aphasia and artistic realization. Brain* 71, 229.

Allen, G. 1878. Note-deafness. *Mind* 3, 157–67.

Arbib, M. A. 2002. The mirror system, imitation and the evolution of language. In *Imitation in Animals and Artifacts* (ed. C. Nehaniv and K. Dautenhahn), pp. 229–80. Cambridge, MA: MIT Press.

Arbib, M. A. 2003. The evolving mirror system: a neural basis for language readiness. In *Language Evolution* (ed. M. H. Christiansen and S. Kirby), pp. 182–200. Oxford: Oxford University Press.

Arbib, M. A. in press. From monkey-like action recognition to human language: An evolutionary framework for neurolinguistics. *Behavioral and Brain Sciences*.

Arensburg, B., Schepartz, L. A., Tillier, A. M., Vandermeersch, B., and Rak, Y. 1990. A reappraisal of the anatomical basis for speech in Middle Palaeolithic hominids. *American Journal of Physical Anthropology* 83, 137–56.

Arensburg, B., Tillier, A. M., Duday, H., Schepartz, L. A., and Rak, Y. 1989. A Middle Palaeolithic human hyoid bone. *Nature* 338, 758–60.

Arribas, A., and Palmqvist, P. 1999. On the ecological connection between sabre-tooths and hominids: faunal dispersal events in the Lower Pleistocene and a review of the evidence for the first human arrival in Europe. *Journal of Archaeological Science* 26, 571–85.

Arsuaga, J. L. 2001. *The Neanderthal's Necklace*. New York: Four Walls Eight Windows.

Arsuaga, J. L., Martinez, I., Gracia, A., Carretero, J. M., Lorenzo, C., Garcia, N., and Ortega, A. I. 1997. Sima de los Huesos (Sierra de la Atapuerca, Spain): the site. *Journal of Human Evolution* 33, 219–81.

Ashton, N. M., and McNabb, J. 1994. Bifaces in perspective. In *Stories in Stone* (ed. N. M. Ashton and A. David), pp. 182–91. Lithics Studies Society, Occasional Paper 4.

Auer, P., Couper-Kuhlen, E., and Muller, F. 1999. *Language in Time: The Rhythm and Tempo of Spoken Language*. New York: Oxford University Press.

Axelrod, R. 1984. *The Evolution of Cooperation*. London: Penguin.

Axelrod, R., and Hamilton, W. D. 1981. The evolution of cooperation. *Science* 211, 1390–6.

Ayotte, J., Peretz, I., and Hyde, K. 2002. Congenital amusia. *Brain* 125, 238–51.

Bailey, B. A., and Davidson, J. W. 2003. Perceived holistic health benefits of three levels of music participation. *Proceedings of the 5th Triennial ESCOM conference*, pp. 220–3. Hanover: Hanover University of Music and Drama.

Balter, M. 1996. Cave structure boosts Neanderthal image. *Science* 271, 449.

Bannan, N. 1997. The consequences for singing teaching of an adaptationist approach to vocal development. In *Music in Human Adaptation* (ed. D. J. Schneck and J. K. Schneck), pp. 39–46. Blacksburg, VA: Virginia Polytechnic Institute and State University.

Baron, R. A. 1987. Interviewer's mood and reaction to job applicants. *Journal of Applied Social Psychology* 17, 911–26.

Baron-Cohen, S. 2003. *The Essential Difference*. London: Allen Lane.

Barton, R. N. E., Current, A. P., Fernandez-Jalvo, Y., Finlayson, J. C., Goldberg, P., Macphail, R., Pettitt, P., and Stringer, C. B. 1999. Gibraltar Neanderthals and results of recent excavations in Gorham's, Vanguard and Ibex Caves. *Antiquity* 73, 13–23.

Bar-Yosef, O., Vandermeersch, B., Arensburg, B., Belfer-Cohen, A., Goldberg, P., Laville, H., Meignen, L., Rak, Y., Speth, J. D., Tchernov, E., Tillier, A.-M., and Weiner, S. 1992. The Excavations in Kebara Cave, Mt. Carmel. *Current Anthropology* 33, 497–551.

Batali, J. 1998. Computational simulations of the emergence of grammar. In *Approaches to the Evolution of Langauge: Social and Cognitive Biases* (ed. J. Hurford, M. Studdert-Kennedy and C. Knight), pp. 405–26. Cambridge: Cambridge University Press.

Batali, J. 2002. The negotiation and acquisition of recursive grammars as a result of competition among exemplars. In *Linguistic Evolution through Language Acquisition: Formal and Computational Models* (ed. E. Briscoe), pp. 111–72. Cambridge: Cambridge University Press.

Beattie, G. 2003. *Visible Thought: The New Language or Body Language*. London: Routledge.

Beattie, G., and Shovelton, H. 1999. Do iconic hand gestures really contribute anything to the semantic information conveyed by speech? An experimental investigation. *Semiotica* 123, 1–30.

Beattie, G., and Shovelton, H. 2000. Iconic hand gestures and the predictability of words in context in spontaneous speech. *British Journal of Psychology* 91, 473–92.

Beattie, G., and Shovelton, H. 2002. What properties of talk are associated with the generation of spontaneous iconic hand gestures? *British Journal of Social Psychology* 41, 403–17.

Bednarik, R. 1994. A taphonomy of palaeoart. *Antiquity* 68, 68–74.

Bednarik, R. 1995. Concept-mediated marking in the Lower Palaeolithic. *Current Anthropology* 36, 605–32.

Bellamo, R. V. 1993. A methodological approach for identifying archaeological evidence of fire resulting from human activities. *Journal of Archaeological Science* 20, 525–55.

Benzon, W. L. 2001. *Beethoven's Anvil: Music in Mind and Culture*. Oxford: Oxford University Press.

Berger, D. 2002. *Music Therapy, Sensory Integration and the Autistic Child*. London: Jessica Kingsley Publishers.

Berger, L. R., and Clarke, R. J. 1995. Eagle involvement of the Taung child fauna. *Journal of Human Evolution* 29, 275–99.

Bergman, C. A., and Roberts, M. 1988. Flaking technology at the Acheulian site of Boxgrove, West Sussex (England). *Revue archéologique de Picardie*, 1–2 (numéro spécial), 105–13.

Berk, L. 1994. Why children talk to themselves. *Scientific American* (November), pp. 60–5.

Berlin, B. 1992. *The Principles of Ethnobiological Classification*. Princeton, NJ: Princeton University Press.

Berlin, B. 2005. 'Just another fish story?' Size-symbolic properties of fish names. In *Animal Names* (ed. A. Minelli, G. Ortalli and G. Singa), pp. 9–21. Venezia: Instituto Veneto di Scienze, Lettre ed Arti.

Bermúdez de Castro, J. M., Arsuaga, J. L., Carbonell, E., Rosas, A., Martinez, I., and Mosquera. M. 1997. A hominid from the Lower Pleistocene of Atapuerca, Spain: possible ancestor to Neanderthals and modern humans. *Science* 276, 1392–5.

Bernstein, L. 1976. *The Unanswered Question*. Cambridge, MA: Harvard University Press.

Bickerton, D. 1990. *Language and Species*. Chicago: Chicago University Press.

Bickerton, D. 1995. *Language and Human Behavior*. Seattle: University of Washington Press.

Bickerton, D. 1998. Catastrophic evolution: the case for a single step from protolanguage to full human language. In *Approaches to the Evolution of Language: Social and Cognitive Biases* (ed. J. Hurford, M. Studdert-Kennedy and C. Knight), pp. 341–58. Cambridge: Cambridge University Press.

Bickerton, D. 2000. How protolanguage became language. In *The Evolutionary Emergence of Language* (ed. C. Knight, M. Studdert-Kennedy, and J. Hurford), pp. 264–84. Cambridge: Cambridge University Press.

Bickerton, D. 2003. Symbol and structure: a comprehensive framework for language evolution. In *Language Evolution* (ed. M. H. Christiansen and S. Kirby), pp. 77–93. Oxford: Oxford University Press.

Binford, L. R. 1981. *Bones: Ancient Men and Modern Myths*. New York: Academic Press.

Binford, L. R. 1984. *Faunal Remains from Klasies River Mouth*. Orlando: Academic Press.

Binford, L. R. 1985. Human ancestors: changing views of their behavior. *Journal of Anthropological Archaeology* 3, 235–57.

Binford, L. R. 1986. Comment on 'Systematic butchery by Plio/Pleistocene hominids at Olduvai Gorge' by H. T. Bunn and E. M. Kroll. *Current Anthropology* 27, 444–6.

Binford, L. R. 1987. Were there elephant hunters at Torralba? In *The Evolution of Human Hunting* (ed. M. H. Nitecki and D.V. Nitecki), pp. 47–105. New York: Plenum Press.

Binford, L. 1989. Isolating the transition to cultural adaptations: an organizational approach. In *The Emergence of Modern Humans: Biocultural Adaptations in the Later Pleistocene* (ed. E. Trinkaus), pp. 18–41. Cambridge: Cambridge University Press.

Bishop, D. V. M. 2002. Putting language genes in perspective. *Trends in Genetics* 18, 57–9.

Blacking, J. 1973. *How Musical Is Man?* Seattle: University of Washington Press.

Bocquet-Appel, J. P., and Arsuaga, J. L. 1999. Age distributions of hominid samples at Atapuerca (SH) and Krapina could indicate accumulation by catastrophe. *Journal of Archaeological Science* 26, 327–38.

Boesch, C. 1991. Teaching among wild chimpanzees. *Animal Behavior* 41, 530–2.

Boyd, R., and Silk, J. B. 2000. *How Humans Evolved* (2nd edn). New York: W. W. Norton and Company.

Boyer, P. 2001. *Religion Explained.* New York: Basic Books.

Brain, C. K. 1981. *The Hunters or the Hunted?* Chicago: Chicago University Press.

Bramble, D. M., and Lieberman, D. E. 2004. Endurance running and the evolution of *Homo. Nature* 432, 345–52.

Brighton, H. 2002. Compositional syntax from cultural transmission. *Artificial Life* 8, 25–54.

Brockleman, W. Y. 1984. Social behaviour of gibbons: introduction. In *The Lesser Apes: Evolutionary and Behavioural Biology* (ed. H. Preuschoft, D. J. Chivers, W. Y. Brockleman and N. Creel), pp. 285–90. Edinburgh: Edinburgh University Press.

Brown, P., Sutikna, T., Morwood, M. J., Soejono, R. P., Jatmiko, Wayhu Saptomo, E., and Rokus Awe Due. 2004. A new small-bodied hominid from the Late Pleistocene of Flores, Indonesia. *Nature* 431, 1055–61.

Brown, S. 2000. The 'musilanguage' model of human evolution. In *The Origins of Music* (ed. N. L. Wallin, B. Merker and S. Brown), pp. 271–300. Cambridge, MA: Massachusetts Institute of Technology.

Brunet, M. et al. (37 authors). 2002. A new hominid from the Upper Miocene of Chad, Central Africa. *Nature* 418, 145–8.

Bunn, H. T. 1981. Archaeological evidence for meat eating by Plio-Pleistocene hominids from Koobi Fora and Olduvai Gorge. *Nature* 291, 574–7.

Bunn, H. T., Harris, J. W. K., Kaufulu, Z., Kroll, E., Schick, K., Toth, N., and Behrensmeyer, A. K. 1980. FxJj50: an Early Pleistocene site in northern Kenya. *World Archaeology* 12, 109–36.

Bunn, H. T., and Kroll, E. M. 1986. Systematic butchery by Plio-Pleistocene hominids at Olduvai Gorge. *Current Anthropology* 27, 431–52.

Bunt, L., and Pavlicevic, P. 2001. Music and emotion: perspectives from music therapy. In *Music and Emotion: Theory and Research* (ed. P. N. Juslin and J. A. Sloboda), pp. 181–201. Oxford: Oxford University Press.

Burr, B. D. 1976. Neanderthal vocal tract reconstructions: a critical appraisal. *Journal of Human Evolution* 5, 285–90.

Byrne, R. W., and Whiten, A. (eds). 1988. *Machiavellian Intelligence: Social Expertise and the Evolution of Intellect in Monkeys, Apes and Humans*. Oxford: Clarendon Press.

Byrne, R. W., and Whiten, A. 1992. Cognitive evolution in primates: evidence from tactical deception. *Man* NS 27, 609–27.

Byron, R. (ed.). 1995. *Music, Culture and Experience: Selected Papers of John Blacking*. Chicago: Chicago University Press.

Callow, P., and Cornford, J. M. (eds). 1986. *La Cotte de St. Brelade 1961–1978: excavations by C. B. M. McBurney*. Norwich: Geo Books.

Calvin, W. H., and Bickerton, D. 2000. *Lingua ex Machina: Reconciling Darwin and Chomsky with the Human Brain*. Cambridge, MA: MIT Press.

Cann, R. L., Stoneking, M., and Wilson, A. 1987. Mitochondrial DNA and human evolution. *Nature* 325, 32–6.

Caporael, L., Dawes, R. M., Orbell, J. M., and van de Kragt, A. J. C. 1989. Selfishness examined: cooperation in the absence of egoistic incentives. *Behavioral and Brain Sciences* 12, 683–739.

Carbonell, E., Bermúdez de Castro, J. M., Arsuaga, J. L., Diez, J. C., Rosas, A., Cuenca-Bescós, G., Sala, R., Mosquera, M., and Rodríguez, X. P. 1995. Lower Pleistocene hominids and artifacts from Atapuerca-TD6. *Science* 269, 826–30.

Carbonell, E., Esteban, M., Martín, A., Martina, N., Xosé, M., Rodríquez, P., Ollé, A., Sala, R., Vergès, J. M., Bermúdez de Castro, J. M., and Ortega, A. 1999. The Pleistocene site of Gran Dolina, Sierra de Atapuerca, Spain: a history of the archaeological investigations. *Journal of Human Evolution* 37, 313–24.

Carruthers, P. 2002. The cognitive functions of language. *Brain and Behavioral Sciences* 25, 657–726.

Carruthers, P., and Smith, P. (eds). 1996. *Theories of Theories of Minds*. Cambridge: Cambridge University Press.

Carter, R. 1998. *Mapping the Mind*. London: Weidenfeld and Nicolson.

Chan, A. S., Ho, Y.-C., and Cheung, M.-C. 1998. Music training improves verbal memory. *Nature* 396, 128.

Changeux, J.-P. 1985. *Neuronal Man: The Biology of Mind*. Princeton, NJ: Princeton University Press.

Chase, P. 1986. *The Hunters of Combe Grenal: Approaches to Middle Palaeolithic*

Subsistence in Europe. Oxford: British Archaeological Reports, International Series, S286.
Cheney, D. L., and Seyfarth. R. S. 1990. *How Monkeys See the World*. Chicago: Chicago University Press.
Christiansen, M. H., and Kirby, S. (eds). 2003. *Language Evolution*. Oxford: Oxford University Press.
Churchill, S. E. 1998. Cold adaptation, heterochrony, and Neanderthals. *Evolutionary Anthropology*, 7, 46–60.
Clegg, M. 2001. *The Comparative Anatomy and Evolution of the Human Vocal Tract*. Unpublished thesis, University of London.
Collard, M., and Aiello, L. C. 2000. From forelimbs to two legs. *Nature* 404, 339–40.
Cooke, D. 1959. *The Language of Music*. Oxford: Oxford University Press.
Corballis, M. 2002. *From Hand to Mouth: The Origins of Language*. Princeton, NJ: Princeton University Press.
Cordes, I. 2003. Melodic contours as a connecting link between primate communication and human singing. *Proceedings of the 5th Triennial ESCOM conference*, pp. 349–52. Hanover: Hanover University of Music and Drama.
Cowlishaw, G. 1992. Song function in gibbons. *Behaviour* 121, 131–53.
Crockford, C., Herbinger, I., Vigilant, L., and Boesch, C. 2004. Wild chimpanzees produce group-specific calls: a case for vocal learning. *Ethology* 110, 221–43.
Cross, I. 1999. Is music the most important thing we ever did? Music, development and evolution. In *Music, Mind and Science* (ed. Suk Won Yi), pp. 10–39. Seoul: Seoul National University Press.
Cross, I. 2001. Music, mind and evolution. *Psychology of Music* 29, 95–102.
Cross, I. in press. Music and meaning, ambiguity and evolution. In *Musical Communication* (ed. D. Miell, R. MacDonald and D. Hargreaves). Oxford: Oxford University Press.
Curtis, V., and Biran, A. 2001. Dirt, disgust and disease. *Perspectives in Biology and Medicine* 44, 17–31.
Dagusta, D., Gilbert, W. H., and Turner, S. P. 1999. Hypoglossal canal size and hominid speech. *Proceedings of the National Academy of Sciences* 96, 1800–4.
Damasio, A. 1994. *Descartes' Error; Emotion, Reason and the Human Brain*. New York: Putnam.
Damasio, A. 2003. *Looking for Spinoza: Joy, Sorrow and the Feeling Brain*. London: William Heinemann.
Damasio, H., and Damasio, A. 1989. *Lesion Analysis in Neuropsychology*. Oxford: Oxford University Press.
D'Amato, M. R. 1988. A search for tonal pattern perception in cebus monkeys: why monkeys can't hum a tune. *Music Perception* 4, 453–80.
Dams, L. 1985. Palaeolithic lithophones: Description and comparisons. *Oxford Journal of Archaeology* 4, 31–46.
Darwin, C. 1871. *The Descent of Man, and Selection in Relation to Sex* (2 vols). London: John Murray.

Darwin, C. 1872. *The Expression of Emotion in Man and Animals*. London: John Murray.

Davidson, I. 2003. The archaeological evidence of language origins: states of art. In *Language Evolution* (ed. M. H. Christiansen and S. Kirby), pp. 140–57. Oxford: Oxford University Press.

Davila, J. M. et al. 1986. Relaxing effects of music in dentristy for mentally handicapped patients. *Special Care Dentist* 6, 18–21.

Day, M. 2004. Religion, off-line cognition and the extended mind. *Journal of Cognition and Culture* 4, 101–21.

De Sousa, R. 1987. *The Rationality of Emotions*. Cambridge, MA: MIT Press.

De Waal, F. 1982. *Chimpanzee Politics*. New York: Harper and Row.

Deacon, T. 1997. *The Symbolic Species: The Co-Evolution of Language and the Human Brain*. London: Allen Lane.

DeMenocal, P. B. 1995. Plio-Pleistocene African climate. *Science* 270, 53–9.

Dempster, D. 1998. Is there even a grammar of music? *Musicae Scientiae* 2, 55–65.

Dennell, R. W. 1983. *European Economic Prehistory*. London: Academic Press.

Dennell, R. W. 1997. The world's oldest spears. *Nature* 385, 767–8.

Dennell, R. W., Rendell, H., and Hailwood, E. 1988. Early tool making in Asia: two-million-year-old artefacts in Pakistan. *Antiquity* 62, 98–106.

D'Errico, F., Henshilwood, C., Lawson, G., Vanhaeren, M., Tillier, A.-M., Soressi, M., Bresson, F., Maureille, B., Nowell, A., Lakarra, J., Backwell, L., and Julien, M. 2003. Archaeological evidence for the emergence of language, symbolism, and music – an alternative inter-disciplinary perspective. *Journal of World Prehistory* 17, 1–70.

D'Errico, F., and Nowell, A. 2000. A new look at the Berekhat Ram figurine: implications for the origins of symbolism. *Cambridge Archaeological Journal* 10, 123–67.

D'Errico, F., and Sanchez Goni, M. F. 2003. Neanderthal extinction and the millennial scale climatic variability of OIS 3. *Quaternary Science Reviews* 22, 769–88.

D'Errico, F., Villa, P., Pinto Llona, A. C., and Ruiz Idarraga, R. 1998. A Middle Palaeolithic origin of music? Using cave-bear bone accumulations to assess the Divje Babe I bone 'flute'. *Antiquity* 72, 65–79.

D'Errico, F., Zilhao, J., Julian, M., Baffier, D., and Pelegrin, J. 1998. Neanderthal acculturation in Western Europe. *Current Anthropology* 39, S1–S44.

Diaz, R. M., and Berk, L. E. 1992. *Private Speech: From Social Interaction to Self-Regulation*. Hillsdale, NJ: Lawrence Erlbaum Associates.

Dibben, N. 2003. Heightened arousal intensifies emotional experience within music. *Proceedings of the 5th Triennial ESCOM conference*, pp. 269–72. Hanover: Hanover University of Music and Drama.

Dibble, H., and Bar-Yosef, O. 1995. *The Definition and Interpretation of Levallois Technology*. Madison: Prehistory Press.

Dissanayake, E. 2000. Antecedants of the temporal arts in early mother–infant interaction. In *The Origins of Music* (ed. N. L. Wallin, B. Merker, and S. Brown), pp. 389–410. Cambridge, MA: Massachusetts Institute of Technology.

Donald, M. 1991. *Origins of the Modern Mind*. Cambridge, MA: Harvard University Press.

Drayna, D., Manichaikul, A., de Lange, M., Snieder, H., and Spector, T. 2001. Genetic correlates of musical pitch in humans. *Science* 291, 1969–72.

Duchin, L. 1990. The evolution of articulated speech: comparative anatomy of the oral cavity in *Pan* and *Homo*. *Journal of Human Evolution* 19, 687–97.

Dunbar, R. I. M. 1988. *Primate Societies*. London: Chapman and Hall.

Dunbar, R. I. M. 1991. Functional significance of social grooming in primates. *Folia Primatologica* 57, 121–31.

Dunbar, R. I. M. 1992. Neocortex size as a constraint on group size in primates. *Journal of Human Evolution* 20, 469–93.

Dunbar, R. I. M. 1993. Coevolution of neocortical size, group size and language in humans. *Behavioral and Brain Sciences* 16, 681–735.

Dunbar, R. I. M. 1996. *Gossip, Grooming and the Evolution of Language*. London: Faber and Faber.

Dunbar, R. I. M. 1998. Theory of mind and the evolution of language. In *Approaches to the Evolution of Language* (ed. J. R. Hurford, M. Studdert-Kennedy and C. Knight), pp. 92–110. Cambidge: Cambridge University Press.

Dunbar, R. I. M. 2003. The origin and subsequent evolution of language. In *Language Evolution* (ed. M. H. Christiansen and S. Kirby), pp. 219–34. Oxford: Oxford University Press.

Dunbar, R. I. M. 2004. *The Human Story*. London: Faber and Faber.

Dunbar, R. I. M., and Dunbar, P. 1975. *Social Dynamics of Gelada Baboons*. Basel: S. Karger.

Edelman, G. 1987. *Neural Darwinism: The Theory of Neuronal Group Selection*. New York: Basic Books.

Edelman, G. 1992. *Bright Air, Brilliant Fire: On the Matter of the Mind*. London: Penguin.

Ekman, P. 1985. *Telling Lies*. New York: Norton.

Ekman, P. 1992. An argument for basic emotions. *Cognition and Emotion* 6, 169–200.

Ekman, P. 2003. *Emotions Revealed*. London: Weidenfeld and Nicolson.

Ekman, P., and Davidson, R. J. 1995. *The Nature of Emotions: Fundamental Questions*. Oxford: Oxford University Press.

Ellefson, J. O. 1974. Anatomical history of white-handed gibbons in the Malayan peninsular. In *Gibbon & Siamang* (ed. D. M. Rumbaugh), pp. 1–136. Basel: S. Karger.

Enard, W., Przeworski, M., Fisher, S. E., Lai, C. S., Wiebe, V., Kitano, T., Monaco, A. P., and Paabo, S. 2002. Molecular evolution of FOXP2, a gene involved in speech and language. *Nature* 418, 869–72.

Evans, D. 2001. *Emotion: The Science of Sentiment*. Oxford: Oxford University Press.

Evans-Pritchard, E. E. 1937. *Witchcraft, Oracles and Magic Among the Azande*. London: Faber and Faber.

Falk, D. 1983. Cerebral cortices of East African early hominids. *Science* 221, 1072–4.

Falk, D. 2000. Hominid brain evolution and the origins of music. In *The Origins of Music* (ed. N. L. Wallin, B. Merker and S. Brown), pp. 197–216. Cambridge, MA: Massachusetts Institute of Technology.

Falk, D. in press. Prelinguistic evolution in early hominids: whence motherese? *Behavioral and Brain Sciences*.

Farizy, C., David, J., and Jaubert, J. 1994. *Hommes et Bisons du Paléolithique Moyen à Mauran (Haute-Garonne)*. Paris: CNRS (*Gallia-Préhistoire* Supplement 30).

Felis, T., Lohmann, G., Kuhnert, H., Lorenze, S. J., Scholz, D., Pätzold, J., Al-Rousan, S., and Al-Moghrabl, S. M. 2004. Increased seasonality in Middle East temperatures during the last interglacial period. *Nature* 429, 164–8.

Fernald, A. 1989. Intonation and communicative intent in mother's speech to infants: is the melody the message? *Child Development* 60, 1497–510.

Fernald, A. 1991. Prosody in speech to children: prelinguistic and linguistic functions. *Annals of Child Development* 8, 43–80.

Fernald, A. 1992. Meaningful melodies in mothers' speech. In *Nonverbal Vocal Communication: Comparative and Developmental Perspectives* (ed. H. Papoušek, U. Jürgens and M. Papoušek), pp. 262–82. Cambridge: Cambridge University Press.

Fernald, A., and Mazzie, C. 1991. Prosody and focus in speech to infants and adults. *Developmental Psychology* 27, 209–21.

Fernald, A., Taeschner, T., Dunn, J., Papoušek, M., de Boysson-Bardies, B., and Fukui, I. 1989. A cross-language study of prosodic modifications in mothers' and fathers' speech to preverbal infants. *Journal of Child Language* 16, 477–501.

Foley, R., and Lahr, M. M. 1997. Mode 3 technologies and the evolution of modern humans. *Cambridge Archaeological Journal* 7, 3–36.

Forgas, J. P., and Moylan, S. 1987. After the movies: the effect of mood on social judgements. *Personality and Social Psychology Bulletin* 13, 465–77.

Frank, R. H. 1988. *Passions Within Reason: The Strategic Role of the Emotions*. New York: W. W. Norton and Co.

Freeman, W. 2000. A neurobiological role for music in social bonding. In *The Origins of Music* (ed. N. L Wallin, B. Merker and S. Brown), pp. 411–24. Cambidge, MA: MIT Press.

Friberg, A., and Sundberg, J. 1999. Does music performance allude to locomotion? A model of final ritardandi derived from measurements of stopping runners. *Journal of the Acoustical Society of America* 105, 1469–84.

Fried, R., and Berkowitz, L. 1979. Music hath charms ... and can influence helpfulness. *Journal of Applied Social Psychology* 9, 199–208.

Friedson, S. M. 1996. *Dancing Prophets: Musical Experience in Tumbuka Healing*. Chicago: University of Chicago Press.

Fry, D. 1948. An experimental study of tone deafness. *Speech* 1948, 1–7.

Gabunia, L., Vekua, A., Lordkipanidze, D., Ferring, R., Justus, A., Majsuradze, G., Mouskhelishvili, A., Noiradze, M., Sologashvili, D., Swisher C. C. III., and Tvalchrelidze, M. 2000. Current research on the hominid site of Dmanisi. In

Early Humans at the Gates of Europe (ed. D. Lordkipanidze, O. Bar-Yosef and M. Otte), pp. 13–27. Liège: ERAUL 92.

Gabunia, L., Vekua, A., Lordkipanidze, D., Swisher, C. C. III., Ferring, R., Justus, A., Nioradze, M., Tvalchrelidze, M., Antón, S. C., Bosinski, G., Jöris, O., de Lumley, M.-A., Majsuardze, G., and Mouskhelishvili, A. 2000. Earliest Pleistocene hominid cranial remains from Dmanisi, Republic of Georgia: taxonomy, geological setting and age. *Science* 288, 1019–25.

Gamble, C. 1987. Man the shoveler: alternative models for Middle Pleistocene colonization and occupation in northern latitudes. In *The Pleistocene Old World* (ed. O. Soffer), pp. 81–98. New York: Plenum Press.

Gamble, C. 1999. *The Palaeolithic Societies of Europe*. Cambridge: Cambridge University Press.

Gardner, R. A., Gardner, B. T., and van Canfort, T. E. (eds). 1989. *Teaching Sign Languages to Chimpanzees*. New York: State University of New York Press.

Gargett, R. H. 1989. Grave shortcomings: the evidence for Neanderthal burial. *Current Anthropology* 30, 157–90.

Gargett, R. H. 1999. Middle Palaeolithic burial is not a dead issue: the view from Qafzeh, Saint-Césaire, Kebara, Amud, and Dederiyeh. *Journal of Human Evolution* 37, 27–90.

Geissmann, T. 1984. Inheritance of song parameters in the gibbon song, analysed in two hybrid gibbons (*Hylobates pileatus x Hylobates lar*). *Folia Primatologica* 42, 216–35.

Geissmann, T. 2000. Gibbon songs and human music from an evolutionary perspective. In *The Origins of Music* (ed. N. L. Wallin, B. Merker and S. Brown), pp. 103–24. Cambridge, MA: Massachusetts Institute of Technology.

Gibbons, A. 1997. A new face for human ancestors. *Science* 276, 1331–3.

Gibert, J., Campillo, D., Arques, J. M., Garcia-Olivares, E., Borja, C., and Lowenstein, J. 1998. Hominid status of the Orce cranial fragment reasserted. *Journal of Human Evolution* 34, 203–17.

Gibert, J., Gibert, Ll., Iglesais, A., and Maestro, E. 1998. Two 'Oldowan' assemblages in the Plio-Pleistocene deposits of the Orce region, south east Spain. *Antiquity* 72, 17–25.

Gibson, K. R., and Ingold, T. (eds). 1993. *Tools, Language and Cognition in Human Evolution*. Cambridge: Cambridge Univerity Press.

Godefroy, O., Leys, D., Furby, A., De Reuck, J., Daems, C., Rondepierre, P., Debachy, B., Deleume, J-F., and Desaulty, A. 1995. Psychoacoustical deficits related to bilateral subcortical hemorrhages, a case with apperceptive auditory agnosia. *Cortex* 31, 149–59.

Goodall, J. 1986. *The Chimpanzees of Gombe*. Cambridge, MA: Harvard University Press.

Gouk, P. 2000. *Music Healing in Cultural Contexts*. Aldershot: Ashgate.

Greenfield, P. M., and Savage-Rumbaugh, E. S. 1990. Grammatical combination in *Pan paniscus*: processes of learning and invention in the evolution and development of language. In *'Language' and Intelligence in Monkeys and Apes:*

Comparative Developmental Perspectives (ed. S. T. Parker and K. R. Gibson), pp. 540–74. Cambridge: Cambridge University Press.

Greenough, W., Black, J., and Wallace, C. 1987. Experience and brain development. *Child Development* 58, 539–59.

Gregersen, P., Kowalsky, E., Kohn, N., and Marvin, E. 2000. Early childhood music education and predisposition to absolute pitch: teasing apart genes and environment. *American Journal of Genetics* 98, 280–2.

Grieser, D. L., and Kuhl, P. K. 1988. Maternal speech to infants in a tonal language: support for universal prosodic features in motherese. *Developmental Psychology* 24, 14–20.

Hauser, M. 1997. *The Evolution of Communication*. Cambridge, MA: MIT Press.

Hauser, M. 2000. The sound and the fury: primate vocalizations as reflections of emotion and thought. In *The Origins of Music* (ed. N. L. Wallin, B. Merker and S. Brown), pp.77–102. Cambridge, MA: Massachusetts Institute of Technology.

Hauser, M., Chomsky, N., and Fitch, W. T. 2002. The faculty of language: what is it, who has it, and how did it evolve? *Science* 298, 1569–79.

Hauser, M., and McDermott, J. 2003. The evolution of the music faculty: a comparative perspective. *Nature Neuroscience* 6, 663–8.

Hawkes, K., O'Connell, J. F., and Blurton-Jones, N. G. 1997. Hadza women's time allocation, offspring provisioning, and the evolution of long post-menopausal life-spans. *Current Anthropology* 38, 551–78.

Heaton, P., Hermelin, B., and Pring, L. 1998. Autism and pitch processing: a precursor for savant musical ability. *Music Perception* 15, 291–305.

Henshilwood, C. S., D'Errico, F., Yates, R., Jacobs, Z., Tribolo, C., Duller, G. A. T., Mercier, N., Sealy, J. C., Valladas, H., Watts, I., and Wintle, A. G. 2002. Emergence of modern human behaviour: Middle Stone Age engravings from South Africa. *Science* 295, 1278–80.

Henshilwood, C. S., D'Errico, F., Vanhaeren, M., van Niekerk, K., and Jacobs, Z. 2004. Middle Stone Age shell beads from South Africa. *Science* 304, 404.

Henshilwood, C. S., and Sealy, J. 1997. Bone artefacts from the Middle Stone Age at Blombos Cave, Southern Cape, South Africa. *Current Anthropology* 38, 890–5.

Henshilwood, C. S., Sealy, J. C., Yates, R., Cruz-Uribe, K., Goldberg, P., Grine, F. E., Klein, R. G., Poggenpoel, C., van Niekerk, K., and Watts, I. 2001. Blombos Cave, Southern Cape, South Africa: Preliminary report on the 1992–1999 excavations of the Middle Stone Age levels. *Journal of Archaeological Science* 28, 421–48.

Henson, R. A. 1988. Maurice Ravel's illness: a tragedy of lost creativity. *British Medical Journal* 296, 1585–8.

Hermelin. B. 2001. *Bright Splinters of the Mind: A Personal Story of Research with Autistic Savants*. London: Jessica Kingsley Publishers.

Hewes, G. 1973. Primate communication and the gestural origin of language. *Current Anthropology* 14, 5–21.

Horden, P. 2000. *Music and Medicine: The History of Music Therapy since Antiquity*. Aldershot: Ashgate.

Hovers, E., Rak, Y., Lavi, R., and Kimbel, W. H. 1995. Hominid remains from Amud cave in the context of the Levantine Middle Palaeolithic. *Paléorient* 21/2, 47–61.

Humphrey, N. 1984. *Consciousness Regained* (chapter 12. The colour currency of nature pp. 146–52). Oxford: Oxford University Press.

Hunt, K. D. 1994. The evolution of human bipedality: ecology and functional morphology. *Journal of Human Evolution* 26, 183–202.

Hurford, J. R. 2000a. The emergence of syntax. In *The Evolutionary Emergence of Language: Social Function and the Origins of Linguistic Form* (ed. C. Knight, M. Studdert-Kennedy and J. R. Hurford) pp. 219–30. Cambridge: Cambridge University Press.

Hurford, J. R. 2000b. Social transmission favours linguistic generalisation. In *The Evolutionary Emergence of Language: Social Function and the Origins of Linguistic Form* (ed. C. Knight, M. Studdert-Kennedy and J. R. Hurford), pp. 324–52. Cambridge: Cambridge University Press.

Hurford, J. R. 2003. The language mosaic and its evolution. In *Language Evolution* (ed. M. H. Christiansen and S. Kirby), pp. 38–57. Oxford: Oxford University Press.

Hurford, J. R., Studdert-Kennedy, M., and Knight, C. (eds). 1998. *Approaches to the Evolution of Language: Social and Cognitive Biases*. Cambridge: Cambridge University Press.

Hurt, C. P., Rice, R. R., McIntosh, G. C., Thaut, M. H. 1998. Rhythmic auditory stimulation in gait training for patients with traumatic brain injury. *Journal of Music Therapy* 35, 228–41.

Ingman, M., Kaessmann, H., Paabo, S., and Gyllensten, U. 2000. Mitochondrial genome variation and the origin of modern humans. *Nature* 408, 708–13.

Isaac, G. (ed.). 1989. *The Archaeology of Human Origins: Papers by Glynn Isaac*. Cambridge: Cambridge University Press.

Isaac, G. 1978. The food-sharing behaviour of proto-human hominids. *Scientific American* 238 (April), 90–108.

Isaac, G. 1983. Bones in contention: competing explanations for the juxtaposition of Early Pleistocene artefacts and faunal remains. In *Animals and Archaeology: Hunters and their Prey* (ed. J. Clutton-Brock and C. Grigson), pp. 3–19. Oxford: British Archaeological Reports, International Series 163.

Isen, A. M. 1970. Success, failure, attention and reactions to others: the warm glow of success. *Journal of Personality and Social Psychology* 15, 294–301.

Isen, A. M., Daubman, K. A., and Nowicki, G. P. 1987. Positive affect facilitates creative problem solving. *Journal of Personality and Social Psychology* 52, 1122–31.

Jackendoff, R. 1999. Possible stages in the evolution of the language faculty. *Trends in Cognitive Sciences* 3, 272–9.

Jackendoff, R. 2000. *Foundations of Language: Brain, Meaning, Grammar, Evolution*. Oxford: Oxford University Press.

Janata, P., and Grafton, S. T. 2003. Swinging in the brain: shared neural substrates for behaviors related to sequencing and music. *Nature Neuroscience* 6, 682–7.

Janzen, J. M. 1992. *Ngoma: Discourses of Healing in Central and Southern Africa.* Berkeley: University of California Press.

Jespersen, O. 1983 [1895]. *Progress in Language.* Amsterdam Classics in Linguistics 17. Amsterdam: John Benjamins Publishing Co.

Jobling, M. A., Hurles, M. E., and Tyler-Smith, C. 2003. *Human Evolutionary Genetics.* New York: Garland Publishing.

Johanson, D., and Edgar, B. 1996. *From Lucy to Language.* London: Weidenfeld and Nicolson.

Juslin, P. N. 1997. Emotional communication in music performance: a functionalist perspective and some data. *Music Perception* 14, 383–418.

Juslin, P. N. 2001. Communication emotion in music performance: a review and a theoretical framework. In *Music and Emotion: Theory and Research* (ed. P. N. Juslin and J. A. Sloboda), pp. 309–37. Oxford: Oxford University Press.

Juslin, P. N., and Sloboda, J. A. (eds). 2001. *Music and Emotion: Theory and Research.* Oxford: Oxford University Press.

Juslin, P. N., and Sloboda, J. A. 2001. Music and emotion: introduction. In *Music and Emotion: Theory and Research* (ed. P. N. Juslin and J. A. Sloboda), pp. 3–20. Oxford: Oxford University Press.

Kahn, P., and Gibbons, A. 1997. DNA from an extinct human. *Science* 277, 176–8.

Kalmus, H., and Fry, D. 1980. On tune deafness (dysmelodia): frequency, development, genetics and musical background. *Annals of Human Genetics* 43, 369–82.

Kappelman, J. 1996. The evolution of body mass and relative brain size in fossil hominids. *Journal of Human Evolution* 30, 243–76.

Kay, R. F., Cartmill, M., and Balow, M. 1998. The hypoglossal canal and the origin of human vocal behaviour. *Proceedings of the National Academy of Sciences* 95, 5417–19.

Kazez, D. 1985. The myth of tone deafness. *Music Education Journal* April 1985, 46–7.

Key, C. A. 1998. *Cooperation, Paternal Care and the Evolution of Hominid Social Groups.* Unpublished PhD thesis, University of London.

Key, C. A., and Aiello, L. C. 1999. The evolution of social organization. In *The Evolution of Culture* (ed. R. Dunbar, C. Knight and C. Power), pp. 15–33. Edinburgh: Edinburgh University Press.

Key, C., and Aiello, L. C. 2000. A Prisoner's Dilemma model for the evolution of paternal care. *Folia Primatologica* 71, 77–92.

Kirby, S. 2000. Syntax without natural selection: how compositionality emerges from vocabulary in a population of learners. In *The Evolutionary Emergence of Language: Social Function and the Origins of Linguistic Form* (ed. C. Knight, M. Studdert-Kennedy and J. R. Hurford), pp. 303–23. Cambridge: Cambridge University Press.

Kirby, S. 2001. Spontaneous evolution of linguistic structure: an iterated learning model of the emergence of regularity and irregularity. *IEEE Journal of Evolutionary Computation* 5, 101–10.

Kirby, S. 2002a. Learning, bottlenecks and the evolution of recursive syntax. In *Linguistic Evolution through Language Acquisition: Formal and Computational Models* (ed. E. Briscoe), pp. 173–204. Cambridge: Cambridge University Press.

Kirby, S. 2002b. The emergence of linguistic structure: an overview of the iterated learning model. In *Simulating the Evolution of Language* (ed. A. Cangelosi and D. Parisi), pp. 121–47. London: Springer.

Kirby, S., and Christiansen, M. H. 2003. From language learning to language evolution. In *Language Evolution* (ed. M. H. Christiansen and S. Kirby), pp. 272–94. Oxford: Oxford University Press.

Kittler, R., Kayser, M., and Stoneking, M. 2003. Molecular evolution of *Pediculus humanus* and the origin of clothing. *Current Biology* 13, 1414–17.

Knight, C., Powers, C., and Watts, I. 1995. The human symbolic revolution: a Darwinian account. *Cambridge Archaeological Journal* 5, 75–114.

Knight, C. 1998. Ritual/speech coevolution: a solution to the problem of deception. In *Approaches to the Evolution of Language* (ed. J. R. Hurford, M. Studdert-Kennedy and C. Knight), pp. 68–91. Cambridge: Cambridge University Press.

Knight, C., Studdert-Kennedy, M., and Hurford, J. (eds). 2000. *The Evolutionary Emergence of Language*. Cambridge: Cambridge University Press.

Knutson, B., Bungdoff, J., and Panksepp, J. 1998. Anticipation of play elicits high frequency ultrasonic vocalisations in young rats. *Journal of Comparative Psychology* 112, 65–73.

Koelsch, S., Gunter, T. C., Yves v. Crammon, D., Zysset, S., Lohmann, G., and Friederici, A. D. 2002. Bach speaks: a cortical 'language-network' serves the processing of music. *NeuroImage* 17, 956–66.

Kohler, W. 1925. *The Mentality of Apes*. London: Routledge and Kegan Paul.

Kohn, M., and Mithen, S. J. 1999. Handaxes: products of sexual selection? *Antiquity* 73, 518–26.

Komarova, N. L., and Nowak, M. 2003. Language learning and evolution. In *Language Evolution* (ed. M. H. Christiansen and S. Kirby), pp. 317–37. Oxford: Oxford University Press.

Krings, M., Stone, A., Schmitz, R. W., Krainitzki, H., Stoneking, M., and Pääbo, S. 1997. Neanderthal DNA sequences and the origin of modern humans. *Cell* 90, 19–30.

Krumhansl, C. 1997. An exploratory study of musical emotions and psychophysiology. *Canadian Journal of Experimental Psychology* 51, 336–52.

Kuhn, S. 1995. *Mousterian Lithic Technology*. Princeton, NJ: Princeton University Press.

Kunej, D., and Turk, I. 2000. New perspectives on the beginnings of music: archaeological and musicological analysis of a Middle Palaeolithic bone 'flute'. In *The Origins of Music* (ed. N. L. Wallin, B. Merker and S. Brown), pp. 235–68. Cambridge, MA: MIT Press.

Laban, R. 1988 [1950]. *The Mastery of Movement* (4th edition). Plymouth: Northcote House.

Lahr, M. M., and Foley, R. 1994. Multiple dispersals and modern human origins. *Evolutionary Anthropology* 3, 48–60.

Lahr, M. M., and Foley, R. 2004. Human evolution writ small. *Nature* 431, 1043–4.

Lai, C. S. L., Fisher, S. E., Hurst, J. A., Vargha-Khadem, F., and Monaco, A. P. 2001. A forkhead-domain gene is mutated in a severe speech and language disorder. *Nature* 413, 519–23.

Langer, S. 1942. *Philosophy in a New Key.* New York: Mentor.

Le Doux, J. 1996. *The Emotional Brain.* New York: Simon and Schuster.

Lee, R. B. 1979. *The !Kung San: Men, Women, and Work in a Foraging Society.* Cambridge: Cambridge University Press.

Leinonen, L., Linnankoski, I., Laakso, M. L., and Aulanko, R. 1991. Vocal communication between species: man and macaque. *Language Communication* 11, 241–62.

Leinonen, L., Laakso, M.-L., Carlson, S., and Linnankoski, I. 2003. Shared means and meanings in vocal expression of man and macaque. *Logoped Phonaiatr Vocol* 28, 53–61.

Lerdahl, F., and Jackendoff, R. 1983. *A Generative Theory of Tonal Music.* Cambridge, MA: MIT Press.

Lewin, R. 1999. *Human Evolution.* London: Blackwell Science.

Lieberman, P. 1992. On Neanderthal speech and Neanderthal extinction. *Current Anthropology* 33, 409–10.

Lieberman, P. 2001. On the neural bases of spoken language. In *In the Mind's Eye: Multidisciplinary Approaches to the Evolution of Human Cognition* (ed. A. Nowell), pp. 172–86. Ann Arbor, MI: International Monographs in Prehistory.

Lieberman, P., and Crelin, E. 1971. On the speech of Neanderthal man. *Linguistic Inquiry* 2, 203–22.

Linnankoski, I., Laakso, M., Aulanko, R., and Leinonen, L. 1994. Recognition of emotions in macaque vocalisations by adults and children. *Language Communication* 14, 183–92.

Lordkipanidze, D., Bar-Yosef, O., and Otte, M. 2000. *Early Humans at the Gates of Europe.* Liège: ERAUL 92.

Luria, A. R., Tsvetkova, L. S., and Futer, D. S. 1965. Aphasia in a composer. *Journal of Neurological Science* 2, 288–92.

McBrearty, S., and Brooks, A. 2000. The revolution that wasn't: a new interpretation of the origin of modern human behavior. *Journal of Human Evolution* 38, 453–563.

McDougall, I., Brown, F. H., and Fleagle, J. G. 2005. Stratigraphic placement and age of modern humans from Kibish, Ethiopia. *Nature* 433, 733–6.

McGrew, W. C., Marchant, L. F., and Nishida, T. 1996. *Great Ape Societies.* Cambridge: Cambridge University Press.

Mâche, F-B. 2000. The necessity of and problem with a universal musicology. In *The Origins of Music* (ed. N. L. Wallin, B. Merker and S. Brown), pp. 473–80. Cambridge, MA: Massachusetts Institute of Technology.

McHenry, H. M. 1996. Sexual dimorphism in fossil hominids and its sociological implications. In *The Archaaeology of Human Ancestry* (ed. J. Steele and S. Shennan), pp. 91–109. London: Routledge.

MacLarnon, A., and Hewitt, G. P. 1999. The evolution of human speech: the role of enhanced breathing control. *American Journal of Physical Anthropology* 109, 341–3.

McNeill, D. 1992. *Hand and Mind: what Gestures Reveal about Thought*. Chicago: Chicago University Press.

McNeill, D. 2000. *Language and Gesture*. Cambridge: Cambridge University Press.

McNeill, W. H. 1995. *Keeping Together in Time: Dance and Drill in Human History*. Cambridge, MA: Harvard University Press.

Maess, B., Koelsch, S., Gunter, T. C., and Friederici, A. D. 2001. Musical syntax is processed in Broca's area: an MEG Study. *Nature Neuroscience* 4, 540–5.

Malloch, S. N. 1999. Mothers and infants and communicative musicality. *Musicae Scientiae* Special Issue 1999–2000, 29–57.

Mandel, S. E. 1996. Music for wellness: music therapy for stress management in a rehabilitation program. *Music Therapy Perspectives* 14, 38–43.

Mania, D., and Mania, U. 1988. Deliberate engravings on bone artefacts of *Homo erectus*. *Rock Art Research* 5, 91–107.

Manning, J. T., and Hartley, M. A. 1991. Symmetry and ornamentation are correlated in the peacock's train. *Animal Behaviour* 42, 1020–1.

Marler, P. 2000. Origins of speech and music: insights from animals. In *The Origins of Music* (ed. N. L. Wallin, B. Merker and S. Brown), pp. 49–64. Cambridge, MA: Massachusetts Institute of Technology.

Marler, P., and Tenaza, R. 1977. Signaling behavior of apes with special reference to vocalization. In *How Animals Communicate* (ed. T. Sebok) pp. 965–1003. Bloomington: Indiana University Press.

Marshack, A. 1997. The Berekhat Ram figurine: a late Acheulian carving from the Middle East. *Antiquity* 71, 327–37.

Marshall, L. 1976. *The !Kung of Nyae Nyae*. Cambridge: Cambridge University Press.

Marshall, L. 2001. *The Body Speaks, Performance and Expression*. London: Methuen.

Martínez, I., Rosa, M., Arsuaga, J.-L., Jarabo, P., Quam, R., Lorenzo, C., Gracia, A., Carretero, J.-M., Bermúdez de Castro, J.-M., and Carbonell, E. 2004. Auditory capacities in Middle Pleistocene humans from the Sierra de Atapuerca in Spain. *Proceedings of the National Academy of Sciences* 101, 9976–81.

Mellars, P. 1989. Major issues in the emergence of modern humans. *Current Anthropology* 30, 349–85.

Mellars, P. 1996. *The Neanderthal Legacy*. Princeton, NJ: Princeton University Press.

Mellars, P., and Gibson, K. R. (eds). 1996. *Modelling the Early Human Mind*. Cambridge: McDonald Institute for Archaeological Research.

Mendez, M. F. 2001. Generalized auditory agnosia with spared music recognition in a left-hander: analysis of a case with a right temporal stroke. *Cortex* 37, 139–50.

Merker, B. 1999. Synchronous chorusing and the origins of music. *Musicae Scientiae* Special Issue 1999–2000, 59–73.

Merker, B. 2000. Synchronous chorusing and human origins. In *The Origins of Music* (ed. N. L. Wallin, B. Merker and S. Brown), pp. 315–28. Cambridge, MA: Massachusetts Institute of Technology.

Merriam, A. P. 1964. *The Anthropology of Music*. Evantson, IL: Northwestern University Press.

Metz-Lutz, M.-N., and Dahl, E. 1984. Analysis of word comprehension in a case of pure word deafness. *Brain and Language* 23, 13–25.

Meyer, L. 1956. *Emotion and Meaning in Music*. Chicago: Chicago University Press.

Meyer, L. B. 2001. Music and emotion: distinctions and uncertainties. In *Music and Emotion* (ed. P. N. Juslin and J. A. Sloboda) pp. 341–60. Oxford: Oxford University Press.

Miller, G. 1997. How mate choice shaped human nature: a review of sexual selection and human evolution. In *Handbook of Evolutionary Psychology: Ideas, Issues and Applications* (ed. C. Crawford and D. L. Krebs), pp. 87–129. Mahwah, NJ: Lawrence Erlbaum Associates.

Miller, G. 2000. Evolution of human music through sexual selection. In *The Origins of Music* (ed. N. L. Wallin, B. Merker and S. Brown), pp. 329–60. Cambridge, MA: Massachusetts Institute of Technology.

Miller, L. K. 1989. *Musical Savants: Exceptional Skill in the Mentally Retarded*. Hillsdale, NJ: Lawrence Erlbaum.

Mitani, J. C. 1985. Gibbon song duets and intergroup spacing. *Behaviour* 92, 59–96.

Mitani, J. C. 1996. Comparative studies of African ape vocal behavior. In *Great Ape Societies* (ed. W. C. McGrew, L. F. Marchant and T. Nishida). Cambridge: Cambridge University Press.

Mitani, J. C., and Brandt, K. L. 1994. Social factors influence the acoustic variability in the long-distance calls of male chimpanzees. *Ethology* 96, 233–52.

Mitani, J. C., and Gros-Louis, J. 1995. Species and sex differences in the screams of chimpanzees and bonobos. *International Journal of Primatology* 16, 393–411.

Mitani, J. C., Hasegwa, T., Gros-Louis, J., Marler, P., and Byrne, R. 1992. Dialects in wild chimpanzees? *American Journal of Primatology* 27, 233–43.

Mitani, J. C., and Marler, P. 1989. A phonological analysis of male gibbon singing behaviour. *Behaviour* 109, 20–45.

Mithen, S. J. 1996. On Early Palaeolithic 'concept-mediated' marks, mental modularity and the origins of art. *Current Anthropology* 37, 666–70.

Mithen, S. J. 1996/1998. *The Prehistory of the Mind: A Search for the Origin of Art, Science and Religion*. London: Thames and Hudson/Orion.

Mithen, S. J. 1998. The supernatural beings of prehistory and the external storage of religious ideas. In *Cognition and Material Culture: The Archaeology of Symbolic Storage* (ed. C. Renfrew and C. Scarre), pp. 97–106. Cambridge: McDonald Institute of Archaeological Research.

Mithen, S. J. 2000. Palaeoanthropological perspectives on the theory of mind. In

Understanding other Minds (ed. S. Baron-Cohen, H. Tager-Flusberg and D. J. Cohen) pp. 488–502. Oxford: Oxford University Press.

Mithen, S. J. 2003. *After the Ice: A Global Human History, 20,000–15,000 BC*. London: Weidenfeld and Nicolson.

Mithen, S. J., and Reed, M. 2002. Stepping out: a computer simulation of hominid dispersal from Africa. *Journal of Human Evolution* 43, 433–62.

Moggi-Cecchi, J., and Collard, M. 2002. A fossil stapes from Stekfontein, South Africa, and the hearing capabilities of early hominids. *Journal of Human Evolution* 42, 259–65.

Møller, A. P. 1990. Fluctuating asymmetry in male sexual ornaments may reliably reveal male quality. *Animal Behaviour* 40, 1185–7.

Monnot, M. 1999. Function of infant-directed speech. *Human Nature* 10, 415–43.

Morwood, M. J., O'Sullivan, P. B., Aziz, F., and Raza, A. 1998. Fission-track ages of stone tools and fossils on the east Indonesian island of Flores. *Nature* 392, 173–6.

Morwood, M. J., Soejono, R. P., Roberts, R. G., Sutikana, T., Turney, C. S. M., Westaway, K. E., Rink, W. J., Zhao, J.-X., van den Bergh, G. D., Rokus Awe Due, Hobbs, D. R., Moore, M. W., Bird, M. I., and Fifield, L. K. 2004. Archaeology and the age of a new hominid from Flores in eastern Indonesia. *Nature* 431, 1087–91.

Mottron, I., Peretz, I., Belleville, S., and Rouleau, N. 1999. Absolute pitch in autism: a case study. *Neurocase* 5, 485–501.

Moya-Sola, S., and Kohler, M. 1997. The Orce skull: anatomy of a mistake. *Journal of Human Evolution* 33, 91–7.

Nelson, D. G. K., Hirsh-Pasek, K., Jusczyx, P. W., and Cassidy, K. W. 1989. How the prosodic cues in motherese might assist language learning. *Journal of Child Language* 16, 55–68.

Nettl, B. 1983. *The Study of Ethnomusicology: Twenty-Nine Issues and Concepts*. Urbana, IL: University of Illinois Press.

Nettle, D. 1999. *Linguistic Diversity*. Oxford: Oxford University Press.

Niedenthal, P. M., and Setterlund, M. B. 1994. Emotion congruence in perception. *Personality and Social Psychology Bulletin*. 20, 401–11.

Noble, W., and Davidson, I. 1996. *Human Evolution, Language and Mind: A Psychology and Archaeological Inquiry*. Cambridge: Cambridge University Press.

Nowak, M. A., and Komarova, N. L. 2001a. Towards an evolutionary theory of language. *Trends in Cognitive Sciences* 5, 288–95.

Nowak, M. A., and Komarova, N. L. 2001b. Evolution of universal grammar. *Science* 291, 114–18.

Nowak, M. A., Komarova, N. L., and Krakauer, D. 1999. The evolutionary language game. *Journal of Theoretical Biology* 200, 147–62.

Nowell, A. (ed.). 2001. *In the Mind's Eye: Multidisciplinary Approaches to the Evolution of Human Cognition*. Ann Arbor, MI: International Monographs in Prehistory.

Oatley, K., and Jenkins, J. J. 1996. *Understanding Emotions*. Oxford: Blackwell.

Oatley, K., and Johnson-Laird, P. N. 1987. Towards a theory of emotions. *Cognition and Emotion* 1, 29–50.

O'Connell, J. F., Hawkes, K., and Blurton-Jones, N. G. 1999. Grandmothering and the evolution of *Homo erectus*. *Journal of Human Evolution* 36, 461–85.

Oelman, H., and Lœng, B. 2003. A validation of the emotional meaning of single intervals according to classical Indian music theory. *Proceedings of the 5th Triennial ESCOM conference*, pp. 393–6. Hanover: Hanover University of Music and Drama.

Oka, T., and Takenaka, O. 2001. Wild gibbons' parentage tested by non-invasive DNA sampling and PCR-amplified polymorphic microsatellites. *Primates* 42, 67–73.

Pagel, M., and Bodmer, W. 2003. A naked ape would have fewer parasites. *Proceedings of the Royal Society of London B, Supplement 1, Biology letters* 270, 5117–19.

Palmqvist, P. 1997. A critical re-evaluation of the evidence for the presence of hominids in Lower Pleistocene times at Venta Micena, southern Spain. *Journal of Human Evolution* 33, 83–9.

Papousek, M., Papousek, H., and Symmes, D. 1991. The meanings of melodies in motherese in tone and stress languages. *Infant Behavior and Development* 14, 415–40.

Parfitt, S., and Roberts, M. 1999. Human modification of faunal remains. In *Boxgrove, A Middle Pleistocene Hominid Site at Eartham Quarry, Boxgrove, West Sussex* (ed. M. B. Roberts and S. Parfitt), pp. 395–415. London: English Heritage.

Parsons, L. 2003. Exploring the functional neuroanatomy of music performance, perception and comprehension. In *The Cognitive Neuroscience of Music* (ed. I. Peretz and R. Zatorre), pp. 247–68. Oxford: Oxford University Press.

Patel, A., Peretz, I., Tramo, M., and Labreque, R. 1998. Processing prosodic and musical patterns: a neuropsychological investigation. *Brain and Language* 61, 123–44.

Patel, A. D. 2003. Language, music, syntax and the brain. *Nature Neuroscience* 6, 674–81.

Payne, K. 2000. The progressively changing songs of humpback whales: a window on the creative process in a wild animal. In *The Origins of Music* (ed. N. L. Wallin, B. Merker, and S. Brown), pp. 135–50. Cambridge, MA: MIT Press.

Pease, A. 1984. *Body Language: How to Read Others' Thoughts by their Gestures*. London: Sheldon Press.

Pelegrin, J. 1993. A framework for analysing prehistoric stone tool manufacture and a tentative application to some early stone industries. In *The Use of Tools by Human and Non-Human Primates* (ed. A. Berthelet and J. Chavaillon), pp. 302–14. Oxford: Clarendon Press.

Peretz, I. 1993. Auditory atonalia for melodies. *Cognitive Neuropsychology* 10, 21–56.

Peretz, I. 2003. Brain specialization for music: new evidence from congenital amusia. In *The Cognitive Neuroscience of Music* (ed. I. Peretz and R. Zatorre), pp. 247–68. Oxford: Oxford University Press.

Peretz, I., Ayotte, J., Zatorre, R. J., Mehler, J., Ahad, P., Penhune, B., and Jutras, B. 2002. Congenital amusia: a disorder of fine-grained pitch discrimination. *Neuron* 33, 185–91.

Peretz, I., and Coltheart, M. 2003. Modularity of music processing. *Nature Neuroscience* 6, 688–91.

Peretz, I., Kolinsky, R., Tramo, M., Labrecque, R., Hublet, C., Demeurisse, G., and Belleville, S. 1994. Functional dissociations following bilateral lesions of auditory cortex. *Brain* 117, 1283–301.

Peretz, I., and Zatorre, R. (ed). 2003 *The Cognitive Neuroscience of Music*. Oxford: Oxford University Press.

Piccirilli, M., Sciarma, T., and Luzzi, S. 2000. Modularity of music: evidence from a case of pure amusia. *Journal of Neurology, Neurosurgery and Psychiatry* 69, 541–5.

Pinker, S. 1994. *The Language Instinct*. New York: William Morrow.

Pinker, S. 1997. *How the Mind Works*. New York: Norton.

Pinker, S. 2003. Language as an adaptation to the cognitive niche. In *Language Evolution* (ed. M. H. Christiansen and S. Kirby), pp. 16–37. Oxford: Oxford University Press.

Pinker, S., and Bloom, P. 1995. Natural language and natural selection. *Behavioral and Brain Sciences* 13, 707–84.

Plavcan, J. M., and van Schaik, C. P. 1992. Intrasexual competition and canine dimorphism in anthropoid primates. *American Journal of Physical Anthropology* 87, 461–77.

Potts, R. 1988. *Early Hominid Activities at Olduvai Gorge*. New York: Aldine de Gruyter.

Potts, R., and Shipman, P. 1981. Cutmarks made by stone tools on bones from Olduvai Gorge, Tanzania. *Nature* 29, 577–80.

Povinelli, D. J. 1993. Reconstructing the evolution of the mind. *American Psychologist* 48, 493–509.

Povinelli, D. J. 1999. *Folk Physics for Apes*. Oxford: Oxford University Press.

Power, C. 1998. Old wives' tale: the gossip hypothesis and the reliability of cheap signals. In *Approaches to the Evolution of Language* (ed. J. R. Hurford, M. Studdert-Kennedy and C. Knight), pp. 111–29. Cambridge: Cambridge University Press.

Premack, A. J., and Premack, D. 1972. Teaching language to an ape. *Scientific American* 227, 92–9.

Rak, Y., Kimbel, W. H., and Hovers, E. 1994. A Neanderthal infant from Amud Cave, Israel. *Journal of Human Evolution* 26, 313–24.

Ramirez Rozzi F. V., and Bermúdez de Castro, J. M. 2004. Surprisingly rapid growth in Neanderthals. *Nature* 428, 936–9.

Reichard, U. 1995. Extra-pair copulations in monogamous gibbons (*Hylobates lar*). *Ethology* 100, 99–112.

Renfrew, A. C., and Zubrow, E. (eds). 1994. *The Ancient Mind*. Cambridge: Cambridge University Press.

Renfrew, A. C., and Bellwood, P. (eds). 2002. *Examining the Farming/Language Dispersal Hypothesis*. Cambridge: MacDonald Institute for Archaeological Research.

Richman, B. 1976. Some vocal distinctive features used by gelada monkeys. *Journal of the Acoustic Society of America* 60, 718–24.

Richman, B. 1978. The synchronisation of voices by gelada monkeys. *Primates* 19, 569–81.

Richman, B. 1987. Rhythm and melody in gelada vocal exchanges. *Primates* 28, 199–223.

Richmond, B. G., and Strait, D. S. 2000. Evidence that humans evolved from a knuckle-walking ancestor. *Nature* 404, 382–5.

Rizzolatti, G., and Arbib, M. A. 1998. Language within our grasp. *Trends in Neurosciences* 21, 188–94.

Robb, L. 1999. Emotional musicality in mother–infant vocal affect, and an acoustic study of postnatal depression. *Musicae Scientiae* Special Issue 1999–2000, 123–54.

Robb, S. L. (ed). 2003. *Music Therapy in Paediatric Healthcare: Research and Evidence-Based Practice*. Silver Spring, MD: American Music Therapy Association.

Robb, S. L., Nicholas, R. J., Rutan, R. L., Bishop, B. L., and Parker, J. C. 1995. The effects of music assisted relaxation on preoperative anxiety. *Journal of Music Therapy* 32, 2–21.

Roberts, M. B., and Parfitt, S. A. 1999. *Boxgrove, A Middle Pleistocene Hominid Site at Eartham Quarry, Boxgrove, West Sussex*. London: English Heritage Archaeological Report no. 17.

Roberts, M. B., Stringer, C. B., and Parfitt, S. A. 1994. A hominid tibia from Middle Pleistocene sediments at Boxgrove, U.K. *Nature* 369, 311–13.

Roe, D. A. 1994. A metrical analysis of selected sets of handaxes and cleavers from Olduvai Gorge. In *Olduvai Gorge 5: Excavations in Beds III–IV and the Masele Beds 1968–71* (ed. M. Leakey and D. Roe), pp. 146–235. Cambridge: Cambridge University Press.

Roe, D. A. 1995. The Orce Basin (Andalucía, Spain) and the initial Palaeolithic of Europe. *Oxford Journal of Archaeology* 14, 1–12.

Rolland, N. 1998. The Lower Palaeolithic settlement of Eurasia, with specific reference to Europe. In *Early Human Behaviour in a Global Context* (ed. M. D. Petraglia and R. Korisettat), pp. 187–220. London: Routledge.

Roseman, M. 1991. *Healing sounds from the Malaysian Rainforest: Temiar Music and Medicine*. Berkeley: University of California Press.

Rosenzweig, M., Leiman, A., and Breedlove, S. 1999. *Biological Psychology: An Introduction to Behavioral, Cognitive and Clinical Neuroscience*. Sunderland, MA: Sinauer Associates.

Rowell, T. E., and Chism, J. 1986. Sexual dimorphism and mating systems: jumping to conclusions. In *Sexual Dimorphism in Living and Fossil Primates* (ed. M. Pickford and B. Chiarelli), pp. 107–11. Firenze: Il Sedicesimo.

Rozin, P., Haidt, J., and McCauley, C. R. 1993. Disgust. In *Handbook of Emotions* (ed. M. Lewis and J. M. Haviland). New York: Guildford Press.

Ruff, C. B., Trinkaus, E., and Holliday, T. W. 1997. Body mass and encephalization in pleistocene *Homo*. *Nature* 387, 173–6.

Russon, A. E., Bard, K. A., and Parker, S. T. 1996. *Reaching into Thought: The Minds of the Great Apes*. Cambridge: Cambridge University Press.

Saffran, J. R. 2003. Absolute pitch in infancy and adulthood: the role of tonal structure. *Developmental Science* 6 35–47.

Saffran, J. R., Aslin, R. N., and Newport, E. L. 1996. Statistical learning by 8-month old infants. *Science* 274, 1926–8.

Saffran, J. R., and Griepentrog, G. J. 2001. Absolute pitch in infant auditory learning: evidence for developmental re-organization. *Developmental Psychology* 37, 74–85.

Saffran, J. R., Johnson, E. K., Aslin, R. N., and Newport, E. L. 1999. Statistical learning of tone sequences by human infants and adults. *Cognition* 70, 27–52.

Savage-Rumbaugh, E. S., and Rumbaugh, D. M. 1993. The emergence of language. In *Tools, Language and Cognition in Human Evolution* (ed. K. R. Gibson and T. Ingold), pp. 86–108. Cambridge: Cambridge University Press.

Savage-Rumbaugh, E. S., Wilkerson, B., and Bakeman, R. 1977. Spontaneous gestural communication among conspecifics in the pygmy chimpanzee (*Pan paniscus*). In *Progress in Ape Research* (ed. G. Bourne), pp. 97–116. New York: Academic Press.

Schaik, C. P. van, and Dunbar, R. I. M. 1990. The evolution of monogamy in large primates: a new hypothesis and some crucial tests. *Behaviour* 113, 30–62.

Scherer, K. R. 1995. Expression of emotion in voice and music. *Journal of Voice* 9, 235–48.

Scherer, K. R., and Zentner, M. R. 2001. Emotional effects of music: production rules. In *Music and Emotion: Theory and Research* (ed. P. N. Juslin and J. A. Sloboda), pp. 361–92. Oxford: Oxford University Press.

Schlaug, G., Jäncke, L., Huang, Y., and Steinmetz, H. 1995. In vivo evidence of structural brain asymmetry in musicians. *Science* 268, 699–701.

Schullian, D., and Schoen, M. 1948. *Music as Medicine*. New York: Henry Schuman.

Scott, K. 1980. Two hunting episodes of Middle Palaeolithic age at La Cotte de St Brelade, Jersey (Channel Islands). *World Archaeology* 12, 137–52.

Sémah, F., Saleki, H., and Falgueres, C. 2000. Did early man reach Java during the Late Pliocene? *Journal of Archaeological Science* 27, 763–9.

Sergent, J. 1993. Music, the brain and Ravel. *Trends in Neuroscience* 16, 168–72.

Sergent, J., Zuck, E., Terriah, S., and MacDonald, B. 1992. Distributed neural network underlying musical sight reading and keyboard performance. *Science* 257, 106–9.

Seyfarth, R. M. in press. Continuities in vocal communication argue against a gestural origin of language (comment on From monkey-like action recognition to human language: an evolutionary framework for neurolinguistics by M. Arbib). *Behavioral and Brain Sciences*.

Seyfarth, R. M., Cheney, D. L., Harcourt, A. H., and Stewart, K. J. 1994. The acoustic features of gorilla double grunts and their relation to behavior. *American Journal of Primatology* 33, 31–50.

Shennan, S. J. 2000. Population, culture history and the dynamics of culture change. *Current Anthropology* 41, 811–35.

Shove, P., and Repp, B. H. 1995. Musical motion and performance: theoretical and empirical perspectives. In *The Practice of Performance: Studies in Musical*

Interpretation (ed. J. Rink), pp. 55–83. Cambidge: Cambridge University Press.

Slater, P. J. B. 2000. Birdsong repertoires: their origins and use. In *The Origins of Music* (ed. N. L. Wallin, B. Merker and S. Brown), pp. 49–64. Cambridge, MA: MIT Press.

Sloboda, J. A. 1985. *The Musical Mind. The Cognitive Psychology of Music*. Oxford: Clarendon Press.

Sloboda, J. A. 1998. Does music mean anything? *Musicae Scientiae* 2, 21–31.

Sperber, D. 1996. *Explaining Culture*. Oxford: Blackwell.

Spoor, F., Wood, B., and Zonneveld, F. 1994. Implications of early hominid labyrinthine morphology for evolution of human bipedal locomotion. *Nature* 369, 645–8.

Spoor, F., and Zonneveld, F. 1998. A comparative review of the human bony labyrinth. *Yearbook of Physical Anthropology* 41, 211–51.

Staal, F. 1988. *Rules Without Meaning. Essays on Ritual, Mantras and the Science of Man*. New York: Peter Lang.

Staal, F. 1989. Vedic mantras. In *Mantra* (ed. H. P. Alper), pp. 48–95. Albany, NY: State University of New York.

Standley, J. M. 1995. Music as a therapeutic intervention in medical and dental treatment: research and clinical applications. In *The Art and Science of Music Therapy: A Handbook* (ed. T. Wigram, B. Saperson and R. West), pp. 3–33. Amsterdam: Harwood Academic Publications.

Standley, J. M. 1998. The effect of music and multimodal stimulation on physiologic and developmental responses of premature infants in neonatal intensive care. *Pediatric Nursing* 24, 532–8.

Standley, J. M. 2000. The effect of contingent music to increase non-nutritive sucking of premature infants. *Pediatric Nursing* 26, 493.

Standley, J. M., and Moore, R. S. 1995. Therapeutic effect of music and mother's voice on premature infants. *Pediatric Nursing* 21, 509–12.

Stanford, C. B. 1999. *The Hunting Apes: Meat Eating and the Origins of Human Behavior*. Princeton, NJ: Princeton University Press.

Steinke, W. R., Cuddy, L. L., and Jakobson, L. S. 2001. Dissociations among functional subsystems governing melody recognition after right-hemisphere damage. *Cognitive Neuropsychology* 18, 411–37.

Straus, L. G., and Bar-Yosef, O. (eds). 2001. Out of Africa in the Pleistocene. *Quaternary International* 75, 1–3.

Street, A., Young, S., Tafuri, J., and Ilari, B. 2003. Mothers' attitudes to singing to their infants. *Proceedings of the 5th Triennial ESCOM conference* (ed. R. Kopiez, A. C. Lehmann, I. Wolther and C. Wolf), pp. 628–31. Hanover: Hanover University of Music and Drama.

Stringer, C. B. 2003. Out of Ethiopia. *Nature* 423, 692–5.

Stringer, C. B., Barton, R. N. E., and Finlayson, J. C. (eds). 2000. *Neanderthals on the Edge*. Oxford: Oxbow Books.

Stringer, C. B., and Gamble, C. 1993. *In Search of the Neanderthals*. London: Thames and Hudson.

Struhsaker, T. 1967. Auditory communication among vervet monkeys

(*Cercopithecus aethiops*). In *Social Communication Among Primates* (ed. A. Altmann), pp. 281–324. Chicago: Chicago University Press.

Studdert-Kennedy, M. 1998. The particulate origins of language generativity: from syllable to gesture. In *Approaches to the Evolution of Language: Social and Cognitive Biases* (ed. J. Hurford, M. Studdert-Kennedy and C. Knight), pp. 202–21. Cambridge: Cambridge University Press.

Studdert-Kennedy, M. 2000. Evolutionary implications of the particulate principle: imitation and the dissociation of phonetic form from semantic function. In *The Evolutionary Emergence of Language* (ed. C. Knight, M. Studdert-Kennedy and J. Hurford), pp. 161–76. Cambridge: Cambridge University Press.

Studdert-Kennedy, M., and Goldstein, L. 2003. Launching language: the gestural origin of discrete infinity. In *Language Evolution* (ed. M. H. Christiansen and S. Kirby), pp. 235–54. Oxford: Oxford University Press.

Sundberg, J., Friberg, A., and Fryden, L. 1992. Music and locomotion: a study of the perception of tones with level envelopes replicating force patterns of walking. *KTH Speech Transmission Laboratory Quarterly Progress and Status Report* 4/1992, 109–22.

Swisher, C. C. III., Curtis, G. H., Jacob, T., Getty, A. G., Suprijo, A., and Widiasmoro. 1994. Age of earliest known hominids in Java, Indonesia. *Science* 262, 1118–21.

Takahashi, N., Kawamura, M., Shinotou, H., Hirayama, K., Kaga, K., and Shindo, M. 1992. Pure word deafness due to left hemisphere damage. *Cortex* 28, 295–303.

Takeuchi, A., and Hulse S. H. 1993. Absolute pitch. *Psychological Bulletin* 113, 345–61.

Tallerman, M. in press. Did our ancestors speak a holistic proto-language? To appear in *Lingua* special issue on language evolution edited by Andrew Carstairs-McCarthy.

Tanner, J. E., and Byrne, R. W. 1996. Representation of action through iconic gesture in a captive lowland gorilla. *Current Anthropology* 37, 162–73.

Tchernov, E. 1995. The hominids in the southern Levant. In *The Hominids and their Environments During the Lower and Middle Pleistocene of Eurasia* (ed. J. Gibert et al.), pp. 389–406. Proceedings of the International Conference of Human Palaeontology. Orce: Museo de Prehistoria y Palaeontology.

Tenaza, R. R. 1976. Songs, choruses and countersinging of Kloss' gibbons (*Hylobates klossii*) in Sinernit Island, Indonesia. *Zeitschrift für Tierpsychologie* 40, 37–52.

Terrace, H. S. 1979. *Nim*. New York: Knopf.

Terrace, H. S., Pettito, L. A., Saunders, R. J., and Bever, T. G. 1979. Can an ape create a sentence? *Science* 206, 891–902.

Thaut, M. H., McIntosh, G. C., Rice, R. R. 1997. Rhythmic facilitation of gait training in hemiparetic stroke rehabilitation. *Journal of Neurological Sciences* 151, 207–12.

Thaut, M. H., McIntosh, K. W., McIntosh, G. C., and Hoemberg, V. 2001. Auditory rhythmicity enhances movement and speech motor control in patients with Parkinson's disease. *Functional Neurology* 16, 163–172.

Thelen, E. 1981. Rhythmical behavior in infancy: an ethological perspective. *Developmental Psychology* 17, 237–57.

Thieme, H. 1997. Lower Palaeolithic hunting spears from Germany. *Nature* 385, 807–10.

Thomas, D. A. 1995. *Music and the Origins of Language: Theories from the French Enlightenment.* Cambridge: Cambridge University Press.

Thornhill, R., and Gangstead, S. 1996. The evolution of human sexuality. *Trends in Ecology and Evolution* 11, 98–102.

Tobias, P. 1981. The emergence of man in Africa and beyond. *Philosophical Transactions of the Royal Society of London, B. Biological Sciences* 292, 43–56.

Tobias, P. 1987. The brain of *Homo habilis*: a new level of organisation in cerebral evolution. *Journal of Human Evolution* 16, 741–61.

Tolbert, E. 2001. Music and meaning: an evolutionary story. *Psychology of Music* 29, 84–94.

Tomasello, M. 2003. On the different origins of symbols and grammar. In *Language Evolution* (ed. M. H. Christiansen and S. Kirby), pp. 94–110. Oxford: Oxford University Press.

Tomasello. M., Call, J., and Hare, B. 2003. Chimpanzees understand psychological states – the question is which ones and to what extent. *Trends in Cognitive Sciences* 7, 153–6.

Tomasello, M., George, B. L., Kruger, A. C., Farrar, M. J., and Evans, A. 1985. The development of gestural communication in young chimpanzees. *Journal of Human Evolution* 14, 175–86.

Tomasello, M., Gust, D., and Frost, G. T. 1989. A longitudinal investigation of gestural communication in young chimpanzees. *Primates* 30, 35–50.

Tomasello, M., Call, J., Nagell, K., Olguin, R., and Carpenter, M. 1994. The learning and use of gestural signals by young chimpanzees: a transgenerational study. *Primates* 35, 137–54.

Tonkes, B., and Wiles, J. 2002. Methodological issues in simulating the emergence of language. In *The Transition to Language* (ed. A. Wray), pp. 226–51. Oxford: Oxford University Press.

Toth, N. 1985. The Oldowan re-assessed: a close look at early stone artefacts. *Journal of Archaeological Science* 12, 101–20.

Toth, N., Schick, K. D., Savage-Rumbaugh, E. S., Sevcik, R. A., and Rumbaugh, D. M. 1993. *Pan* the tool-maker: investigations into the stone tool-making and tool-using capabilities of a bonobo (*Pan paniscus*). *Journal of Archaeological Science* 20, 81–91.

Trehub, S. E. 2003. Musical predispositions in infancy: an update. In *The Cognitive Neuroscience of Music* (ed. I. Peretz and R. Zatorre), pp. 3–20. Oxford: Oxford University Press.

Trehub, S. E., and Schellenberg, E. G. 1995. Music: its relevance to infants. *Annals of Child Development* 11, 1–24.

Trehub, S. E., Schellenberg, E. G., and Hill, D. 1997. The origins of music perception and cognition: a developmental perspective. In *Perception and Cognition of Music* (ed. I. Deliege and J. A. Sloboda), pp. 103–28. Hove: Psychology Press.

Trehub, S. E., and Nakata, T. 2001. Emotion and music in infancy. *Musicae Scientiae* Special Issue 2001–02, 37–59.

Trehub, S. E., and Trainor, L. J. 1998. Singing to infants: lullabies and playsongs. *Advances in Infancy Research* 12, 43–77.

Trehub, S. E., Unyk, A. M., and Trainor, L. J. 1993. Adults identify infant-directed music across cultures. *Infant Behaviour and Development* 16, 193–211.

Trehub, S. E., Unyk, A. M., Kamenetsky, S. B., Hill, D. S., Trainor, L. J., Henderson, M., and Saraza M. 1997. Mothers' and fathers' singing to infants. *Developmental Psychology* 33, 500–7.

Trevarthen, C. 1999. Musicality and the intrinsic motive pulse: Evidence from human pyschobiology and infant communication. *Musicae Scientiae* Special Issue 1999–2000, 155–215.

Trinkaus, E. 1983. *The Shanidar Neanderthals*. New York: Academic Press.

Trinkaus, E. 1985. Pathology and posture of the La-Chapelle-aux-Saints Neanderthal. *American Journal of Physical Anthropology* 67, 19–41.

Trinkaus, E. 1995. Neanderthal mortality patterns. *Journal of Archaeological Science* 22, 121–42.

Trinkaus, E., and Shipman, P. 1995. *The Neanderthals*. New York: Alfred A. Knopf.

Turk, I. (ed.). 1997. *Mousterian 'Bone Flute', and Other Finds from Divje Babe 1 Cave Site in Slovenia*. Ljubljana: Založba ZRC.

Turner, A. 1992. Large carnivores and earliest European hominids: changing determinations of resource availability during the Lower and Middle Pleistocene. *Journal of Human Evolution* 22, 109–26.

Unkefer, R. (ed.). 1990. *Music Therapy in the Treatment of Adults with Mental Disorders*. New York: Schirmer Books.

Unyk, A. M., Trehub, S. E., Trainor, L. J., and Schellenberg, E. G. 1992. Lullabies and simplicity: a cross-cultural perspective. *Psychology of Music* 20, 15–28.

Van Peer, P. 1992. *The Levallois Reduction Strategy*. Madison: Prehistory Press.

Vaneechoutte, M., and Skoyles, J. R. 1998. The memetic origin of language: modern humans as musical primates. *Journal of Memetics* 2 ‹http://jom-emit.cfpm.org/1998/vol2›.

Vieillard, S., Bigand, E., Madurell, F., and Marozeau, J. 2003. The temporal processing of musical emotion in a free categorisation task. *Proceedings of the 5th Triennial ESCOM conference*, pp. 234–7. Hanover: Hanover University of Music and Drama.

Wallin, N. L. 1991. *Biomusicology: Neurophysiological, Neuropsychological and Evolutionary Perspectives on the Origins and Purposes of Music*. Stuyvesant, NY: Pendragon Press.

Wallin, N. L., Merker, B., and Brown, S. (eds). 2000. *The Origins of Music*. Cambridge, MA: Massachusetts Institute of Technology.

Wanpo, H., Ciochon, R., Yumin, G., Larick, R., Qiren, F., Schwarcz, H., Yonge, C., de Vos, J., and Rink, W. 1995. Early *Homo* and associated artefacts from China. *Nature* 378, 275–8.

Ward, R., and Stringer, C. 1997. A molecular handle on the Neanderthals. *Nature* 388, 225–6.

Watt, R. J., and Ash, R. L. 1998. A psychological investigation of meaning in music. *Musicae Scientiae* 2, 33–53.

Weisman, R. G., Njegovan, M. G., Williams, M. T., Cohen, J. S., and Sturdy, C. B. 2004. A behavior analysis of absolute pitch: sex, experience, and species. *Behavioural Processes* 66, 289–307.

Wenban-Smith, F. 1999. Knapping technology. In *Boxgrove, A Middle Pleistocene Hominid Site at Eartham Quarry, Boxgrove, West Sussex* (ed. M. B. Roberts and S. Parfitt), pp. 384–95. London: English Heritage.

Wheeler, P. 1984. The evolution of bipedality and the loss of functional body hair in hominids. *Journal of Human Evolution* 13, 91–8.

Wheeler, P. 1988. Stand tall and stay cool. *New Scientist* 12, 60–5.

Wheeler, P. 1991. The influence of bipedalism on the energy and water budgets of early hominids. *Journal of Human Evolution* 21, 107–36.

Wheeler, P. 1992. The influence of the loss of functional body hair on hominid energy and water budgets. *Journal of Human Evolution* 23, 379–88.

Wheeler, P. 1994. The thermoregulatory advantages of heat storage and shade seeking behaviour to hominids foraging in equatorial savannah environments. *Journal of Human Evolution* 21, 107–36.

White, M. J. 1998. On the significance of Acheulian biface variability in southern Britain. *Proceedings of the Prehistoric Society* 64, 15–44.

White, R. 1993. A social and technological view of Aurignacian and Castelperronian personal ornaments in S.W. Europe. In *El Origin del Hombre Moderno en el Suroeste de Europa* (ed. V. Cabrera Valdés), pp. 327–57. Madrid: Ministerio des Educacion y Ciencia.

White, T. D., Asfaw, B., Dagusta, D., Gilbert, H., Richards, G. D., Suwa, G., and Clark Howell, F. 2003. Pleistocene *Homo sapiens* from Middle Awash, Ethiopia. *Nature* 423, 742–7.

Wilson, S. J., and Pressing, J. 1999. Neuropsychological assessment and modeling of musical deficits. *MusicMedicine* 3, 47–74.

Wood, B. 1992. Origin and evolution of the genus *Homo*. *Nature* 355, 783–90.

Wood, B. 2002. Hominid reveleations from Chad. *Nature* 418, 133–5.

Wood, B., and Collard, M. 1999. The human genus. *Science* 284, 65–71.

Wray, A. 1998. Protolanguage as a holistic system for social interaction. *Language and Communication* 18, 47–67.

Wray, A. 2000. Holistic utterances in protolanguage: the link from primates to humans. In *The Evolutionary Emergence of Language: Social Function and the Origins of Linguistic Form* (ed. C. Knight, M. Studdert-Kennedy and J. R. Hurford), pp. 285–302. Cambridge: Cambridge University Press.

Wray, A., 2002a. *Formulaic Language and the Lexicon*. Cambridge: Cambridge University Press.

Wray, A. 2002b. (ed.). *The Transition to Language*. Oxford: Oxford University Press.

Wray, A. 2002c. Dual processing in protolanguage. In *The Transition to Language* (ed. A. Wray), pp. 113–37. Oxford: Oxford University Press.

Wray, A., and Grace, G. W. in press. The consequences of talking to strangers: evolutionary corollaries of socio-cultural influences on linguistic form. *Lingua*.

Wright, A. A., Rivera, J. J., Hulse, S. H., Shyan, M., and Neiworth, J. J. 2000. Music perception and octave generalization in rhesus monkeys. *Experimental Psychology: General* 29, 291–307.

Wynn, T. 1993. Two developments in the mind of early *Homo*. *Journal of Anthropological Archaeology* 12, 299–322.

Wynn, T. 1995. Handaxe enigmas. *World Archaeology* 27, 10–23.

Wynn, T., and Coolidge, F. L. 2004. The expert Neanderthal mind. *Journal of Human Evolution*. 46, 467–87.

Yaqub, B. A., Gascon, G. G., Al-Nosha, M., Whitaker, H. 1988. Pure word deafness (acquired verbal auditory agnosia) in an Arabic speaking patient. *Brain* 111, 457–66.

Yellen, J. E., Brooks, A. S., Cornelissen, E., Mehlman, M. H., and Stewart, K. 1995. A Middle Stone Age worked bone industry from Katanda, Upper Semliki Valley, Zaire. *Science* 268, 553–6.

Zahavi, A. 1975. Mate selection: a selection of handicap. *Journal of Theoretical Biology* 53, 205–14.

Zahavi, A., and Zahavi, A. 1997. *The Handicap Principle*. New York, NY: Oxford University Press.

Zatorre. R. J. 2003. Absolute pitch: a model for understanding the influence of genes and development on neural and cognitive function. *Nature Neuroscience* 6, 692–5.

Zhu, R. X., Potts, R., Xie, F., Hoffman, K. A., Deng, C. L., Shi, C. D., Pan, Y. X., Wang, H. Q., Shi, R. P., Wang, Y. C., Shi, G. H., and Wu, N. Q. 2004. New evidence on the earliest human presence at high northern latitudes in northeast Asia. *Nature* 431, 559–62.

Zuberbühler, K. 2003. Referential signalling in non-human primates: cognitive precursors and limitations for the evolution of language. *Advances in the Study of Behaviour* 33, 265–307.

Zubrow, E. 1989. The demographic modelling of Neanderthal extinction. In *The Human Revolution: Behavioural and Biological Perspectives on the Origins of Modern Humans* (ed. P. Mellars and C. B. Stringer), pp. 212–32. Edinburgh: Edinburgh University Press.

Picture acknowledgements

I am grateful to Margaret Mathews for drawing figures 7, 8, 13–17 and 20, and to Jane Burrell for figure 18.

Figure 1. Adapted from Lahr & Foley (2004).
Figures 2–4. From Carter (1998).
Figure 5. From Peretz & Coltheart 2003, reproduced with permission of Isabelle Peretz.
Figure 6. After Jobling *et al.* (2004).
Figures 7, 8, 13, 14, after Johnson & Edgar (1996).
Figure 9. Reproduced from Lewin (1999) with permission of John Fleagle.
Figure 10. Drawn on basis of data from Aiello & Dunbar (1993).
Figure 12. Reproduced from Roberts & Parfitt (1999), permission sought from English Heritage.
Figure 15. After Turk (1997).
Figure 16. After White *et al.* (2003).
Figure 17. After Henshilwood *et al.* (2002).
Figure 20. After D'Errico *et al.* (2003).

Index